# Progress in Botany

Volume 86

**Series Editors**

Ulrich Lüttge
Department of Biology
Technische Universität Darmstadt
Darmstadt, Germany

Francisco M. Cánovas
Department of Molecular Biology and Biochemistry
University of Málaga
Málaga, Spain

Hans Pretzsch
Chair for Forest Growth and Yield Science
School of Life Sciences Weihenstephan
Technical University of Munich
Freising, Germany

María Carmen Risueño Almeida
Pollen Biotechnology of Crop Plants Group
Centro de Investigaciones Biologicas "Margarita Salas" (CIB) CSIC Madrid
Madrid, Spain

Christoph Leuschner
Pflanzenökologie
Universität Göttingen
Göttingen, Germany

Progress in Botany is devoted to all the colourful aspects of plant biology. The annual volumes consist of invited reviews spanning the fields of molecular genetics, cell biology, physiology, comparative morphology, systematics, ecology, biotechnology and vegetation science, and combine the depth of the frontiers of research with considerable breadth of view. Thus, they establish unique links in a world of increasing specialization. Progress in Botany is engaged in fostering the progression from broad information to advanced knowledge and finally deepened understanding. All chapters are thoroughly peer-reviewed.

Ulrich Lüttge • Francisco M. Cánovas •
María Carmen Risueño Almeida •
Christoph Leuschner • Hans Pretzsch
Editors

# Progress in Botany Vol. 86

Springer

*Editors*
Ulrich Lüttge
Department of Biology
Technische Universität Darmstadt
Darmstadt, Germany

María Carmen Risueño Almeida
Pollen Biotechnology of Crop Plants Group
Centro de Investigaciones Biologicas
"Margarita Salas" (CIB) CSIC Madrid
Madrid, Spain

Hans Pretzsch
Chair for Forest Growth and Yield Science
School of Life Sciences Weihenstephan
Technical University of Munich
Freising, Germany

Francisco M. Cánovas
Department of Molecular Biology
and Biochemistry
University of Málaga
Málaga, Spain

Christoph Leuschner
Pflanzenökologie
Universität Göttingen
Göttingen, Germany

ISSN 0340-4773 ISSN 2197-8492 (electronic)
Progress in Botany
ISBN 978-3-032-20082-2 ISBN 978-3-032-20083-9 (eBook)
https://doi.org/10.1007/978-3-032-20083-9



This Springer imprint is published by the registered company Springer Nature Switzerland AG
The registered company address is: Gewerbestrasse 11, 6330 Cham, Switzerland

# Contents

# Contributors

**Mutez Ahmed** Root-Soil Interaction, TUM School of Life Sciences, Technical University of Munich, Freising, Germany

**Lorena Almagro** Department of Plant Biology, Faculty of Biology, University of Murcia, Campus of Espinardo, Murcia, Spain

**Jörg D. Becker** Instituto de Tecnologia Química e Biológica António Xavier, Universidade Nova de Lisboa, Oeiras, Portugal

**Sabine Braun** Institute for Applied Plant Biology, Witterswil, Switzerland

**Brendan Choat** Hawkesbury Institute for the Environment, Western Sydney University, Richmond, NSW, Australia

**Danielle R. Cooney** Department of Crop and Soil Sciences, North Carolina State University, Raleigh, NC, USA

**Sylvain Delzon** University of Bordeaux, INRAE, BIOGECO, Pessac, France

**Hannah E. R. Frank** Department of Crop and Soil Sciences, North Carolina State University, Raleigh, NC, USA

**Kevin Garcia** Department of Crop and Soil Sciences, North Carolina State University, Raleigh, NC, USA

**Sven-Eric Hopf** Institute for Applied Plant Biology, Witterswil, Switzerland

**Pierre L. Ibisch** Centre for Econics and Ecosystem Management at Eberswalde University for Sustainable Development (HNEE), Eberswalde, Germany

**Steven Jansen** Institute of Botany, Ulm University, Ulm, Germany

**Alena P. Jones** Department of Crop and Soil Sciences, North Carolina State University, Raleigh, NC, USA

**Arjun Kafle** Department of Crop and Soil Sciences, North Carolina State University, Raleigh, NC, USA
Agriculture Research Service, Plant Science Research Unit, USDA, Raleigh, NC, USA

**Gitanjali Nanda Kafle** Department of Crop and Soil Sciences, North Carolina State University, Raleigh, NC, USA

**Martyna M. Kotowska** School of Natural Sciences, Macquarie University, Sydney, NSW, Australia

**Laurent J. Lamarque** Natural Resources Canada, Canadian Forest Service, Laurentian Forestry Centre, Québec, QC, Canada
Département des Sciences de l'environnement, Université du Québec à Trois-Rivières, Trois-Rivières, QC, Canada

**Christoph Leuschner** Plant Ecology, Albrecht von Haller Institute for Plant Sciences, University of Goettingen, Goettingen, Germany

**Torben Lübbe** Plant Ecology, Albrecht von Haller Institute for Plant Sciences, University of Goettingen, Goettingen, Germany

**Diogo A. Lucas** Instituto de Tecnologia Química e Biológica António Xavier, Universidade Nova de Lisboa, Oeiras, Portugal
Gulbenkian Institute for Molecular Medicine, Oeiras, Portugal

**José Manuel Martí-Guillén** Department of Plant Biology, Faculty of Biology, University of Murcia, Campus of Espinardo, Murcia, Spain

**Sara Esperanza Martínez-Lorente** Department of Plant Biology, Faculty of Biology, University of Murcia, Campus of Espinardo, Murcia, Spain

**Scott McAdam** Department of Botany and Plant Pathology, Purdue University, West Lafayette, IN, USA

**María Ángeles Pedreño** Department of Plant Biology, Faculty of Biology, University of Murcia, Campus of Espinardo, Murcia, Spain

**Peter Petrik** Institute of Forest Botany and Forest Zoology, Technical University of Dresden (TUD), Tharandt, Germany

**Andreas Roloff** Institute of Forest Botany and Forest Zoology, Department of Forestry Science, Faculty of Environment, Dresden University of Technology (TU Dresden), Tharandt, Germany

**Benjamin D. Rose** Department of Crop and Soil Sciences, North Carolina State University, Raleigh, NC, USA

**Ana Belén Sabater-Jara** Department of Plant Biology, Faculty of Biology, University of Murcia, Campus of Espinardo, Murcia, Spain

**Christian Schindler** University of Basel, Swiss Tropical and Public Health Institute, Basel, Switzerland

**Bernhard Schuldt** Institute of Forest Botany and Forest Zoology, Technical University of Dresden (TUD), Tharandt, Germany

**José M. Torres-Ruiz** Instituto de Recursos Naturales y Agrobiologıa (IRNAS), Consejo Superior de Investigaciones Cientıficas (CSIC), Seville, Spain

**Simon Tresch** Institute for Applied Plant Biology, Witterswil, Switzerland

**Julia Unkelbach** Department of Palynology and Climate Dynamics, Albrecht-von-Haller-Institute for Plant Sciences, University of Göttingen, Göttingen, Germany

# Curriculum Vitae

Prof. Dr. **Andreas Roloff**
Forest Scientist, Tree Biologist
70 years
born 1955.05.15 in Bremen (NW Germany)
living near Meißen, Saxony (SE Germany)
Dresden University of Technology (TU Dresden)
Institute of Forest Botany and Forest Zoology, Pienner Str. 7, D-01737 Tharandt, Germany
Chairman German Tree Institute

| | |
|---|---|
| Since 2022 | Senior Professorship of Research and Science Communication on Tree Biology |
| Since 2019 | Chairman of the Board 'National Heritage Trees' |
| **1994–2022** | Chair of Forest Botany, Dept. of Forest Sciences, Faculty of Environmental Sciences |

| | |
|---|---|
| | **Director Institute of Forest Botany and Forest Zoology and Tharandt Botanic Garden/Saxonian State Arboretum, TU Dresden** |
| 2016–2020 | Chairman of the German committee for Tree Assessment: revision of the Guidelines for 'Tree Safety Inspection' |
| Since 2016 | Head of German Tree Institute, TU Dresden |
| 2009–2015 | Chairman of the board 'Tree of the Year' in Germany |
| 1994–2003 | Study Dean of the Department of Forest Sciences, TU Dresden |
| Before | |
| 1990–93 | Professorship of Forest Botany/Dendrology at University Goettingen, Germany |
| 1988–1990 | Assistant Professor at Institute of Silviculture, Univ. Goettingen, responsible for teaching in Forest Vegetation Science (Geobotany) |
| 1989 | Habilitation on tree crown development, title of the thesis: *Crown development and vitality assessment of selected tree species of the temperate zone. (in German).* Sauerländer's, Frankfurt/M.: 258 pp. |
| 1987–88 | Scientist at Institute of Silviculture, University of Goettingen |
| 1986 | PhD Forest Botany, title of thesis*: Morphology of crown development in European beech with regard to possibly new modifications. (in German)* Ber. Forschungsz. Waldökosysteme Göttingen 18: 1–177 |
| 1984–1987 | Scientist at Institute of Forest Botany, University of Goettingen |
| 1979–84 | Study Course Forestry Science Univ. Goettingen |
| 1976–79 | Study Course Psychology Univ. Goettingen |

# Forty Years of Tree Biology Research and Method Development for Tree Management in Forests, Cities and Parks (1984–2024)

**Andreas Roloff**

## Contents



Communicated by Ulrich Lüttge and Hans Petzsch

A. Roloff (✉)
Institute of Forest Botany and Forest Zoology, Department of Forestry Science, Faculty of Environment, Dresden University of Technology (TU Dresden), Tharandt, Germany
e-mail: andreas.roloff@tu-dresden.de

 
Progress in Botany (2026) 86: 1–81, https://doi.org/10.1007/124_2025_89,
Published online: 11 June 2025

**Abstract** My professional career began in 1979 with the study course in forestry sciences at Goettingen university (Germany), which I completed with my diploma thesis in 1984 and then continued as a research assistant in forest botany with a forest decline-project. For this purpose, investigations on the branching of *Fagus sylvatica* (European beech) were carried out as part of my PhD thesis with a number of new, surprising and significant results on bud and branching development. These soon led to the development of a completely new inspection key for assessing the vitality of beech trees. Immediately after its first publications in 1985, the method was applied in forestry practice. Based on this, the method was transferred to other tree species as part of my habilitation thesis in 1987, adapted and finally generally modified for deciduous trees of the temperate zone. The method has already been tested in the USA and China and also works there and is established and widely used in Central Europe to assess the condition of trees ('Roloff score').

After a short 2-year interim period at the Institute of Silviculture in Goettingen in the field of Forest Vegetation Science, I was appointed to the Chair of Forest Botany/ Dendrology at the University of Goettingen in 1990 and from there to the Chair of Forest Botany at TU Dresden in 1994. There, the main areas of work were tree-ring research, scientific species identification, tree species monographs, tree water balance and drought stress, urban tree management, tree species selection for climate change, and tree aging. Finally, in 2019, the initiative for a 10-year project with ancient trees in Germany was launched, in which 100 very old trees are searched for, selected and proclaimed as 'National Heritage Trees' in order to protect them permanently and so that they can reach an age of 1,000 years with optimal support. This project runs until 2029 and is then intended to be transferred to a new permanent nature conservation category in Germany.

**Keywords** Bud development, Crown architecture, Ecophysiology, Forest Botanic Garden, Long-lived tree species, Species determination, Tree ageing, Tree species selection, Urban tree management, Vitality assessment

## 1 Introduction: Start of my Professional Career and Motivation for a Science Communication Strategy

I began to study psychology in Goettingen (1976) but soon realized that this was not the optimal orientation. So after 3 years I switched to forestry sciences because I had started working with wood, building furniture and got very interested in the development of wood and trees. This field of interest has inspired me from the beginning and to this day. I was particularly interested in forest botany right from the beginning, as this subject was represented in the first semesters with two extensive lecture series as well as a large number of species determination exercises and excursions. So I started with botanical drawings (bark, trees, 'pointer plants') until the dendrology professor in charge, Horst Bartels, saw some of my drawings and asked me (in the second semester) if I could draw on dendrological topics for his lecture scripts and for his new dendrology textbook (Bartels 1993). As a result, I became scientific assistant with him and relatively soon it was also clear that I would write my diploma thesis with his supervision. After I finished my studies, he asked me to become doctorate with him on the forest decline situation ('Waldsterben') that was current at the time, and it became a research project.

This is how my forest botanical career began, which continues to these days. From the very beginning, it was a strong concern of mine to work on issues that are important and useful for tree management practice. Through my familiarization with the branching development and its change with decreasing vigor status of the tree (investigating European beech as the most important natural and forestry deciduous tree species in Central Europe), it soon became my goal: to develop an inspection aid for beech trees, how to assess the vigor of trees relatively easy and practicable on the basis of their branching pattern. After the publication of the resulting 'vitality stages key' of my PhD thesis (Roloff 1986a), the practice immediately gratefully took up this method and incorporated it into its tree condition surveys. It has proven itself as one of two methods used until today (in addition to foliage density) in Central Europe is now widely established for most tree inspections, also in the urban area (Roloff 1999, 2017d, 2018a; Stribley and Ashmore 2002; Dobbertin 2005).

The aim and important strategy of my science communication of research results was therefore from the very beginning to implement them in practice: i.e. method development and recommendations for action, based on tree biological scientific investigations, surveys and experiments after discussion with representatives from practice. As an essential medium, I have written a number of specialist books for those interested in trees and users and have promoted their popularity through further practical publications as well as with many lectures/workshops and information events (Roloff 2001, 2004, 2010, 2013, 2016a, 2018a, 2022a, 2023a; Roloff and Bärtels 2018).

From all this, the sequence of findings and their implementation and application in practice has emerged, as it is presented in detail below.

## 2 Crown Architecture and Vitality Stages (Mainly 1983–2018)

*For this first major topic of my professional life (during the 1980s), the great forest decline at the time (called 'Waldsterben') and an approved federal project in Goettingen forest botany with Prof. Bartels were decisive. He knew my ability to draw (see Sect. 1), explicitly encouraged it and thus engaged me for his lecture notes and dendrology books.*

*At that time, there was only the method of foliage density (UN/ECE* 1989*) to assess the condition of trees: the leaf loss in % in four damage levels (crown transparency). I immediately noticed that this probably cannot be enough to* *objectively assess the condition of trees. While drawing hundreds of branches for my diploma thesis and PhD thesis, there was suddenly a breakthrough: that I discovered a significant change in the branching structure with decreasing shoot length, which can also be distinguished in four gradations and could be a completely different, more well-founded basis for the condition assessment of beech trees. I then illustrated this with concise drawings and converted it into the vitality key for European beech.*

## 2.1 Bud Development, Sprouting, Shoot Length's Growth of European Beech (Diploma Thesis 1983)

Extensive investigations on European beech buds over the course of the year showed that all primordia for the sprouting of the coming spring (shoots, leaves and lateral buds) were already fully developed in early summer of the previous year and have been ready in the bud in miniature since then (Fig. 1, Roloff 1983). This was relatively unknown until then, and even the significance of it for the interpretation of the annual shoot length was not known. This is because the shoot lengths are also significantly predetermined and are only less influenced by the temperature and water supply conditions during sprouting in spring. In particular, the distinction between short and long shoots is irreversibly predetermined. This means that the length of the spring shoots in European beech is determined in the previous year from 75 up to 100% and that drought damage or particularly favourable conditions of a growing season, for example, are therefore usually only reflected in the shoot lengths with a 1-year delay (Roloff 1987).

Syllepsis and prolepsis (lammas shoots) as a simultaneous sprouting of lateral branches with the sprouting main axis in spring or a second sprouting at the beginning of summer were investigated in European beech stands of different ages, exposed and recorded in relation to shoot length growth. According to this, they represent a possibility of faster additional conquest of airspace if optimal water, light and nutrient conditions are given, especially in the youth of tree development (Fig. 2, Roloff 1983, 1987).

The length development of short and long shoots of European beech during sprouting was recorded in detail, with the important result that budding and length growth (determined by the previous year's bud development, see above) is already completed after 2 weeks by mid-May for short shoots and after 4 weeks by the end of May for long shoots. After that, only lammas shoots are possible as a second flush in summer (see before, Roloff 1987). During sprouting of long shoots in European beech, it is important that they initially grow horizontally (as an adaptation to shade conditions) and erect themselves after 6 weeks with better light conditions but remain horizontally oriented (plagiotropy) in the shade for years or even decades (Fig. 3, Roloff 1983, 1987).

This 'waiting stage' in the deep shade of young European beech trees, which can last for several years or even decades, is impressive. In this way, young beeches of

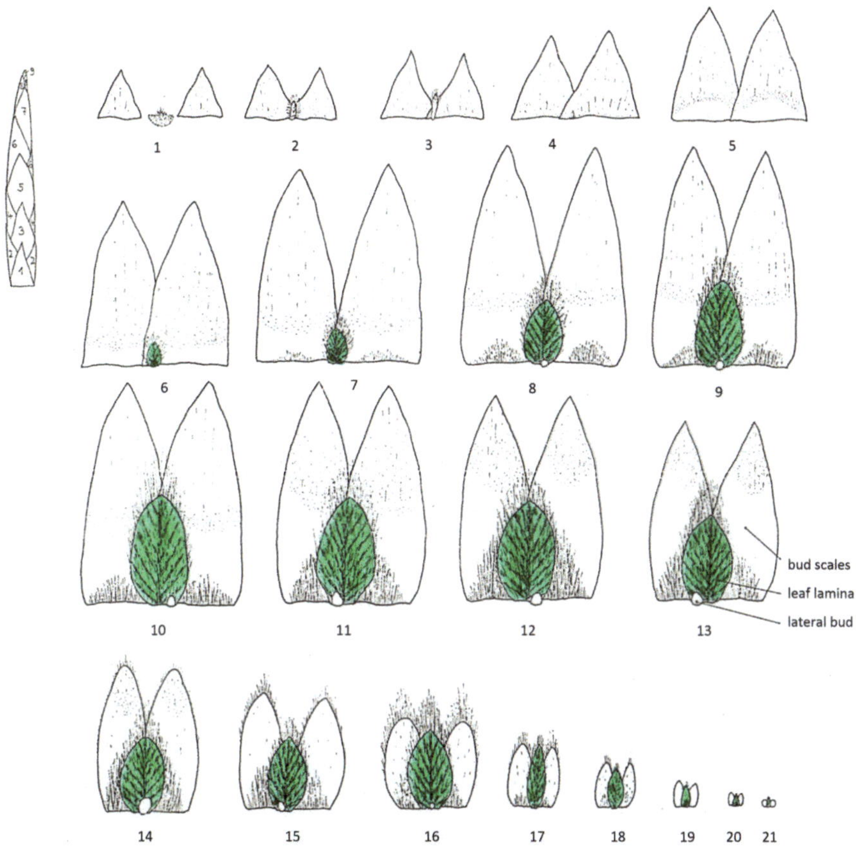


**Fig. 1** Components of a European beech terminal bud in January (exterior view top left), free-preparation from below/outside to up/inside; always belonging to one unit: two stipules (bud scales), leaflet lamina (green), lateral bud primordium (for its sprouting 15 months later)

50 cm in size can be more than 50 years old (Fig. 4, Roloff 1983), in extreme cases more than 100 years. This is well known from primeval forest research and in silviculture of mixed mountain forests (Mayer 1999), there named as 'sit and wait state' (Mitscherlich 1970). It otherwise only occurs in young *Abies alba* and *Taxus baccata*.

Short shoot chains of European beech can live a maximum span of 15–20 years, then they die – if for 1 or 2 years long shoots are not inserted. This means that *Fagus sylvatica* has the longest short-shoot lifespan of all the tree species investigated and shows one of the best shade adaptations (Fig. 5, Roloff 1987, 1989b, c, 2001). According to this, there are deciduous tree species that do not develop short shoots at all (*Salix alba* and *Robinia pseudoacacia*, e.g.), other tree species with a short shoot lifespan of about 5 years (e.g. *Betula pendula*, *Quercus robur*, *Fraxinus excelsior*),

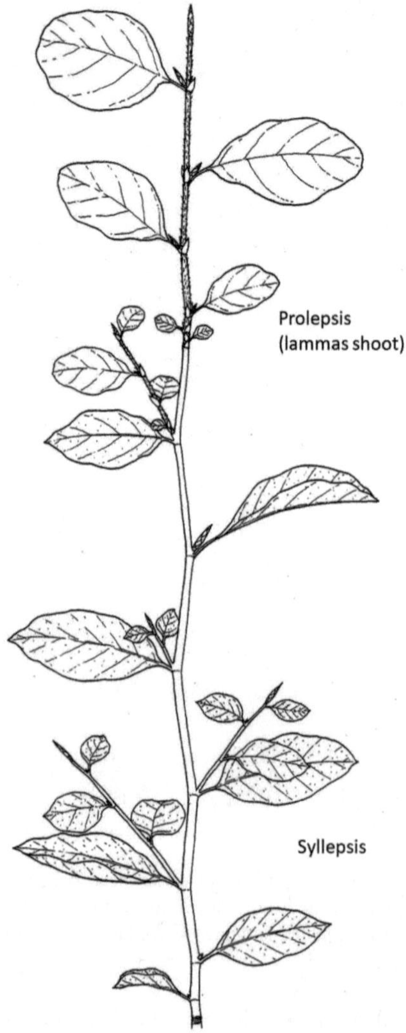


**Fig. 2** Additional shoot development (European beech terminal shoot) under optimal conditions: syllepsis (lateral shoots at the bottom) and prolepsis (lateral shoots and terminal extension of the leader)

about 10 years (e.g. *Acer pseudoplatanus*, *Tilia cordata*, *Carpinus betulus*) or up to 15 years (e.g. *Aesculus hippocastanum*, *Acer saccharum*, *Fagus sylvatica*).

## 2.2 Shoot Base Scars, Bud Cycle, Branching Pattern and Vitality Stages of European Beech (PhD Thesis 1986)

It was also relatively unknown until these investigations and their publication that the bud scales leave scars on the shoot surface and fall off when they sprout and are

**Fig. 3** Sprouting development from the terminal bud (left), elongation in length (middle) and straightening up (right) of a European beech terminal shoot with good light exposure (dates in bottom line)

referred to by the (new) term 'shoot base scars' (Fig. 6, Roloff 1986a). In the case of smooth-barked tree species such as *Fagus sylvatica*, they are preserved and visible for up to 20 (40) years and can be used, for example, for retrospective shoot length measurements. So they make it possible to measure the annual shoot lengths back at least 10 years for many tree species and up to 40 years for European beech (Roloff 1986a). This also gives you a non-destructive way of determining the age of any branch (if no lammas shoots have occurred, which can occur especially with oaks and on young, well-exposed trees of other tree species). This method can be used to trace the reactions of trees to drought stress back over the past 10 (to 20) years (Roloff 1986a, 1989a, b, c).

Investigations into the so-called morphogenetic cycle of European beech (Fig. 7) showed that bud development begins as early as 1.5 years before sprouting, with the creation of secondary leaves as bud scales, and thus continuously invisible from the outside in the buds, the sprouting for the next spring is prepared (Roloff 1986a, 1987).

Furthermore, investigations were carried out on the formation and anatomy of the leaf abscission zone in the petiole base of European beech trees, with the result that it is already anatomically prepared by thinner cell walls and smaller cells when the

**Fig. 4** 37-year-old European beech with horizontal shoot orientation ('waiting state') in shading conditions

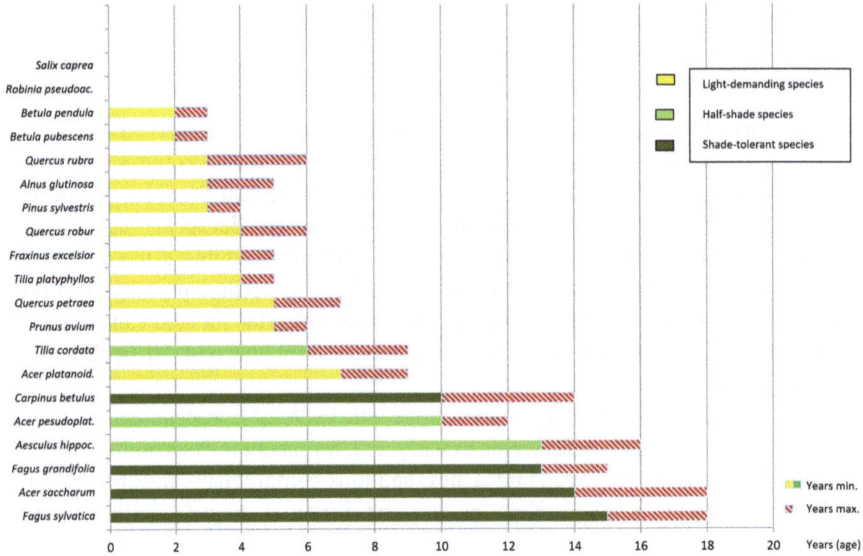


**Fig. 5** Maximum short shoot lifespan of 20 tree species

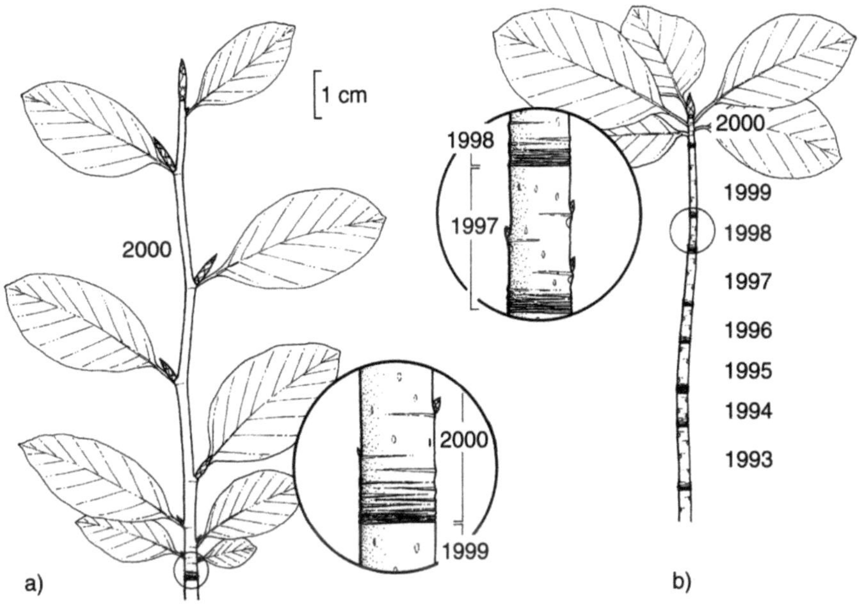


**Fig. 6** Shoot base scars in European beech (**a**) on a long shoot (left at bottom), (**b**) on a short shoot chain (right)

leaves are unfolded in April (Fig. 8, Roloff 1986a, 1987). It is then activated only 6 months later by dissolving the middle lamellae shortly before the leaves abscission.

If young European beech trees in unshaded environments have regular lammas shoots (second flush at the end of June, see Sect. 2.1), they can often not fully develop their terminal buds by autumn due to unexpected weather influences, and the sprouting remains 'stuck' in a variable state of development. Then growth anomalies can occur in the following spring with frequent bifurcation of the top shoot (Fig. 9, Roloff 1986a, 1987), which is undesirable from a forestry point of view, as the later trunk forks. It turned out that the phenomenon only occurs on young European beech trees in the open field at full light, not in the shade of older trees (Fig. 10, Roloff 1983, 1986a, 1987).

By measuring the length of the annual shoots on the main branches of European beech crowns 20 years back with the help of the base scars of the shoots (see above), interesting crown development details could be determined and crown growth patterns created (Fig. 11, Roloff 1986a). These patterns clearly show the influence of the competition situation and drought years on crown development (cf. Pretzsch et al. 2011; Dieler and Pretzsch 2013; Pretzsch 2019).

Based on the above-mentioned and intensive studies on the branching pattern of European beech, it was found that the branching structure changes markedly with decreasing annual shoot lengths. This makes it possible to distinguish four different structures that can be interpreted as a measure of the tree vitality – if well-exposed branches of the upper crown are used. In lower crown areas, the shading of the sun

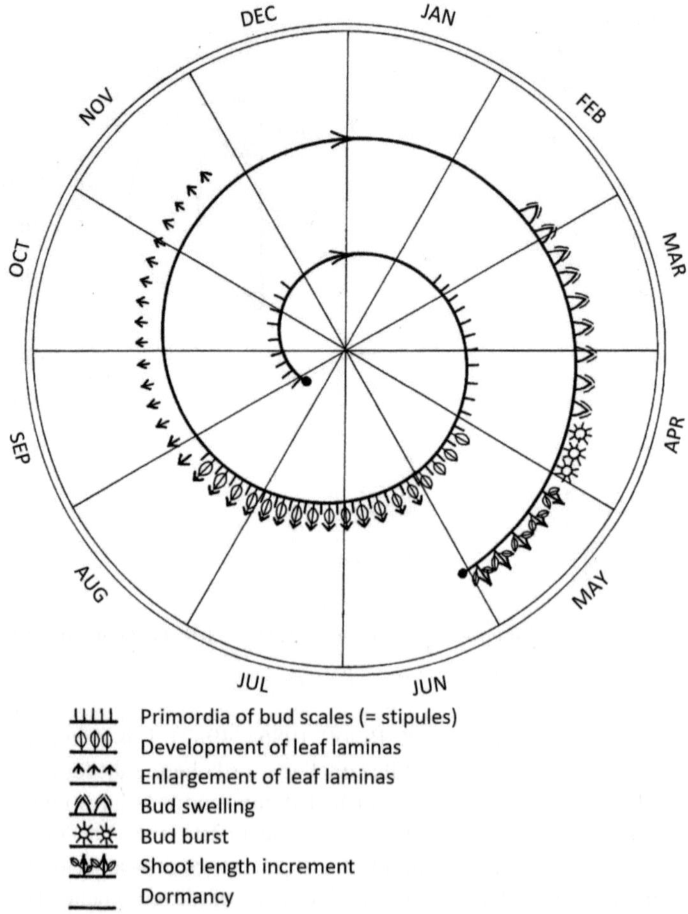


**Fig. 7** Morphogenetic cycle of European beech bud development, process course from inner spiral (first secondary stipule origins in August 2 seasons before) to middle spiral (leaf lamina development, last year's season) and outer spiral (sprouting and longitudinal growth, this spring)

crown above is too strong, and therefore, the branches there are not suitable for an interpretation of the tree's vitality. Of course, the age of the tree also has an important influence on the shoot lengths, but it can also be interpreted as a factor of the tree vigour. A concept of four different branching structures (modes) was developed (Roloff 1986a, 1988a, b, 1989a, b, c, 2001, 2016d, 2018a), designated with new terms and these are briefly presented here (Fig. 12):

**Long-Shoot Mode (Exploration)**

In the so-called 'long-shoot mode', a branching pattern is created, as is typical for the tree species and in which the long-shoots visually dominate. At each annual shoot increment, the lateral shoot lengths decrease from top to bottom (so-called acrotony); only at the bottom there are some short shoots and dormant buds. This creates storey-

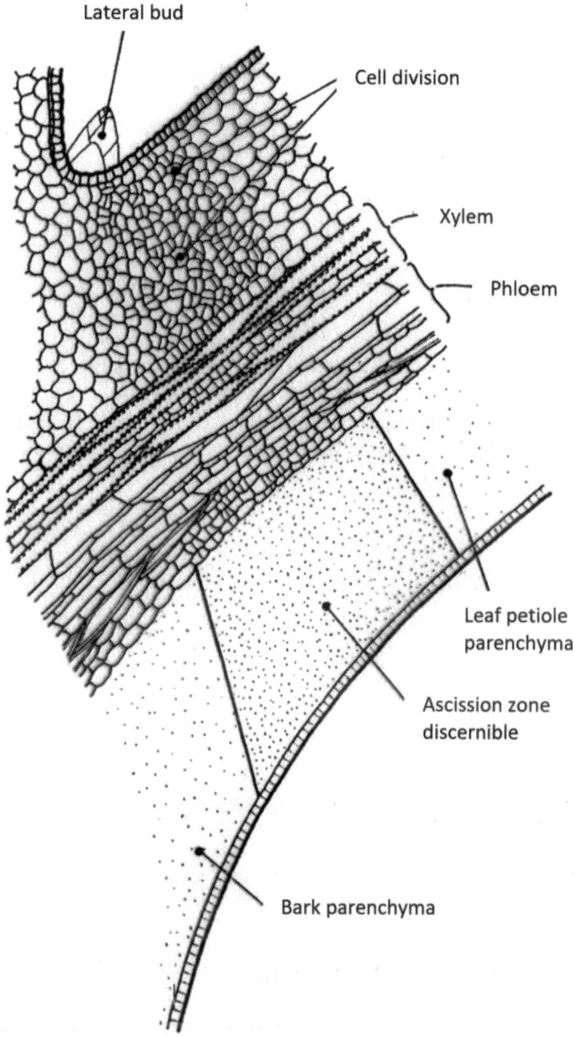


**Fig. 8** Development status of the abscission zone in May shortly after sprouting: recognizable by increased cell divisions and thus smaller cells in the petiole base (longitudinal section, in the upper part anatomical drawing, at bottom schematic sketch)

like ledges on the branch. That long-shoot mode is the normal case for the leader shoots of vigorous trees during young and middle age. Only in this way the treetop shoots can fulfil their main task for the tree and constantly conquer new air space with long shoots, fill it with lateral branching and prevail against competitors.

**Brush Mode (Degeneration)**

In the so-called 'brush mode', the terminal bud still forms long shoots every year – albeit shorter ones – but from all lateral buds, i.e., also from the uppermost ones, short shoots (or short shoot chains) emerge almost without exception. As a result, the branching is clearly impoverished, and elongated, bottlebrush-like structures are formed, which protrude from the periphery of the crown and on which the leaves

**Fig. 9** Growth anomaly due to lammas shoots: several leader shoots after sprouting from lammas shoot buds of the previous year, which lead to multiple bifurcation of the trunk after straightened up

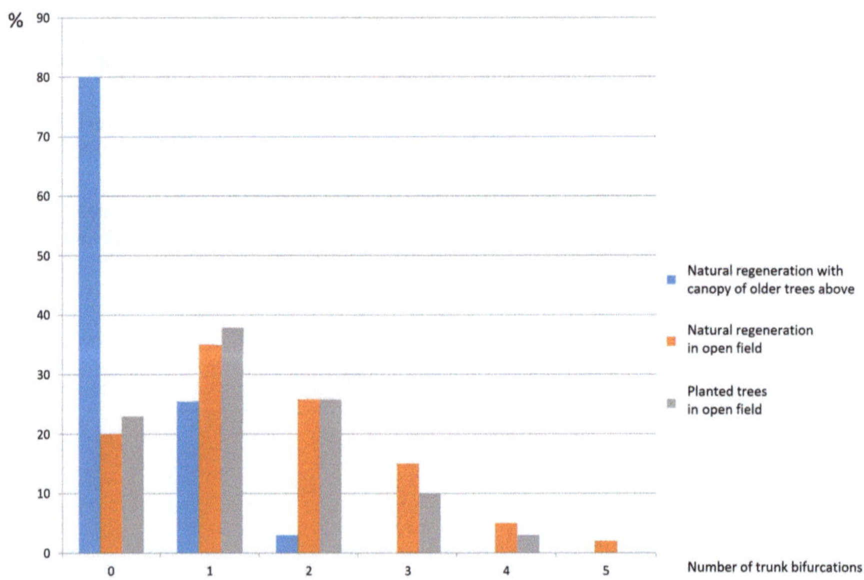


**Fig. 10** Frequency of bifurcation of the top shoot after lammas shoots: accumulation of bifurcations in open field, absence in the shade under old forest canopy

are densely arranged all around (at the end of the lateral short shoots or short shoot chains).

**Short-Shoot Mode (Stagnation)**
With a significant reduction in vitality, the terminal buds of the leader-shoots themselves start to form short shoots. So there develop no more lateral branches on them, because short shoots do not branch. Due to the short shoot lengths, the length increment of the branch or the height gain of the tree stagnates, hence the term

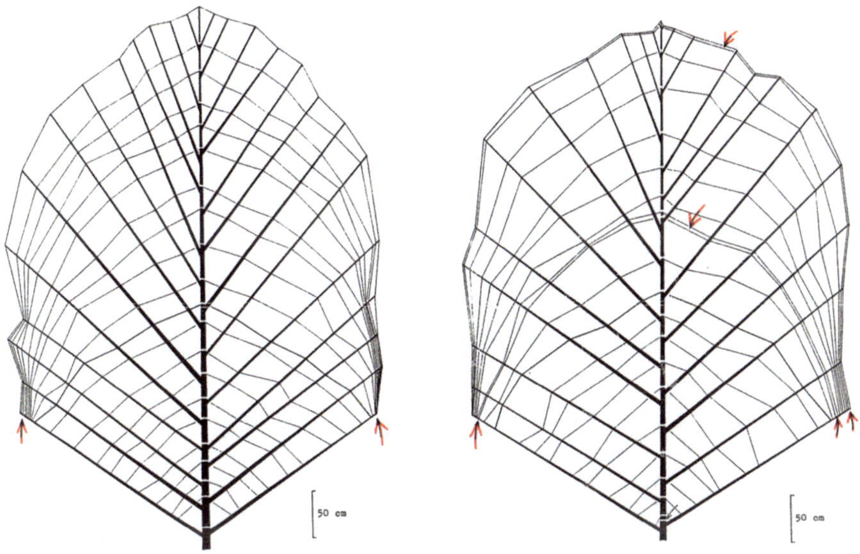


**Fig. 11** Crown growth patterns measuring the length of the annual shoots on the main branches of European beech crowns over 20 years back with the help of the shoot base scars (arrows left figure: competition effect of neighbouring trees, arrows in right figure: short shoots as effect of drought years last season and 8 years ago)

'short-shoot mode'. This is the 'claw-foot or crow's foot stage', as the short shoots in the crown periphery become longer and predominate and stretch towards the light in a claw-like manner.

**Retreating Mode (Retraction)**
If the short shoot mode lasts for a longer period of time (i.e. it is not only temporary), the branch or – if it is the leader-shoot – the entire top shoot begins to die, the 'retraction mode' is reached. Due to their unfavourable mechanical-static properties (dense tufts of leaves at the end of very tender shoots), the short shoot chains in the upper crown area exposed to the wind cannot become arbitrarily long or old (for more information, see short shoot lifespan Fig. 5). The exact time of death is now decided by secondary factors.

Based on these growth phases, an inspection tool for the vitality assessment of European beech was developed, which distinguishes between four vitality stages (Roloff 1986a, 1988a, b, 1989a, 1991b, 1992a, 1994, 1999, 2001, 2014b, 2018a; Rust and Roloff 2002). It has been widely used since the 1990s for the assessment of forest and urban trees and has even established itself as a decisive criterion for assessment in urban areas for the standard inspection of trees. It is also presented here for European beech, as the branching pattern modes (Fig. 12) on this tree species have been explained beforehand. Forest damage in European beech trees is currently widespread again because of severe drought years.

As another symptom of the tree condition, which was already known and used, there was previously only the foliage density available (called 'leaf loss'), which is

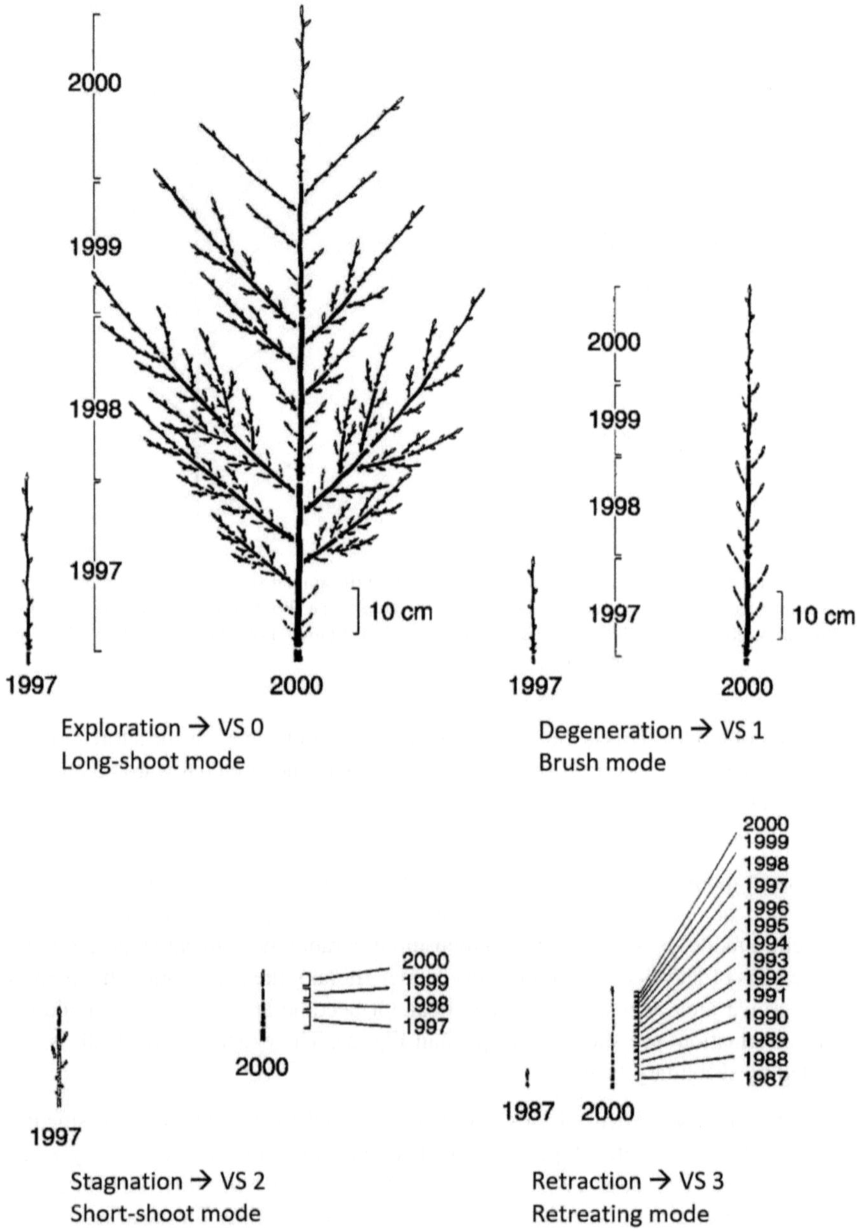


**Fig. 12** Four different branching structures and modes of European beech based on decreasing shoot lengths (the initial state on the left and the development 3 years later on the right)

widely referred to as damage levels (UN/ECE 1989). However, this has the disadvantage that the method cannot be used in the winter months for deciduous tree species.

With decreasing shoot lengths, the branching structures of the crown change so significantly that this can be used for a vitality assessment by distinguishing four vitality stages (VS) (assessment of the upper crown, Fig. 13):

- *VS 0 (vitality stage 0)*: vigorous trees without impairment of growth in long-shoot mode (Fig. 12): branching net-like, long shoots also dominate the overall impression on side axes, mostly the foliage without large crown gaps.
- *VS 1* (trees with slightly reduced vitality in brush mode): spike-like to elongated brush-like structures protrude from the upper crown, which are caused by reduced longitudinal growth of the main axes with laterally (almost) only short shoots or short shoot chains, the crown appears frayed on the outside.
- *VS 2* (trees with significantly reduced vitality in short shoot-mode): as a result of short shoot formation now also on the main axes, claw-like outer branch structures form in winter with tufted foliage clusters in summer and inner crown gaps; this condition can also occur due to older age and last for a long time without the tree having to die soon: for old trees of a long-lived tree species, this can be the 'normal state' without cause for concern or need for action.
- *VS 3* (trees with greatly reduced vitality and dying main axes in retraction mode): branching like VS 2, but also with as a clear warning signal by dying main axes, the crown disintegrates into partial crowns. These trees are often to be classified as problematic in terms of safety risks because of the dead branches in the treetop area. After professional pruning measures, if necessary, the tree can live for a longer time with a reduced crown size (depending on the tree species), which trees in undisturbed locations also do by themselves (see secondary crown Fig. 16 below).

The vitality level key presented above is of additional importance due to its applicability in aerial photo interpretation. This is because the changing, described structures of the upper crown are very easy to recognize in the leafy state due to the also changing arrangement of the leaves (Runkel and Roloff 1985; Roloff 1986a, 2001, 2018a; VDI 1993). Figure 14 shows four European beech crowns of different vitality from a bird's eye view in a vitality level key for aerial photographs. The stereoscopic view and the resulting spatial viewing possibility of aerial photographs make differentiation even easier: the fanned, closed upper crown of a vigorous tree of VS 0 differs significantly from the bottle brush structures of VS 1. You can recognize a tree of the VS 2 by the noticeably loosened crown with cloud-like structures and the dissolved, skeletonized crowns with dead leader shoots of the VS 3.

With the help of aerial photo interpretation, large-scale vitality studies of entire urban districts or forest areas are possible. This can also help to find the causes of locally or regionally occurring problems with trees. In forests, vitality monitoring of trees in the leafy state (terrestrial and by aerial photographs) can be obstructed and difficult due to visual impairment of trees and crown parts, e.g., due to stratification

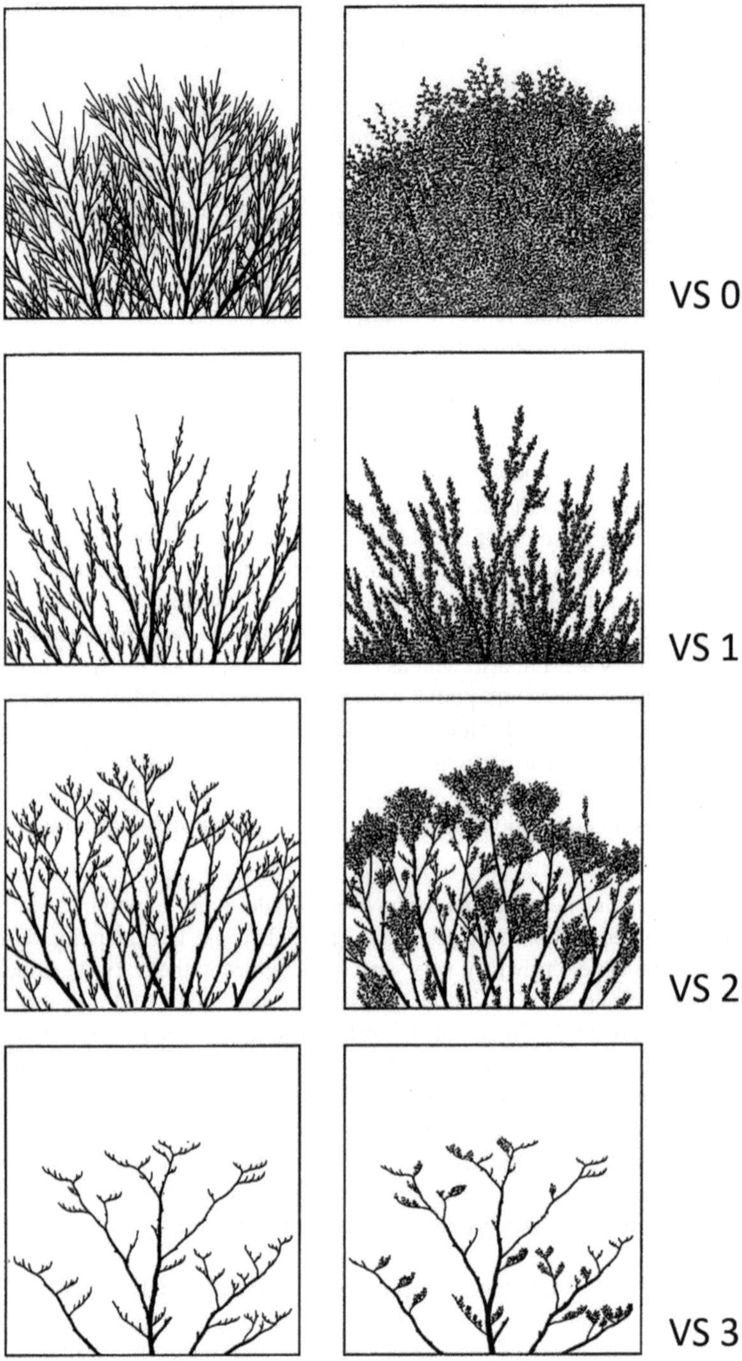


**Fig. 13** Four vitality stages (VS) 0, 1, 2 and 3 of European beech, based on the four different branching structures/modes in Fig. 12

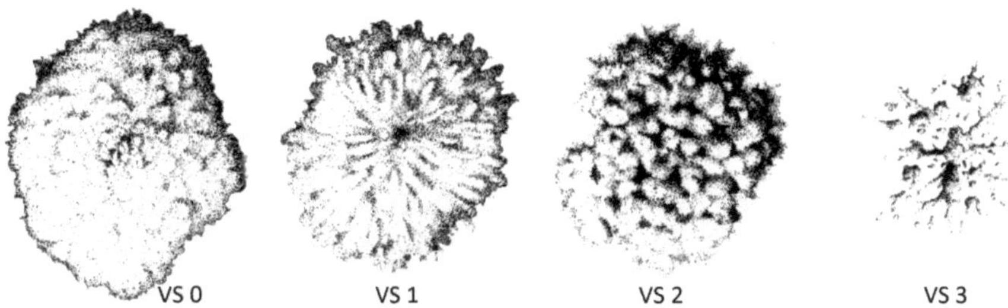


**Fig. 14** The four vitality stages (VS) in aerial view for European beech (from left to right: vitality stages VS 0–3)

of stand structures (Pretzsch et al. 2011). In urban areas, it is usually easier with many trees standing free/solitary, e.g., aside roads and in parks.

## 2.3 *Vitality Assessment in Deciduous and Coniferous Tree Species (Habilitation Thesis 1989)*

The method of vitality assessment by branching pattern then was transferred to and tested on other 20 tree species (including the tree species *Acer saccharum*, which is very important in Canada). Depending on the tree species, there are modifications, but they are only so moderate that the 4 vitality levels can be distinguished as well (Roloff 1989b, 1991b, 1992a, 1994, 1999, 2001, 2014b, 2018a). From this, a species-independent VS key was developed (Fig. 15), which can also be applied to other tree species in the temperate zone (e.g., in North America or temperate Asia). For conifer species, a fundamentally different approach is necessary due to their strongly different crown architecture (e.g., *Picea abies*, *Abies alba* and *Pseudotsuga menziesii*, see Roloff 2018a).

Although dead branches in the treetop area have an impact on safety risks of old trees, they do not have to be a sign of the imminent death of the entire tree but can also represent an adaptation process as a crown retreat. Weaker and coarser branch areas can be abandoned in the course of crown modification, and a secondary crown can develop deeper on the trunk (Fig. 16, Roloff 1989c, 2001, 2018a). The variety of processes and further development is variable and difficult to schematize or describe in fixed sub-steps. Long-lived tree species such as oaks and limes cannot grow continuously in height for over 500 years but have often reached their maximum tree height by the time they are 150–200 years old (see Fig. 60, Sect. 11).

Roots are a very important and interesting part of trees, but much of it is still unknown or uncertain (like a black box) because root investigations are laborious, time-consuming, challenging and complicated (Kozlowski et al. 1991; Kutschera and Lichtenegger 2002; Tyree and Zimmermann 2002; Larcher 2003). During excavations of 24 European beech trees about 3–4 m in size (age 10–31 years)

**Fig. 15** Species-independent vitality stage (VS) key for deciduous trees

**Fig. 16** Secondary crown of *Aesculus hippocastanum* 30 years after topping

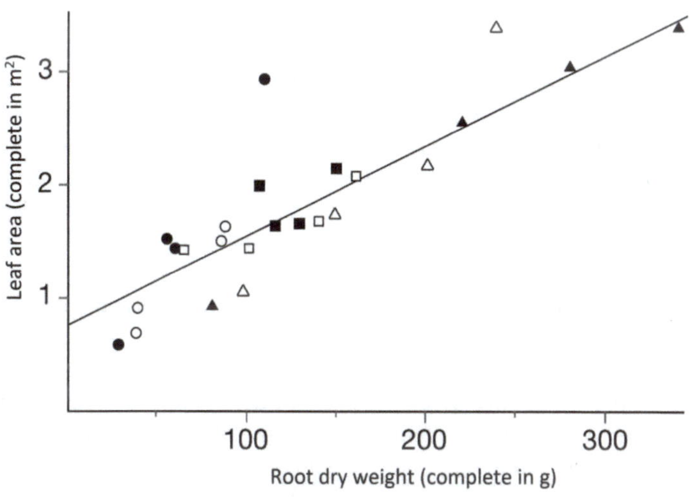


**Fig. 17** Relation between root biomass and leaf area in 24 completely excavated European beech trees (24 trees, height: 3–4 m, age 10–31 years; three symbol types: three different sites; black symbols: trees with syllepsis; white symbols: trees without syllepsis)

down to the fine roots, a close and significant relationship between all investigated 10 root and crown parameters was found (Roloff and Römer 1989; Roloff 1989c, 1991a, 2001, 2018a). This was particularly impressive between root biomass and leaf area (Fig. 17), so that based on this relationship, it can be assumed: if parts of one crown or root parameter are lost or damaged, other root or crown parameters will also be affected. Of course, this makes sense, since the roots supply the leafs with water and nutrients and, conversely, they are supplied with assimilates from the leaves.

Due to the continuing growth of the branches in the crown periphery, the areas entering the interior are increasingly shaded, so that the branches there finally die by lack of light. A process of branch removing takes place (Fig. 18a, b, Roloff 1986a, 2001, 2018a, 2022a), in the case of some tree species already in advance by so-called branch drops (Fig. 35 below). Dead branches will be attacked by fungi and break off sooner or later; this can take different time periods depending on the tree species. In most cases, branch removing on 5-year-old branches starts with the death of the lateral short shoots and is completed 10–15 years later with only the survival of the strongest side branches (Fig. 18a, b), if they are still connected to the outer crown and thus to the light due to their own growth.

## 3 Vegetation Science, Forest Floor Vegetation and Silviculture (Mainly 1989–2002)

*After the project in forest botany with the completion of my dissertation, the question of my further professional development was still open. I had good contact with silviculture, as forest botany is used as a basis there. Prof. Gisela Jahn, as a lecturer in forest vegetation science (AKS* 1980*; Fischer* 2003*), retired in 1988, and there was no successor. Therefore, I applied for the position and became her successor (as assistant professor), which was of course a significant reorientation.*

*Vegetation science was not new territory for me, as we carried out botanical identification exercises and excursions in forest botany and the use of herbaceous layer plants as indicators for the forest site interested and fascinated me very much from the beginning.*

Due to the great difficulties in determining forest floor grass and moss species, but their importance as efficient site indicators, an identification key for the 36 most important forest grass species was developed during my student days and published (Roloff 1984, Fig. 19), which quickly found its way into forestry practice and is still widely used today. The determination is made possible solely on the basis of the special features at the leaf basis connected with the stalk, i.e., a characteristic that can be used almost all year round in the open ground and regardless of flowering.

In 1987, an identification key for 21 forest moss species followed, which was based on special features of the growth habit and of culm, leaf and habitus characteristics directly at the site (Roloff 1986b).

During my time as assistant professor for forest vegetation science at the Institute of Silviculture at University of Goettingen, a book for the excursions and exercises in the field was created (Roloff and Peik 1990), in which the natural forest communities in the vicinity of Goettingen and interesting field examples are presented, with

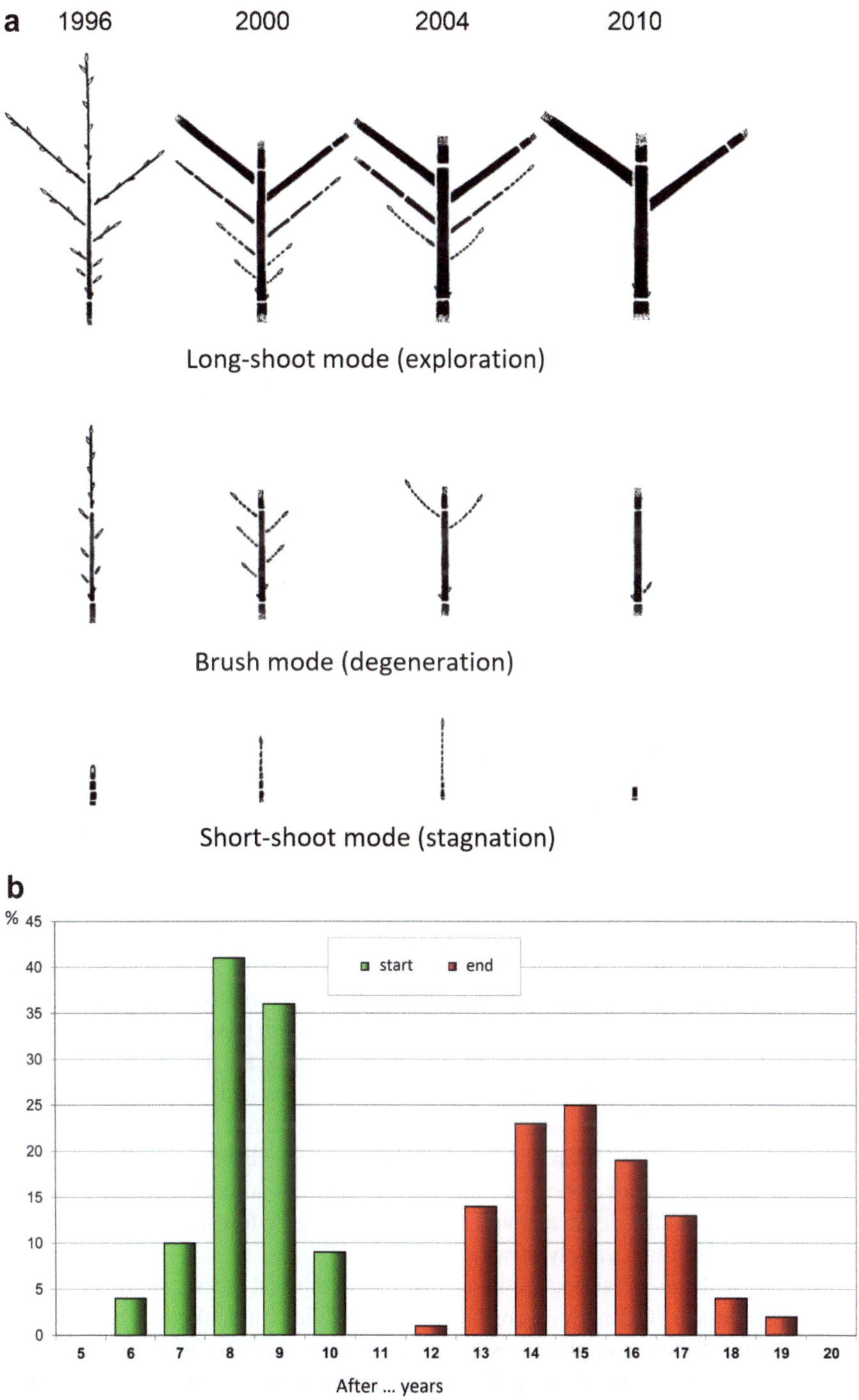


**Fig. 18** (**a**) Course of lateral twig removal: visual impression. (**b**) Course of lateral twig removal: diagram with starting age of the main twig and ending age

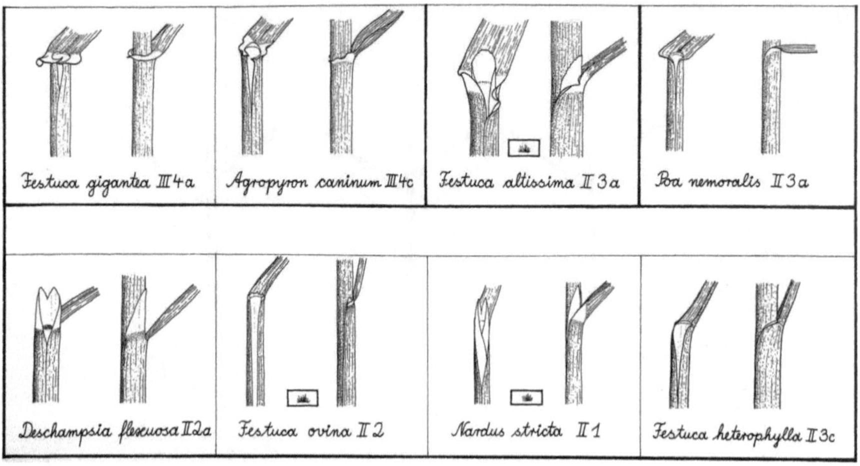


**Fig. 19** Impression of some details from the forest grass species identification key – transition area from stalk to leaf with ears and leaf membranes (tufts: caespitose grass species; numbers: site indicator value of geobotany)

precise location informations and species lists of the tree, shrub and herb layer. It also described the course of natural succession to pure European beech forests and for ten other natural forest types (Fig. 20).

For understandable reasons, biomass surveys in forests have so far mostly concentrated on the tree layer, but there are repercussions of the herb layer on tree growth, e.g., during drought due to water competition, which is known from *Pinus sylvestris* forests with the grass *Calamagrostis epigejos* (Hornschuch et al. 2008). So for 7 years, biomass surveys were carried out for a dissertation (Bolte 1999) on 129 sites with 40 cover-forming soil vegetation species (Fig. 21), and the following parameters were determined during the main development period of the plant species: above-ground biomass, degree of coverage, mean shoot length, element contents (N, K, Ca, Mg, P, S) of the above-ground plant organs. This means that all sites and forest vegetation units as well as forest ecosystem types of the north-eastern German lowlands have been recorded. From this, models for the above-ground dry matter estimation of 31 species or species combinations and 13 groups of similar growth form groups were derived via statistical calculations depending on the coverage ratio and the mean shoot length (Bolte 1999; Bolte et al. 2001, 2002). The estimated tables developed for more than 40 types of ground vegetation enable a rapid derivation of the dry matter and element stock of the forest floor flora and have proven to be used in a variety of ways in practice.

Growth and eco-physiological investigations in alluvial forests along the middle Elbe near Dessau (Germany) were intended to provide an ecological basis for the management of these complex forest ecosystems (Bonn and Roloff 2002; Roloff and Bonn 2002). The studies show that drought stress conditions can occur at irregular intervals on floodplain sites. In dry seasons, oaks consume more water than ash and

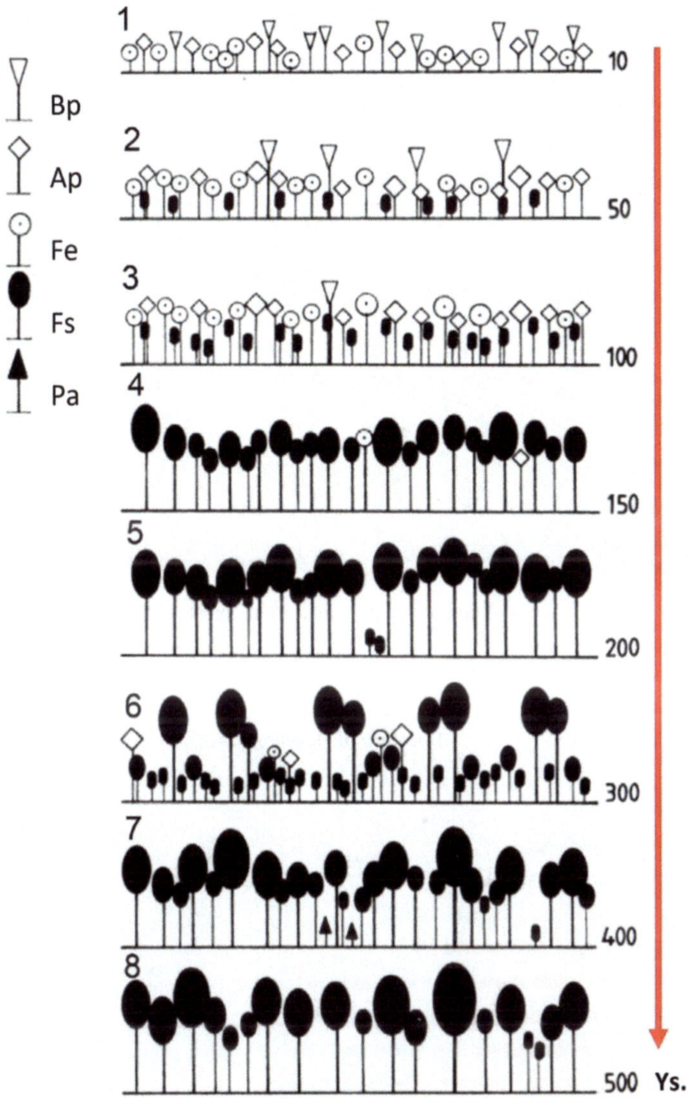


**Fig. 20** Natural succession from any initial state to a pure European beech stand over 500 years (symbols: Bp *Betula pendula*, Ap *Acer pseudoplatanus*, Fe *Fraxinus excelsior*, Fs *Fagus sylvatica*, Pa *Picea abies*)

elm. This can be attributed to their deeper root system, which allows them to take advantage of additional water supplies and therefore hardly experience drought stress. This can lead to competitive shifts between the tree species involved. In addition, the flood tolerance of the tree species also has an effect on this, and oaks also have advantages (Fig. 22). Oaks are expected to benefit most from the expected climate change and are therefore well adapted.

**Fig. 21** Forest floor cover-forming vegetation in a Pine forest with the grass *Calamagrostis epigejos*

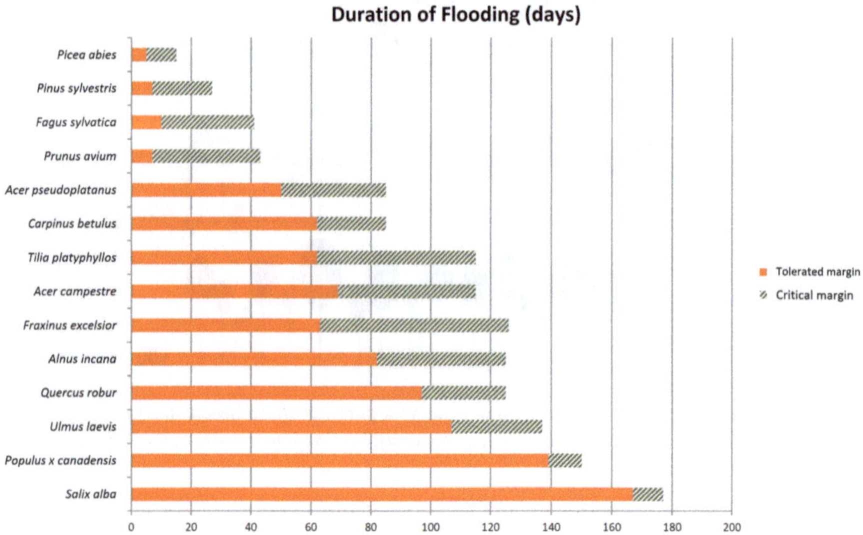


**Fig. 22** Flood tolerance of important riparian forest tree species (according to Späth (1988) and after own studies)

It is assumed that climate change will favour *Fagus sylvatica* to *Picea abies* at its northern range margins and induced disturbance events (Rust and Roloff 2009). An old-growth mixed forest of spruce and European beech, situated near the northern beech margin, was studied to reveal effects of disturbances and response processes on natural forest dynamics, focussing on the understory (Bolte et al. 2010, 2014; Grundmann et al. 2011). We carried out analyses on understory dynamics of European beech and spruce in relation to overstory release. This was done based on a sequence of stand and tree vitality inventories after a series of abiotic and biotic disturbances. It became apparent that European beech (understory) has a larger adaptive capacity to disturbance impacts and overstory release (68% standing volume loss) than spruce. Understory dynamics can play a key role for forest succession from spruce to European beech-dominated forests. Disturbances display an acceleration effect on forest succession in the face of climate change. European beech is poised strategically to replace spruce as the dominant tree species at the study area. Due to an increasing productivity and a lower risk of stand failure, European beech may raise into the focus of forestry in southern Sweden (Bolte et al. 2010, 2014; Grundmann et al. 2011).

*During this period on vegetation science, I have untiringly completed my forest botany habilitation thesis with vitality keys for 20 tree species, which were immediately included in the forest damage surveys. Then in 1990 my doctoral supervisor Horst Bartels retired, so I applied to succeed him and was selected. There I was able to intensively continue and expand my tree research and, of course, above all to make the vitality levels known through many publications and lectures, also abroad,* e.g., *in the UK and the USA as well as Canada. The professorship in Goettingen was associated with the leader position of the Forest Botanic Garden, which I was very enthusiastic about, as it was still relatively young in development.*

## 4 Management and Further Development of Tharandt Forest Botanic Garden, TU Dresden (1994–2023)

*This benefited me when I moved to Tharandt and the TU Dresden, because there the director position of the most beautiful and largest of all Forest Botanic Gardens is also linked to the Chair of Forest Botany. Although this was a big task in addition to the professorship and usually had the scope of another half additional position, it gave me great pleasure and was very important to me: an ingenious opportunity to implement research results from Forest Botany in the same way and to test topics at garden tours and events to see how they were grasped.*

*The many positive feedbacks from the scientific community, the political scene and the public (media interest),* e.g., *with the garden expansion to almost double the area to 35 hectares, were motivation to expand and intensify the very individual scientific path that I had taken. I just felt completely at home in Tharandt (it still is until today), although the change of location with my family to Tharandt in 1994 was*

**Fig. 23** Practical work in the Forest Experience Workshop of Tharandt Forest Botanic Garden

*of course also very exciting, especially with four schoolchildren, two of them in puberty – that of course required some effort.*

Garden guides were designed, written and printed for the Goettingen Forest Botanic Garden (Roloff et al. 1993) and the Tharandt Botanic Garden & Arboretum immediately after starting the duties there, to provide visitors with more information and background information and to enable them more experience and relaxation (Roloff and Pietzarka 1996, 2006).

At the beginning of TU Dresden garden's management period 1994, the so-called 'WaldErlebnisWerkstatt' (Forest Experience Workshop) was founded for environmental education (forest education, especially for children and young people) and has since been fully booked with events every year (Fig. 23). In the meantime, more than 100,000 participants have attended the courses, and you have to register up to 1 year in advance to get an appointment.

During the time of my management of Tharandt Forest Botanic Garden at TU Dresden (= Saxonian State Arboretum), targeted activities and a newly founded support association gave rise to the possibility of expanding the garden to almost twice the area, with a concept for the North American forest types. For this purpose, seed and fruit collection trips were carried out in North America and the woody plants were grown in Tharandt Forest Botanic Garden for a few years until they reached planting size (height 1 m). Previously, the expansion area had been purchased and prepared by creating the North American landscape with the Rocky Mountains, Appalachian Mountains, Great Lakes, prairie and Great Salt Lake in miniature. This has proven to be particularly attractive for visitors, the area was

**Fig. 24** Expansion area 'North American Forests' (Forest Park) with Indian summer in Tharandt Forest Botanic Garden, TU Dresden

opened in 2002 and named 'Tharandt ForestPark', since then it has been continuously planted with more and more trees and shrubs.

In fact, the 43 main forest types of North America have been established and are developing excellently: they now have grown into closed forests on several sub-areas. This 'Little North America' is absolutely unique in Europe and has become the largest Indian Summer area (Fig. 24), with the result that the complete number of the garden visitors per year has doubled, from about 80,000 to 150,000. Furthermore, the Tharandt ForestPark with an area of about 16 hectares has become one of the largest and most attractive outdoor laboratory areas for tree studies, with several hundred tree species growing in small stands (now 15–25 years old). For example, climate adaptation mechanisms can be investigated, such as the reactions of different tree species in height increment to dry years.

## 5 Tree-Ring Research (Mainly 1995–2016)

*Already in Goettingen, tree-ring research played an essential role in my research group (Worbes et al.* 1992, 2003*), and some of my Goettingen team followed me to TU Dresden, so that we were able to continue and expand this there (Grundmann et al.* 2011*; Gillner et al.* 2013b*). In addition, decisive sub-steps of water transport take place in the tree rings, and it takes on essential functions in the event of drought stress. Furthermore, the most important requirement there is that the water flow upwards does not get interrupted by embolism or cavitation (Tyree and Zimmermann* 2002*).*

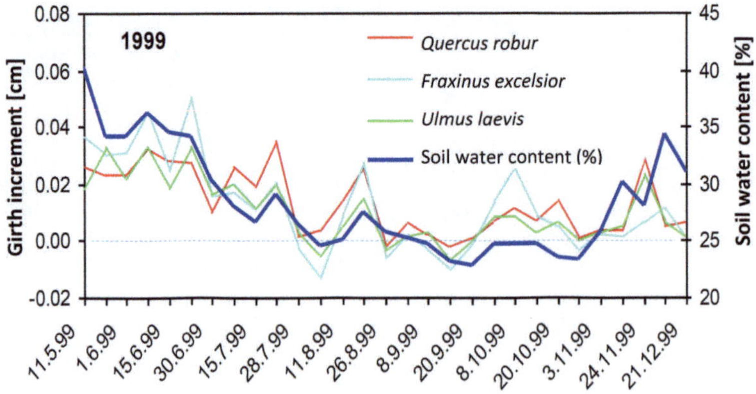


**Fig. 25** Shrinkage of the trunk girth (below line 0.00 cm) due to consumption of the trunk water reservoir (From Bonn and Roloff (2002))

*In stress reactions, the tree rings are always an important indicator for many past events (weather, pests, stress…, Gillner et al.* 2013a*). The basis for this work was always a well-equipped and supervised tree-ring measurement laboratory, which worked out excellently. In particular, the linking of tree rings and shoot length investigations under the influence of environmental effects and ageing processes has led to significant further advances in science (Liu et al.* 2022*).*

The trunk can act as a water reservoir in a number of tree species and thus have a vital function in times of drought stress (Gärtner 1995; Bonn and Roloff 2002; Roloff and Bonn 2002). This characteristic is particularly pronounced in some conifer species (e.g. spruce) and diffuse-porous deciduous tree species (e.g. hornbeam), as many older tree rings are still involved in water transport and storage. In dry periods, therefore, a measurable shrinkage of the trunk can occur if the trunk storage water is temporarily used to supply the leaves (Fig. 25). While it is initially possible to replenish this store overnight every day, this becomes more and more difficult with increasing drought stress and finally impossible after embolisms. For example, the trunk girth can decrease from one year to the next when the same tree is measured repeatedly. In the case of ring-porous trees, there is only a lower storage capacity, as only the outer, youngest annual rings can be used.

Since the antiquity man has cultivated olive trees in the Mediterranean Region. The Olive tree is one of the oldest cultivated plants on earth and can reach high age. A closer look into the data for the oldest olive trees of the world resp. Mediterranean region displays information about height, diameter and external girth. Measurement and gaining this information are quickly and easily accessible; however, they give not insight into the age, rather than information about the enormous dimensions of the trees. The information about the ages of more than 2,000 years is frequently to find for some individual – sometimes up to 5,000 years. However, this information is regularly indicated without any reference or method used for assessing the age such

**Fig. 26** One of the oldest olive trees, the age dated by the annual rings with about 1,120 years on Corfu (declared as nature monument)

as for the supposedly oldest olive tree in the world in Vouves on Crete (Cherubini et al. 2013).

With our study, the age of three olive trees on Corfu/Greece has been investigated with dendrochronological methods for the study (Gillner and Roloff 2016). The dimensions via diameter, height and girth of these trees are comparable to the oldest olive trees in the world (Fig. 26). Linear regressions were calculated on the basis of diameter, radii and the included number of tree rings. Results indicate 735 years for the youngest and 1,120 years for the oldest tree. So this is one of the two or three oldest olive trees of the world, which has been investigated for its age with scientific methods. That is why the tree has after our investigations been designated as nature monument and declared a world heritage.

In the case of very old trees with a trunk girth of more than 400 cm, the annual rings are usually no longer usable for age derivation, as these trees are (almost) always hollow on the inside, and therefore, their early life history can no longer be reconstructed from an increment drill or possibly from a trunk disc after cutting the tree. One could then only use the remaining outer trunk mantle, but extrapolations of the tree age from its annual rings of the last decades or its trunk girth are usually inadmissible or remain estimates with uncertainty range (Roloff 2021b). In very important cases, one can make tree ring measurements on younger and younger trees in the area around, but this is very time-consuming and still does not provide an exact age of the specific tree in question (see previously olive tree, Fig. 26).

If you take several trunk girth measurements in a short time on such bulging trunks of very old trees, as in Fig. 26, you will get different results several times and then can average them. Occasionally, the annual increase in girth is derived from circumference measurements at intervals of 1 year or several years (decades), and then a year-by-year age derivation is derived from this. However, this method is not permissible for the reason just mentioned above (shrinkage of the trunk, Fig. 25), also because the growth process can have changed very significantly over the course of decades or even centuries of tree life. Reasons for this can be temporary competition and shading effects by neighbouring trees, pruning and maintenance measures, changes in local/regional climate, and there is also an age-related growth curve with stronger growth in youth up to a maximum in the 'early average' tree life and then gradual decline again. In managed forest stands, thinning can have an impact.

This will be described here again using an example of an ancient oak. There were current annual circumference measurements for this tree as well as older ones from 10 and 50 years ago. From this, an age of 700 years was then derived (not by us), which is about twice as high as in our own derivation from location, tree shape and competitive situation (oak as a very light-demanding tree species). Thus, the only way to estimate the coarse age of very old trees is often the following (Roloff 2021b): using today's tree shape/sculpture, looking for old photos (there are no earlier than about 1900), paintings (usually from about 1800 or rarely 1500 at the earliest), on which the already impressively large tree can be seen, or old descriptions of the younger tree, e.g. in documents (in extremely rare cases back to 1300, if they were court, thing, or boundary trees).

Unfortunately, over 500 years ago, no one wrote down when a tree was planted with completing a church or castle building – so such a thing can only be assumed. The age of the tree is then benevolently deduced quite often as follows: if the church was built in 1325 and there is a thick lime tree next to it, it is also 700 years old today. However, this remains mere speculation, because it may have been planted much later – e.g. for the construction of the successor church in 1525 and would therefore 'only' be 500 years old. Or it could also be a 400-year-old re-sprouting of the original lime tree, which had died at the time (see Fig. 27 and cf. Schweingruber et al. 2006).

There are many impressive stories in Germany about the age of thick trees with rounding up to preferably 1,000 years from 700, 500 or even 400 years (Roloff 2021b), especially when it comes to the oldest tree in Germany, which still has a single trunk or at least some original trunk remains are still connected to each other at their base. Unfortunately, during closer inspections of many '1,000-year-old' oaks, limes or yews, the age of none of the trees could be proven to be 1,000 years – so today there is no reliably documented tree over 1,000 years old in Germany (Fig. 27). In England, however, according to own enquiries and survey, there are over 100 ancient trees that are more than 1,000 years old. The reasons for this will follow in Sects. 11 and 12. The world's oldest tree is a *Pinus longaeva* tree with around 5,000 years of age in the Great Basin mountains of California (Nevada/ USA), 3,500 m a.s.l. (Schütt and Lang 1996).

**Fig. 27** Supposedly over 1,000-year-old lime tree, whose current four regrowth stems are only max. 200 years old without any remaining parts of the original trunk

# 6 Scientific Species Determination of Over 2000 Central European Woody Plants (Mainly 1995–2018)

*Part of my daily job as forest botanist was and is species identification. This has always given me a lot of pleasure and was very important to me (also without leaves during winter, only with buds). But even when I was a student, I was very surprised and annoyed that I couldn't get along with the only identification book available at that time (*Fitschen 1977*), because it contained a lot of text and only a few illustrations, so that you couldn't compare the leaves outside. As early as 1984, the desire arose to make my own contribution with my skills for drawings. However, it took more than 10 years until the new reference book for the identification of over 2,000 woody species was completed (*Roloff *and* Bärtels 1996*: first edition, today in fifth edition 2018).*

*My specification was to create at least one typical leaf drawing for each tree and shrub species, and for special important other characteristics also further detailed illustrations, so that in some cases, up to ten details per species had to be shown. At times it was a very hard job, but I was also allowed to do this at home in the evening and during weekends at the large kitchen table. As a result, all our four children learned to draw fantastically (not only leaves and trees), because at times they liked to sit with me at the table drawing on their own along on used paper sheets. So by this I was able to spend a lot of time with them, and it was also a good opportunity to talk to each other...*

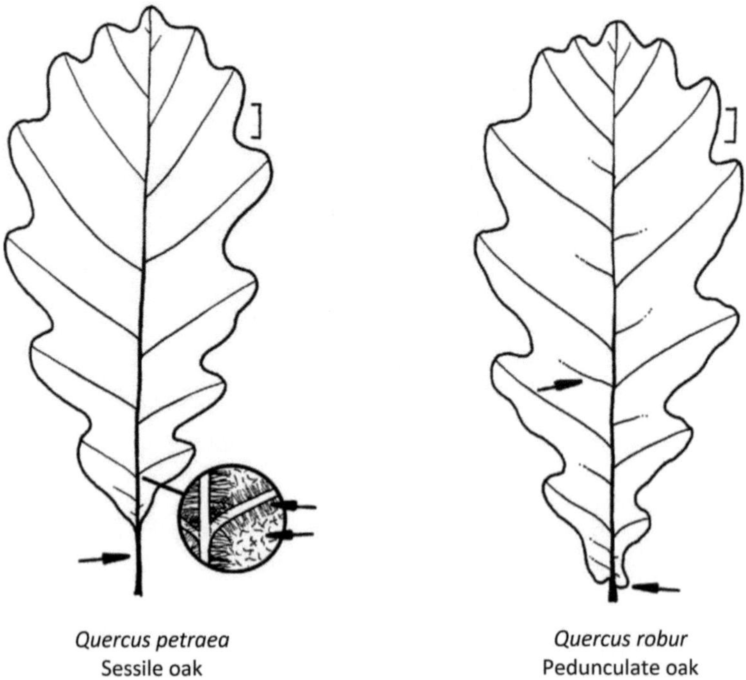


**Fig. 28** Example of leaf illustrations for *Quercus petraea* (left) and *Quercus robur* (right) for the oak determination key. (From Roloff and Bärtels (2018))

In an extensive work over a period of 10 years (1986–95), identification keys were developed for more than 2,000 native and introduced non-native woody plants (trees and shrubs), and the most important biological, ecological, aesthetic and use characteristics were compiled for each species (first edition 1996: 695 pp.). For each species of woody plants, there are detailed drawings with identification of the decisive species characteristics (Fig. 28). It has become the only standard work on determination of trees and shrubs in Central Europe (in German) with these quality characteristics and is now published in its fifth edition (Roloff and Bärtels 2018: 912 pp.). Both newly added species because of global warming or renaming processes were always taken into account.

Digital identification keys with photo support have improved significantly since then but still have major problems with the variability of characteristics (on the photo of one single leaf as a basis, only one characteristic expression occurs) and with the enormous number of species (more than 2,500 woody species). However, further progress through AI can be expected – there are already some systems available for smartphone use: you take a photo of a twig and then the species name is shown. But the results of identification are often wrong and it is difficult (and so far not possible) for more than 500 different tree species.

**Fig. 29** Two different leaf types of *Quercus robur* from the same crown of an 800-year-old oak tree on two different very old branches: upper row pedunculate oak leaves (middle north side of the crown), lower row sessile oak leaves (upper south side of the same crown)

In 1995, for the first time and several times thereafter, we published the appraisal that *Quercus robur* and *Quercus petraea* are not separate tree species, but two ecotypes of the same oak species, with decisive consequences for their use (Kleinschmit et al. 1995; Kleinschmit and Roloff 1998; Roloff 2021a, d). This view is reinforced by our own current findings, as both leaf types have been found on an approximately 800-year-old *Quercus robur* tree: sessile oak leaves on the upper south side of the crown on an approximately 300-year-old branch and pedunculate oak leaves on the north side of the crown of the same tree on a 400-year-old branch. This confirms for us the hypothesis that they are adaptation types (to drought) of the same original leaf type of this tree species (Fig. 29). Due to the hybridization ability of *Q. robur* and *Q. petraea*, their different ecological requirements and the lack of disjoint characteristics of traits between the two types, it can be assumed that there is only one biological species and sessile oak and pedunculate oak are to be interpreted as their ecotypes *Quercus robur* ssp. *robur* (pedunculate, common or English oak) and *Qu. robur* ssp. *sessiliflora* (sessile oak).

For the 'Tree of the Year' (now 36 tree species from 1989 to 2025), which have been proclaimed annually in Germany since 1989 (Fig. 30), the official detailed descriptions of their properties, biology, ecology and use were published for more

**Fig. 30** Proclamation of the Tree of the Year 2010, the wild cherry, in Berlin (Germany)

than 20 years (1996–2016) (www.baum-des-jahres.de, Roloff 1998, 2006, 2015, 2017a; Roloff and Pietzarka 2000). This significantly improved knowledge about these tree species and made the public more aware of their importance.

Another special determination key has been developed for ten species of the genus *Salix* (willows, Fig. 31), which are important but difficult to identify, in order to better distinguish these woody plants for the practice of engineering biology (Dachsel et al. 2018). Together with the description of their characteristics, most of these ten willow species can be used for near-natural bank reinforcement along streams and rivers: they are planted there and thus imitate the natural riparian vegetation if that has been missing so far. Engineering biology is a very important, special and interesting field for application of tree biology, which is based on natural processes of succession in alluvial forests.

## 7 Tree Species Monographs (Editor in Chief Encyclopaedia of Woody Plants, Mainly 1997–2021)

*With the enormous biodiversity of woody plants and many own experiences on trips, hikes, bike tours, the desire grew to describe individual monographs of some species more in detail,* i.e. *an original dendrology work. Together with various co-authors and the many lectures, this resulted in almost 20 such monographs, and I became familiar with the Encyclopaedia of Woody Plants (Roloff et al.* 2021b*). I was soon (from 1997) asked to work on it as co-editor, and eventually became editor-in-chief.*

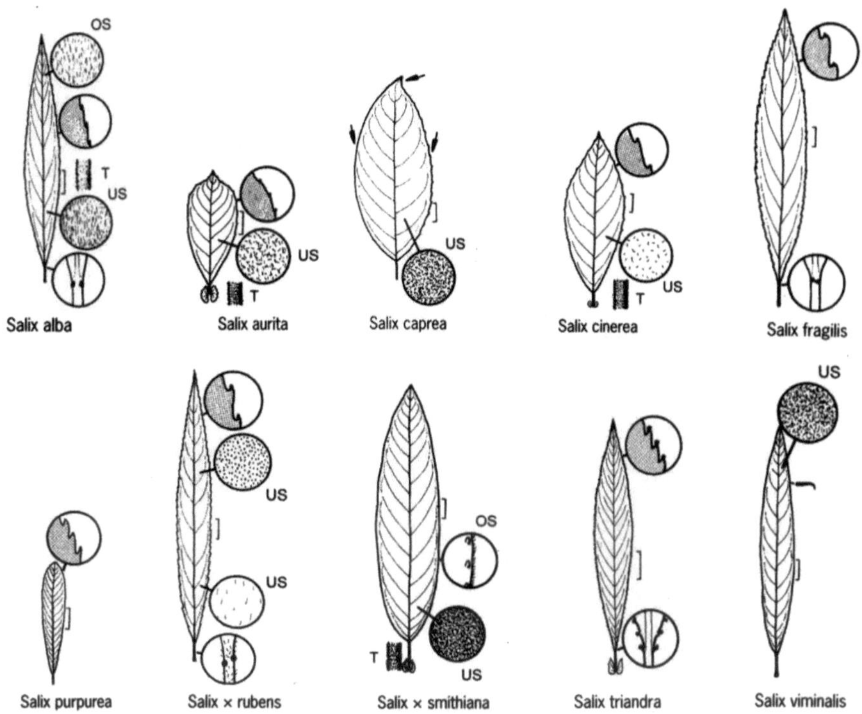


**Fig. 31** Willow determination key for engineering biology (ten important *Salix* species; abbreviations: OS upside, US underside, T shoot)

According to detailed investigations in mixed young *Fraxinus excelsior* and *Fagus sylvatica* stands, the root growth of common ash trees in dense natural regeneration is so intensive and so strongly concentrated on the upper soil horizons (upper 15 cm) that the roots of the European beech trees associated with them are displaced into deeper soil horizons (Fig. 32, Rysavy and Roloff 1994; Roloff et al. 2021a). As a result, the young European beech trees suffer drought damage and eventually die in summer droughts, when the plants have to cover their necessary water from the scant rainfall by showers/thunderstorms and there is not any more infiltration into deeper soil layers – this phenomenon is called 'ashing'. Pure common ash stands that are undesirable for forestry purposes then develop. This can only be prevented if a sufficient canopy by old trees is maintained above the regeneration, which inhibits the growth of the common ash tree and thus also its root growth by shading to such an extent that the European beech can survive.

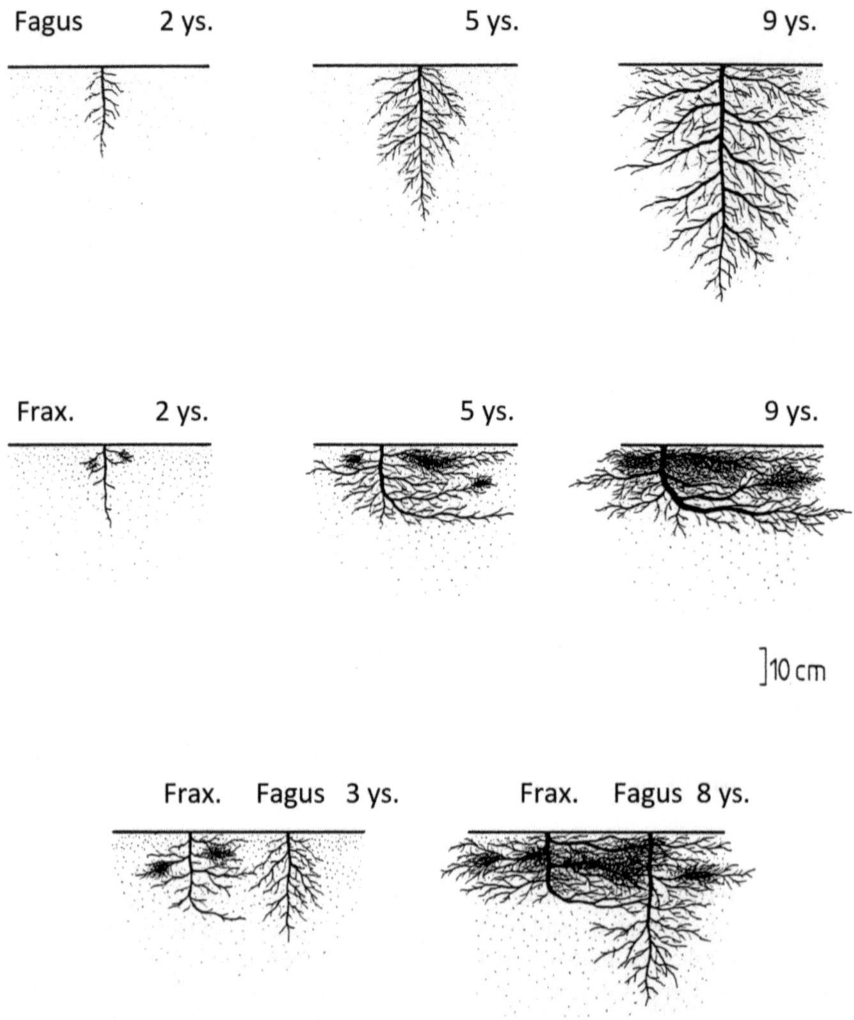


**Fig. 32** Root development of young *Fagus sylvatica* trees (upper row, in pure stand, with age), young *Fraxinus excelsior* trees (middle row, in pure stand, with age) and in a mixed stand of both tree species with root competetion (bottom row, with age) with displacement of the European beech downwards. (From Rysavy and Roloff (1994))

# 8 Ecophysiology – Tree Water Balance and Drought Stress (Mainly 1998–2022)

*During the 1990s, hardly anyone talked about climate change or did any research on it, but we did and published on it even then (Roloff* 1992b*). It quickly became clear that drought stress would become the main requirement for trees in the future (which has now occurred), so that this also became and still is one of our physiology* *focuses (Roloff* 2021b*). On the one hand, this is about how already established older trees cope with such drought periods, but on the other hand, it is also about which tree species – that have received less attention or are known to date – may be able to cope better with them. These are mainly the rarer, sometimes smaller and therefore often overlooked native tree species such as Acer campestre, Malus sylvestris, Prunus avium, Pyrus pyraster Sorbus aria, S. domestica, S. torminalis, but also other non-native tree species.*

There is a lot of publications on ecophysiology and drought stress problems and adaptation mechanisms in trees with background information for science and practice: some fundamental books are Kozlowski et al. (1991), Tyree and Zimmermann (2002), Larcher (2003). In the following, we report the main recent and interesting findings of our own investigations (summarized in Roloff 2021b).

In a 3-year desiccation experiment with 40-year-old spruce trees in the Ore Mountains (Germany), the trees of the trial area were completely cut off from the water supply via precipitation and lateral soil water (Nebe et al. 1998; Rothe and Roloff 1998; Vogel et al. 1998; Rothe et al. 1999, 2002; Roloff et al. 2000; Rothe 2005). Outside the area, trees were measured as a control without intervention. Then the physiological reactions were followed for 3 years (Fig. 33): in the third year, the

**Fig. 33** Investigations during the desiccation experiment of spruce trees with the support of a computer tomography specially converted for tree investigations (for recording the sapwood conductive area)

saturation moisture of the soil was still 22%, and the permanent wilting point was undercut in the summer of the second year. After that, no more sap flow in the trunk was measurable after the trunk water reservoir was also used up.

The shoot growth of the treated trees was mainly stopped in the third year; there were only very short annual shoots on individual branches. Photosynthesis and transpiration largely came to a standstill but at times remained at a low level. Surprisingly, the treated trees showed no signs of dieback until the end of the third year. This could only be explained by water absorption via the needles (from precipitation, dew and fog absorption); in experiments, this could be confirmed on spruce branches (see following paragraph).

When trees are under drought stress because no rain has fallen for a long time and the soil has dried out, it happens that the air is more humid than the soil. If dew falls at night or a shower during thunderstorms, leaves can also absorb water (in small quantities) directly through their surface and thus temporarily supply their leaves (Kramer and Boyer 1995; Roloff and Rothe 2000). This direct leaf water absorption usually plays a subordinate role in terms of quantity (estimated 3–5% of the annual water demand), but during drought stress periods maybe up to 80% per day. So it can be of considerable importance for the tree's survival in times of extreme drought stress. It works particularly well with conifers (studied on *Pinus*, *Picea pine*, *Sequoia*), as they can comb out fog from the air better than deciduous trees (Fig. 34). In extreme situations, drought stress situations can even lead to water transport backwards (downwards) in the branches and in young trees also in the trunk downwards to the roots (Herbst et al. 2002).

**Fig. 34** Dew droplets on spruce needles with the possibility of needle water absorption directly through the surface

**Fig. 35** Branch drops (cladoptosis) of common oak under an old tree in a drought period

The phenomenon of branch drops (cladoptosis) is caused by the active shedding of branches with an abscission zone, especially on *Quercus* (Fig. 35), *Salix* and *Populus* species. It has been studied extensively and continuously on oak (Klugmann and Roloff 1999; Roloff and Klugmann 2000; Rust et al. 2000, 2004; Roloff 2001; Roloff et al. 2001; Rust and Roloff 2002, 2004) to better understand what leads to this phenomenon and what consequences it has for the tree. It became clear that it is primarily a reaction and adaptation to drought stress to rapidly reduce the leaf area (accumulation of branch drops during drought periods) and improve the light conditions inside the crown for these light demanding tree species. It is interesting that only side branches can be separated, for anatomical reasons of the abscission zone, and only branches that are one to a maximum of 8 years old can be shed.

Individual dead branches in the crown (e.g. above a road, Fig. 36) cannot have their cause in a dead main root just below (e.g. under the road) because there is no direct relationship between individual branches and individual roots. Rather, dye experiments showed (Schmidt and Roloff 2000, 2001; Roloff 2022a) that the water absorbed by a root is already distributed around the trunk girth at a short trunk height and, conversely, a branch also supplies many roots with assimilates. If there was such a direct relationship between main roots and main branches, often a different branch would have to die because of the many twisted trunks of trees.

Mistletoe infestation has been increasing in importance for about 20 years, as mistletoe is becoming more competitive – among other things due to warm winters (as it is an evergreen shrub species), weakening of host trees (due to drought stress

**Fig. 36** Dead branch on a Norway maple above the road (see red arrow) – cannot be the consequence of a dead root under the road

and pruning measures) and due to more frequent overwintering of mistletoe-spreading birds (decreasing bird migration, Newton 2020, 2023; Rappole 2022). From our extensive investigations (Grundmann et al. 2012; Roloff et al. 2012, 2013, Fig. 37), recommendations for action were derived as to whether and when mistletoe should be combated or whether and when it should/can be allowed to grow. Since the removal of mistletoe from older trees is very labour-intensive and therefore expensive, sensible and effective measures can be proposed: e.g. up to 10 mistletoes per tree are unproblematic, for significantly more there is a need for action. This is because in water shortage situations, host tree branches occasionally die, as mistletoe, with its stronger water tension, preferentially diverts the water from the branch to itself (Fig. 37), thus increasing drought stress for the continuing tree branch.

Trees have several ways of adaptation to periods of drought and heat (Larcher 2003): high resilience through origin-related genetic adaptations (e.g. sclerophylly, silvery undersides, Fig. 38), through medium-term adaptations that are expressed not until the following year (e.g. smaller leaves with smaller but more stomata, short shoots), and acclimations, i.e. immediate responses to drought stress (e.g. lowering the leaf water potential). Adaptation strategies on all three levels are presented for urban climate tree species from the long-term study 'Urban Green 2021' (Böll et al. 2022) and our own investigations (Meyer et al. 2005; Gillner and Roloff 2013; Gillner et al. 2014, 2015a, 2017; Roloff 2022a, b).

The investigations of the 3-year field experiment 'AdapTree' (Fig. 39, Dreßler et al. 2017; Roloff 2021b) on 90 1- to 2-year-old trees of five urban tree species (*Acer campestre*, *Populus nigra* 'Italica', *Sorbus aucuparia*, *Styphnolobium japonicum*,

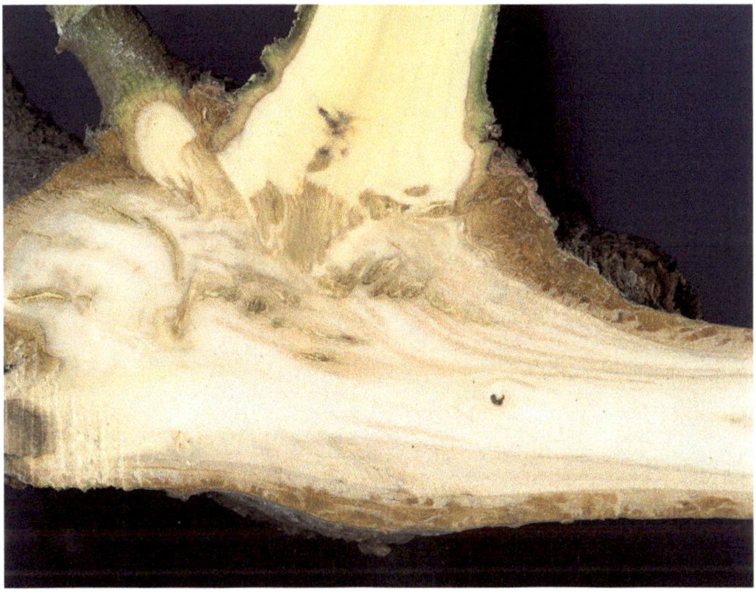

**Fig. 37** Longitudinal section through the branch of a hybrid poplar with mistletoe (above: mistletoe, below: the transition area between mistletoe-xylem and host xylem, at bottom: branch of the poplar)

**Fig. 38** Silver lime tree (*Tilia tomentosa*) with silvery undersides of the leaves (due to dead hair felt) turned outwards during heat periods to reduce leaf heating

**Fig. 39** Experimental area for the drought stress project 'AdapTree' with five young urban tree species

*Tilia platyphyllos*) have shown that a certain degree of adaptation of young woody plants to drought is possible even in a short period of time. From this, it can be deduced that intermittent watering of the young trees at longer intervals would be the optimal option for three of the tree species (variant with moderate drought stress and more intensive watering at monthly intervals). It is precisely this variant that corresponds to the expected climate scenarios. These results show some positive effects of hardening. Depending on the tree species, however, the two irrigation variants have to be considered even more differentiated (Dreßler et al. 2017).

Enduring shade is one of the most important characteristics of tree species that grow naturally in forest stands (which is the case with almost all tree species by nature). Only in this way can natural regeneration survive under the forest canopy of old trees and establish the next generation of stands. Pioneer tree species lack the characteristic of shade tolerance, with the result that they usually colonize the site only once and temporarily by primary succession. In most park and avenue trees, the shade phase is usually eliminated/bypassed by the spacing of the trees. But in dense avenues, old tree groups/forest stands as well as in parks and urban forests, it is an important issue (Table 1). Shade tolerance can be achieved at the leaf level (light/shadow leaf differentiation, compensation point of photosynthesis) or/and by the crown architecture (distichous leaf positioning and horizontal direction of growth with erection of top shoots in light).

Shade tree species are usually also climax tree species, e.g. *Fagus sylvatica*, *Taxus baccata*, *Tsuga heterophylla*. Some tree species (e.g. *Fraxinus excelsior*, *Acer platanoides*, *A. pseudoplatanus*) try to deal with a shady situation by growing

**Table 1** Limit values for shade tolerance of important tree species in youth, as % of outdoor radiation (according to Lerch 1991, Lyr et al. 1992, Matyssek et al. 2010, Roloff 2019a, b and own surveys)

| | % of outdoor radiation |
|---|---|
| *Betula pendula* (white birch) | 10–15 |
| *Larix decidua* (European larch) | 10–20 |
| *Robinia pseudoacacia* (black locust) | 10–12 |
| *Populus tremula* (trembling aspen) | 9–12 |
| *Pinus sylvestris* (Scots pine) | 8–12 |
| *Sorbus aucuparia* (rowan berry) | 4–12 |
| *Alnus glutinosa* (black alder) | 7–9 |
| *Quercus robur* (pedunculate oak) | 2–5 |
| *Picea abies* (Norway spruce) | 2–4 |
| *Fraxinus excelsior* (common ash) | 1–3 |
| *Acer pseudoplatanus* (sycamore) | 1–3 |
| *Ulmus laevis* (European white elm) | 1–3 |
| *Ilex aquifolium* (common holly) | 1–2 |
| *Carpinus betulus* (common hornbeam) | 1–2 |
| *Pseudotsuga menziesii* (Doug fir) | 1–2 |
| *Tilia cordata* (small-leaved lime) | 1–2 |
| *Abies alba* (European silver fir) | 0.5–1 |
| *Buxus sempervirens* (common box) | 0.5–1 |
| *Fagus sylvatica* (European beech) | 0.5 |
| *Taxus baccata* (common yew) | 0.5 |

upwards as quickly as possible and neglecting branching (Roloff 2019a, b, 2022a). However, this can only work if the trees above are not 30 m high, because the young tree cannot make it there and the 'Oskar syndrome' occurs (Fig. 40): the trees stagnate in growth with a height of 1 m (like Oskar in Günter Grass's book 'Tin Drum') and finally die. On the other hand, this strategy is very successful in high herbaceous biotopes in alluvial forests and in hedges, shrub communities or crevices/cracks in the wall, where the tall tree can reach the light after a few metres vertical growth. The European beech tree's strategy of waiting for more light with short shoots is completely different and in other ecological situations successful ('waiting stage', see Sect. 2.1). This is also much seen in tropical rainforests: longevity of seedlings waiting for a gap or clearing by storm-fall or human foresting (Hallé et al. 1978).

In 2010, a 'climate forest' with ten tree species (2,000 trees) was created near Cologne (Germany), which were selected according to our assessments of the tree species with regard to drought stress adaptation (Climate-Species-Matrix, see Sect. 9). After 10 years, intensive investigations were carried out on the growth development (diagrams top shoot lengths: Fig. 41a and tree ring widths: Fig. 41b) of the remaining 1,342 trees with interesting results (Forker et al. 2020; Forker and Roloff 2023). One conclusion from this study is that young trees are well adapted in

**Fig. 40** 'Oskar syndrome' – stagnating height growth of shaded Norway maple and sycamore regeneration

their growth to survive climatic events that deviate from average, such as the dry summers of 2016 and 2018. Of the selected tree species, sweet chestnut stood out in particular, which had its strongest shoot length increment in 2018 (average 140 cm, Fig. 41a).

In their radial growth (Fig. 41b), the tree species react to dry years 2016 and 2018 after different times. Some reduce their growth in the same year, but some of the tree species studied did not react to the summer drought stress until the following year. A significantly lower increment than in the previous year was recorded for *Abies grandis* and *Juglans regia*, *Prunus avium* and, to a lesser extent, *Sorbus aria*. The growth of *Betula pendula* corresponded to the previous year. The delayed reaction of the tree species *Quercus pubescens*, *Castanea sativa*, *Sorbus torminalis* and *Juglans regia* is noticeable not only in the reduced growth of the previous and following year but also in the larger tree-ring widths in the dry year – this increasing growth, however, results from the 'normal' previous year.

There are published in Botany textbooks (e.g. Lüttge et al. 2010; Evert and Eichhorn 2012; Lüttge 2017; Kadereit et al. 2021) over 40 different short-, medium- and long-term adaptation mechanisms to drought stress, of which the 10 most important ones were selected and tested to see how often they are present in almost 100 drought stress-tolerant tree species. Seven of these ten characteristics have proven to be particularly meaningful, in this ranking (Roloff 2021b, 2022b):

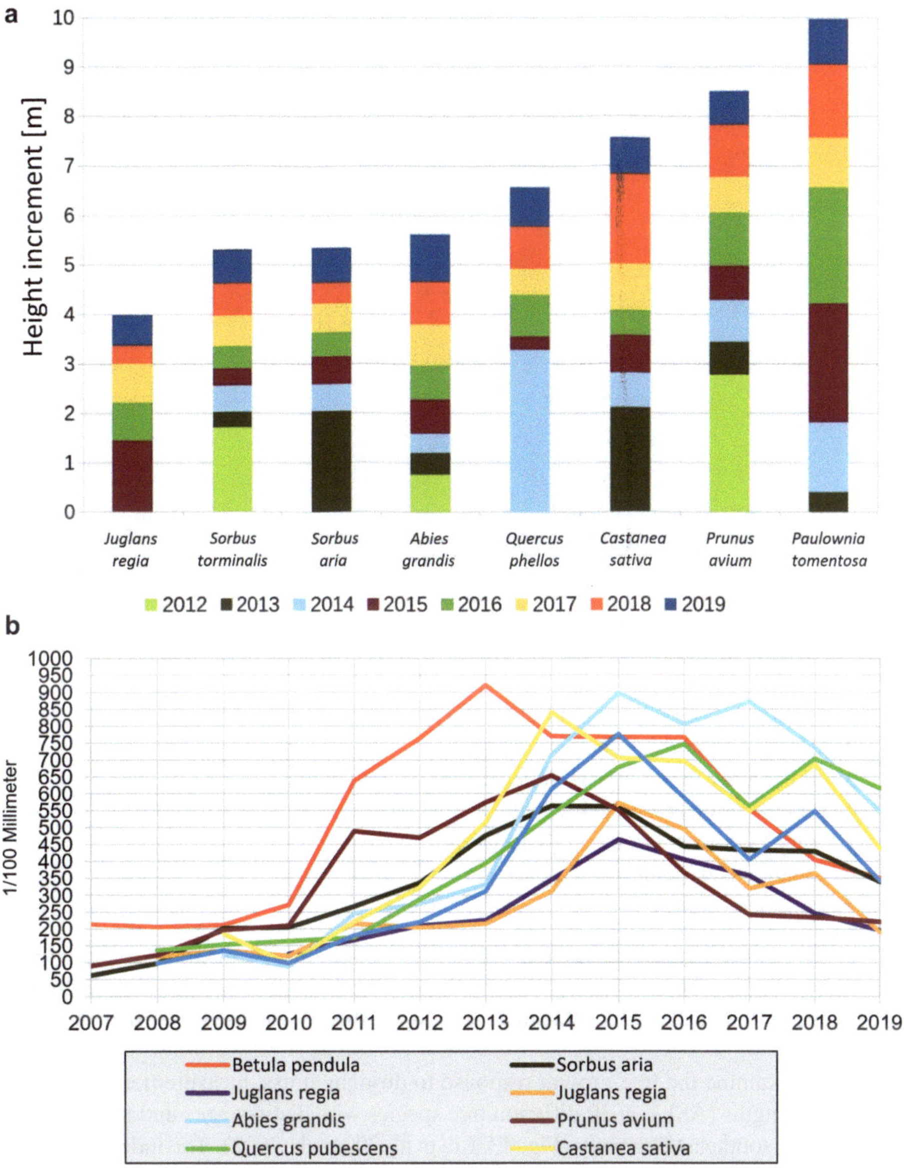


**Fig. 41** (**a**) Annual treetop length growth of eight tree species in the Climate Forest Cologne during the period 2012–2019. (**b**) Tree-ring widths of eight tree species in the Climate Forest Cologne 2009–2019 (From Forker and Roloff (2023))

**Fig. 42** *Styphnolobium japonicum*, a tree species with five of the seven most important drought stress tolerance indicators

1. Natural occurrence on natural sites with frequent water shortage
2. Glossy and/or thick-leathery leaves
3. Bluish/grey or silvery/white leaves on the underside
4. Pinnate or clearly lobed leaves
5. Very high frost tolerance (below −30°C)
6. Small leaves (length < 10 cm)
7. Presence of thorns

From this, it can be deduced for little-known tree species: the more of these seven characteristics they show (Fig. 42), the better should be their adaptability to drought stress, and especially if there are several of these characteristics with a higher ranking.

To examine the tree growth response to drought stress, measurements of annual shoot lengths (ASL) on 30 different tree species with distinct ages and habitats were taken in southeast Germany (Fig. 43, Liu et al. 2021a, b, 2022). The index of relative ASL increase, which is a standardized ASL increase, was calculated to indicate the growth rate of various tree species between the humid years (2015–2017) and the drought period (2018–2020). The results showed the ASLs of most early-young trees (<10 years) significantly showed a decreasing tendency in the drought period, whereas the middle-young (11–15 years) and late-young trees (>16 years) presented a relatively stable trend.

The height increment of early-young road-side trees studied (Fig. 43) shows a differentiated picture: while young *Fraxinus excelsior* and *Quercus robur* were able

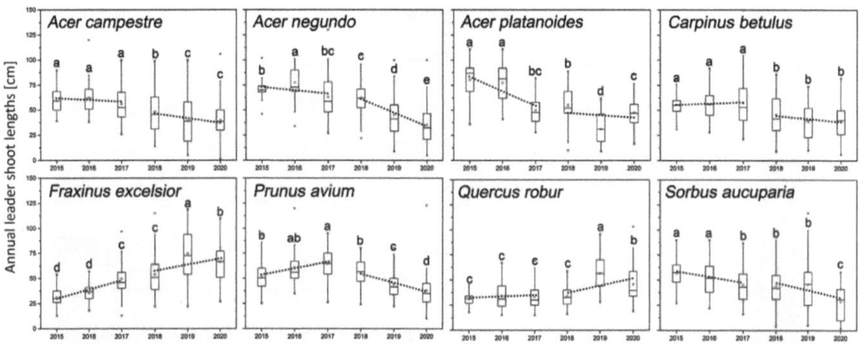


**Fig. 43** Shoot length measurements of the 6 years 2015–2020 on eight tree species (50 trees each) at road-side locations – 2018–2020 was an extreme drought period. (From Liu et al. (2022)). The determined growth values and their distribution are displayed as box plots; each box shows the middle 50% of the data, the line that cuts across the box indicates the median; the 'antennas' at top and bottom describe the scattering of the rest of the data; points outside the antennas are outliers. Different letters show significant differences between the annual increments

to increase their growth during the drought period, significant declines in shoot length growth compared to previous years can be observed for *Acer campestre*, *A. negundo*, *A. platanoides*, *Carpinus betulus*, *Prunus avium*. *Sorbus aucuparia* develops indifferently. However, the box plots also show that the dispersion of the values for almost all tree species increases significantly during the study period – i.e. the individual trees of a tree species behave increasingly inconsistently. ASL measurement can be recommended to be an easy, fast and effective method to detect tree growth rate and its response to environmental stress (Liu et al. 2021b, 2022; Roloff 2021b).

If buds remain closed longer than next spring, they are called dormant buds, also known as preventive buds. They are envisaged for unexpected situations such as pruning or sudden release, because the tree cannot react sufficiently and quickly enough with its 'regular' buds alone. In this respect, dormant buds are of considerable importance for the reaction possibilities and the adaptation potential, i.e. the survival of long-lived trees as stationary organisms (Fig. 44, Roloff 1986a, 2001, 2022a). The lifespan of dormant buds can be a few years; in extreme cases, dormant buds over 100 years old have been found in the bark cracks of old oaks. These buds had to be kept alive for over 100 years and also grow a bit in the bark outwards every year (as the trunk gets thicker every year), otherwise they would have been enclosed in the wood of the trunk in the meantime and lost touch with the light. Dormant buds are also the most important starting point for reiterations (see paragraph below) and for reactions to pruning measures.

When branches occur at locations in the crown and at times that are unpredictable according to the architectural model of the tree species, they are called reiterations (Hallé et al. 1978; Oldeman 2014). They can arise from dormant buds (see previous paragraph) or newly formed adventitious buds and are of enormous importance for

**Fig. 44** Dormant bud on the shoot of a hornbeam in summer (centre)

the responsiveness of trees to changes in their environment and for adaptation (Roloff 1989c, 2001, 2022b). The cause can be injuries (so-called traumatic reiterations, e.g. due to pruning, browsing, late frost, construction work, etc., Fig. 45) or significant improvements in the site situation (adaptive reiterations, e.g. by creating more light or rooting space). Many different types can be found on almost all trees, only some palm species are not able to do so and accordingly have considerable adaptation problems to changes or after damage (see Fig. 52).

The diversity of reiteration types and their frequency were examined and evaluated in detail for different tree species (Fig. 46). Comprehensive information and application examples have been compiled for the different types of reiterations, their occurrence and their significance. In this context, the capability of reiterations refers to the different potential of tree species to react immediately to environmental influences and interventions with reiterations.

## 9 Urban Tree Management (Mainly 2003–2025)

*Since the mid-1990s, we have been increasingly asked questions about urban trees. This is a completely different scientific subject than forest trees, because in cities up to 65 characteristics are important, and in the forest 5 are usually enough. Therefore, this area of urban forestry is very exciting and challenging for tree biology,*

**Fig. 45** Example of traumatic reiteration on common ash tree after vandalism

*also because countless additional stress factors and problems effect the trees in cities: heat, sealing, soil compaction, vandalism, construction sites,* etc.

*To this end, we are constantly challenged and responded to with our tree knowledge. For example, a large number of projects on urban trees have been worked on, and we have intensively integrated the subject into the teaching of the master courses Forestry Sciences and Tropical Forestry. Our graduates are therefore now working on this field in larger numbers, are very sought-after and enthusiastic, as it is a very diverse and exciting job – the need for tree specialists there is immense. We can be a little proud of the fact that we recognized this very early on and reacted immediately to it. Some of our tropical master's graduates started urban tree management for the first time in their tropical home countries, in a larger city there.*

Since 2000, urban tree management has become a current, new and important field of application for forest botany and tree biology, because for their functions of shading, cooling, air moistening, biodiversity, etc. biological features are responsible and improvable by a specific tree management. Through the awareness of this, this subject became even the most important for my own further professional development and career. Here, tree biology is so much in demand for applications and for problem solving and method development that urban tree management is not possible without comprehensive knowledge of tree biology. We have been able to make significant contributions to this, as will be explained below. The ten most important books on urban forestry worldwide are currently: Harris et al. (2004), Konijendijk et al. (2005), Carreiro et al. (2008), Forman (2014), Dover (2015),

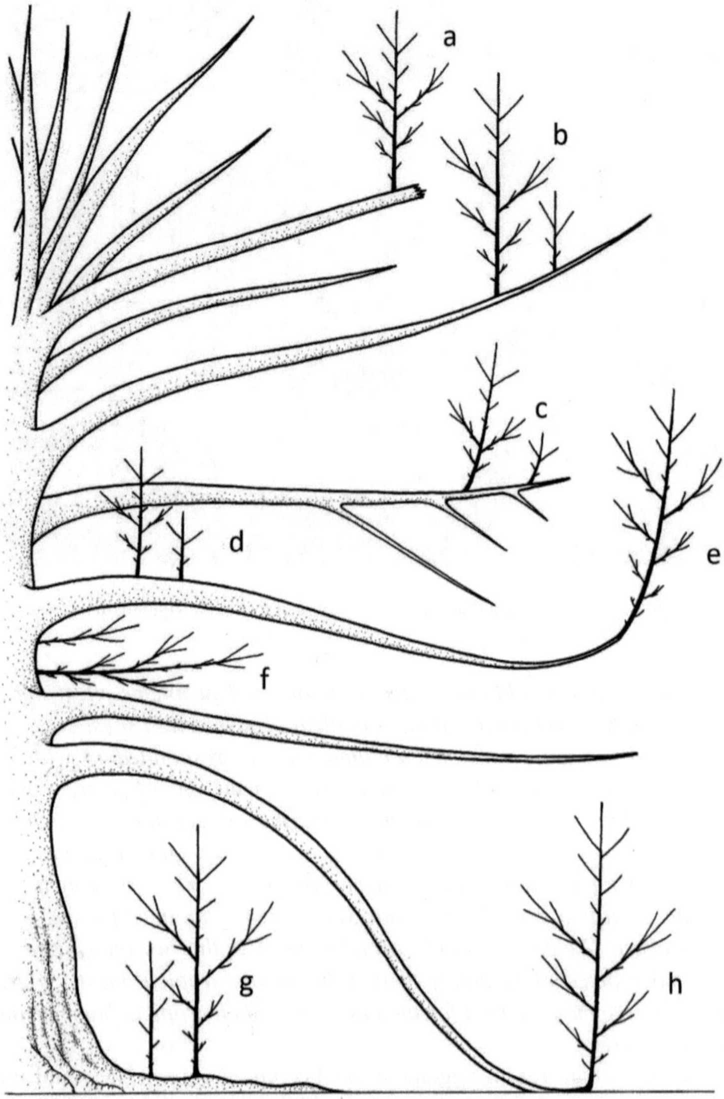


**Fig. 46** Overview of reiteration types, shown as a graphic on one tree; explanation of some examples: (a) traumatic reiteration after branch breakage, (b) adaptive reiteration for light exploitation, (e) straightening up of branches, (g) root suckers, (h) layering branch

Miller et al. (2015), Roloff (2016a), Dujesiefken et al. (2017), Ferrini et al. (2017) and McBride (2017).

From our research described since 1983, a number of methods for tree maintenance and urban tree management have been derived, which are now well established in practice:

- For vitality assessment: vitality stages and regeneration stages (see Sects. 2.2 and 2.3)
- For the identification of tree and shrub species: Book *Flora of Woody Plants* (see Sect. 6)
- For understanding and assessing the reactions of trees to injuries and environmental changes/stress: reiterations (see Sect. 8)
- For interpretation of the 'tree body language' as a sign of their life history, inner trunk defects or safety aspects (Roloff 2016e, 2022b)
- For regular routine inspections of urban trees: Tree Inspection Guidelines (see Sect. 9)
- For the search of suitable tree species on problematic urban locations and by need due to climate change: Climate-Species-Matrix and database Citree (see Sect. 10)
- To derive the biological and ecological value of old trees: evaluation forms (see Sect. 11)
- For the implementation of tree control/investigations: Tharandt Tree Diagnosis (next paragraph)

With the so-called 'Tharandt Tree Diagnosis', a method and software for the detailed assessment of urban and street trees has been developed, which came about through very close contact between science and practice and is continuously being further developed. The method combines the qualified inspection for the evaluation of damage and risk characteristics in trees, the vitality assessment from a biological-botanical point of view and the evaluation of tree statics using different approaches. Tharandt Tree Diagnosis thus tries to bring together the existing knowledge in the field of tree inspection and further tree assessment. In addition to the extensive, visual description of the tree's condition, low-destructive diagnostic procedures are usually used for further investigations (especially a sonic tomography). Only in exceptional cases and in the case of severely damaged trees that are particularly worth the preservation, the results are checked at a few measuring points with the help of drilling methods (Weiß et al. 2004).

Stem injections of systemic insecticides against horse chestnut (*Aesculus hippocastanum*) leaf miner are very successful in vigorous trees (Fig. 47, Schmidt and Roloff 2003, 2004, 2005). Uptake rates are significantly lower at the root collar than at breast height, while efficiency is independent of injection height. For injections prior to flowering, the efficiency is higher than after flowering. When administerd in a adequate dosage, the effects persist until 2 years following the treatment. But in Germany stem injection for this purpose is not legally allowed to use yet.

For the dissemination of new research results and their transfer/application into practice, the Dresden Urban Tree Conference has been established and organized annually in March since 2007. This event is so popular for practitioners, users and planners that it is always fully booked at least 6 months in advance since many years. The interest in the conference is very high and it takes place annually alternating in the lecture hall of the forestry sciences in Tharandt (Fig. 48), in the town hall in

**Fig. 47** Re-sprouting with horse chestnut blossoms in autumn after complete leaf damage due to severe leaf miner infestation in summer

**Fig. 48** Exhibition during annual Dresden Urban Tree Conference (at Tharandt campus)

**Fig. 49** Stress for urban trees due to construction work, soil compaction/sealing, little root space, radiation/heat (*Styphnolobium japonicum*)

Dresden city or in the horticultural college in Dresden-Pillnitz. With a maximum of 150 participants, the organization is carried out by a team from German Tree Institute, the Forest Sciences Department and Forest Botany Institute. Conference proceedings are always published in book form in 'Contributions to Forest Sciences' (initiator and editor of this book series is the author).

Urban trees are exposed to a variety of stress factors due to the situation along roads, in sealed areas, by lack of water, little rooting space, soil compaction, greater heat, construction work (Fig. 49), damage/vandalism, etc. It is all the more surprising that so many tree species and trees are still coping well with this situation. In addition to all the topics mentioned before and below, other questions were also examined (Roloff 2013, 2016b, c, 2023a).

For street trees, site-specific soil and microclimate conditions can restrict growth, health and longevity for several species. Many urban sites are characterized by a high amount of impervious surface and paved areas resulting in a high rate of surface runoff, low infiltration into soil and strong radiation effects. In maintaining the ecologic and economic benefits of street trees through future decades, it is essential to establish species with a high tolerance to stressful urban environments. Our team measured leaf-gas exchange of *Acer platanoides*, *A. pseudoplatanus*, *Platanus* × *hispanica*, *Quercus rubra* and *Tilia platyphyllos* to assess stomatal conductance, transpiration and net photosynthesis at impervious urban sites in the city of Dresden, Germany (Gillner et al. 2015b). The results show significant higher leaf-gas exchange rates for the species *Platanus* × *hispanica* and *Quercus rubra*, compared to the species *Acer platanoides* and *A. pseudoplatanus*. The significant higher mean values of water-use efficiency of *Platanus* × *hispanica* and *Quercus rubra*, and in particular the values during the selected periods with a high vapour pressure deficit, indicate more economical water consumption of these species.

Vegetation and trees in particular play a key role in mitigating thermal load in urban areas. The goals of another study were to analyse species-specific microclimatic effects and to identify the role of leaf-area-density, transpiration, and stomatal conductance for cooling effects at urban sites (Gillner et al. 2015b). Air temperature, relative humidity and surface temperature were used to study the effects of six tree species with similar age structure in a residential area with comparable site conditions. Transpiration, stomatal conductance and leaf-area density were measured during the summer months. Air temperatures and relative humidity were monitored for four trees per street in hourly intervals, and surface temperatures were recorded in the afternoon for three hot days on the shaded and unshaded asphalt surfaces beneath the trees. Species differ significantly in their ability to reduce air and surface temperatures, as well as to increase relative humidity. Trees showing both a high leaf-area density and a high rate of transpiration are more effective in cooling the air temperatures. Differences in the surface temperatures of the tree-shaded areas are more pronounced compared to the air temperatures. Thus, the surface temperatures of the full sun exposed areas are up to 15.2 K warmer than those in the tree-shaded areas. The results provide information to decrease thermal load in urban areas for future tree planting by choosing species with high cooling potential. The highest cooling potential was measured for *Corylus colurna* and *Tilia cordata* and the lowest for *Ulmus* × *hollandica* (Gillner et al. 2015b).

The tree inspection on the basis of the 'Guidelines for Tree Inspections' (Baumkontroll-Richtlinien, Dujesiefken and Roloff 2020; FLL 2020) is structured and gives the inspectors legal and professional security in their work. These guidelines, and especially in connection with the training and certification, thus indirectly serve the preservation of the tree. If abnormalities and defect symptoms are correctly assessed, this leads to fewer 'precautionary' tree fellings or drastic pruning measures in the crown. Impressive tree shapes can thus continue to develop, shape our landscape, make their contribution in times of climate change and are preserved as a testimony to our cultural heritage (see Sect. 12). The interest in the tree control guidelines is now also present in other European countries. With the revision of the tree inspection guidelines in 2020 (chairman of the board was the author) and with the various court rulings on issues of tree control, which now largely refer to this set of rules, this will increase even further (Dujesiefken and Roloff 2020).

Investigations were carried out on various current problems of the damage caused by pathogens to woody plants, e.g. on the status and dieback of *Alnus glutinosa* and *Fraxinus excelsior* (Fig. 50, Schumacher et al. 2006; Volke et al. 2019). Both have a considerable impact on the use of these tree species, and in some cases planting bans were imposed at times.

There are relatively few studies surveying how meteorological and qualitative characteristics of tree sites in urban areas influence *Hymenoscyphus fraxineus*-infection of *Fraxinus excelsior*. Based on the significance of common ash trees in urban areas all over Europe and the frequent use of this species and its varieties, it becomes increasingly important to identify factors that drive the pathogen. We

**Fig. 50** Ash dieback with dying branches of the upper crown due to fungal infection

assessed 401 ash trees along streets and in green spaces distributed across Leipzig (Germany). The assessment included scoring the level of damage by *H. fraxineus*, visually estimating the degree of soil sealing under a tree's crown, measuring tree height, the height of crown base and the diameter at breast height, year of planting and species variety, as well as recording qualitative characteristics (e.g. secondary pathogens, feeding marks). Mean relative air humidity, mean annual temperature and a combined index of meteorological conditions were used as meteorological parameters. In addition, we sampled tissue of 45 ash trees for the purpose of genetically verifying the occurrence of *H. fraxineus*. We analysed the effects of all variables on ash dieback by using linear mixed effect models. Results show an extensive damage by *H. fraxineus* in the whole city area with 70% of trees being infected. Mean relative air humidity and the index of meteorological conditions show a significant positive influence on the severity of ash dieback, whereas the year of planting has a negative effect. Trees with feeding marks at the trunk tend to be more damaged by the fungus. Both meteorological conditions and the susceptibility of species and their varieties should be taken into account when planting ash trees in urban areas (Volke et al. 2019).

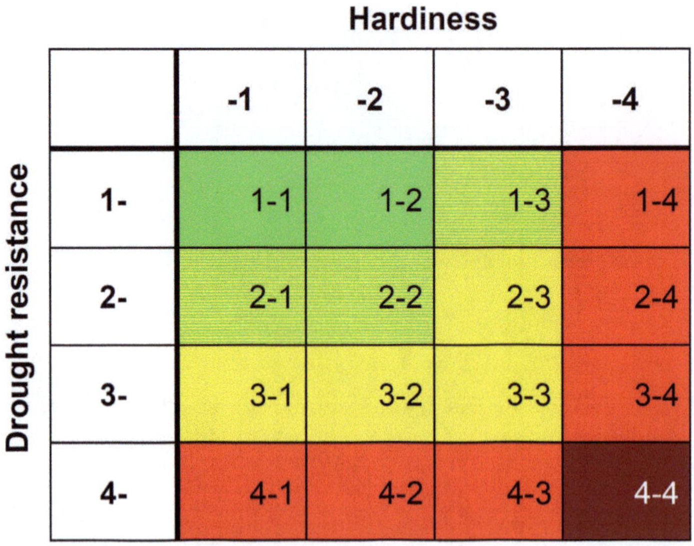


**Fig. 51** Categories of the climate-species matrix with 16 subcategories, based on drought tolerance (first grade) and frost hardiness (second grade). Categories in red, yellow and green: green = 'very suitable': 1–1, 1–2/green/yellow = 'suitable': 1–3, 2–1, 2–2/yellow = 'suitable but occasionally problematic': 2–3, 3–1, 3–2, 3–3/red = 'not very suitable': 1–4, 2–4, 3–4, 4–1, 4–2, 4–3/ purple = 'unsuitable': 4–4

## 10 Tree Species Selection for Urban Forestry Regarding Climate Change (Mainly 2008–2021)

*With the urban tree topics, the desire to investigate and contribute more to the choice of tree species very soon arose. The many trips were also useful for this (especially China, UK, Italy, Spain, Greece), and very important tools for tree species selection were created: Climate-Species-Matrix and the Citree database. They have been used intensively and proved for over 10 years, were immediately taken up by tree nurseries for their offers and have certainly led to a rethink in young tree cultivation.*

An extensive study (Roloff 2014a, 2020b, 2021b; Roloff et al. 2009, 2018, 2022; Liu et al. 2021b) has been based on reliable publications and represents a basic attempt to classify and assess more than 250 urban woody species that can be used in Central European cities (in parks, gardens and as road-side trees) with regards to their usability regarding the expected climate change. The new Climate-Species-Matrix was developed for this purpose. Four degrees of drought resistance and four degrees of winter hardiness are the decisive criteria in a two-dimensional assessment (Fig. 51). The woody species (divided into trees of more than 10 m of potential height and trees of up to 10 m of height) were placed in 16 categories of decreasing tolerance, ranging from 1–1 (very suitable) to 4–4 (very limited usability).

**Fig. 52** Palm trees (*Phoenix canariensis*) as urban and street trees in the Mediterranean region (Mallorca, Spain), left without, right with adequate pruning measures, in the middle a tree after death of the summit bud due to beetle damage

For many of these 250 species, as well as for many others that are not named in this study, not all questions could be answered. Nonetheless this categorization on the basis of drought resistance and winter hardiness provides a sound basis for decisions in planning the use of urban tree species in future. This planning process must of course also include additional criteria like soil parameters, shade resistance, aesthetics, etc., depending on individual requirements. For this purpose, there was developed additionally the data base Citree (see section below).

'New' street tree species are being sought not only in the urban area of the temperate zone, which cope better than native species with drought stress, but also in Mediterranean and subtropical regions. Tree maintenance also plays a role there, and these are also special issues for palm trees. For this purpose, two investigations were carried out and elaborations were developed for practice there. An interesting aspect in urban palm tree management are special pruning methods (Fig. 52, Hosek and Roloff 2015). Overpruning palms can have negative side-effects, which reduce the health and therewith the aesthetical value of the plant. Despite this, it is not uncommon to see palms in cities that had too many of their fronds removed. This study aimed to find relations between various factors, such as 'species' or 'distance to the closest road', and the pruning state of Arecaceae in order to suggest reasons for the continued existence of this practice and approaches to reduce its prevalence. Only two factors, 'height' and 'distance to the closest object', showed the same relations with the pruning state for the three most common species and should therefore be the first to address in order to reduce overpruning. Data analysis proposes three underlying reasons: personal taste in relation with misinformation, improper choice of species for available planting sites and economic factors as most likely to be responsible for widespread overpruning.

Palms (Arecaceae) are part of many urban environments in warmer climatic zones where they are mainly planted for their aesthetical value. Due to physiological and morphological characteristics, they provide many common services at lower levels than trees, putting them in danger of being replaced. To avoid this, it is necessary to adequately match them to available sites so that costs remain small in comparison to their benefits. To determine what factors might be related to palm health, and therefore their aesthetic value, information on the total Arecaceae population in Olhão, Portugal, was collected and statistically analysed for relationships between crown state and variables like palm height, distance to the nearest road or objects located within a species-specific growing space (Hosek and Roloff 2016).

While city-scape-related variables showed no or only very weak correlations, it was found that the three most common species differed significantly from each other as comparatively many *Phoenix canariensis* palms were in poor state due to Red Palm Weevil damage. The results of the study suggest that, in order to reduce financial inputs required to create high levels of desired benefits, a focus should be on macro environmental conditions which cannot easily be manipulated or only with considerable costs associated. In addition to often discussed general climatic suitability of different species, it is also necessary to consider the factor ‘pests and diseases’ when evaluating the appropriateness of certain plant types for designated growing spaces.

For the cities of Jena and Erfurt (Eastern Germany), a working group with us in advisory board has developed handbooks on tree species recommendations due to

**Fig. 53** Jena city green – market square with problematic horse chestnut tree species in the sealed inner city area

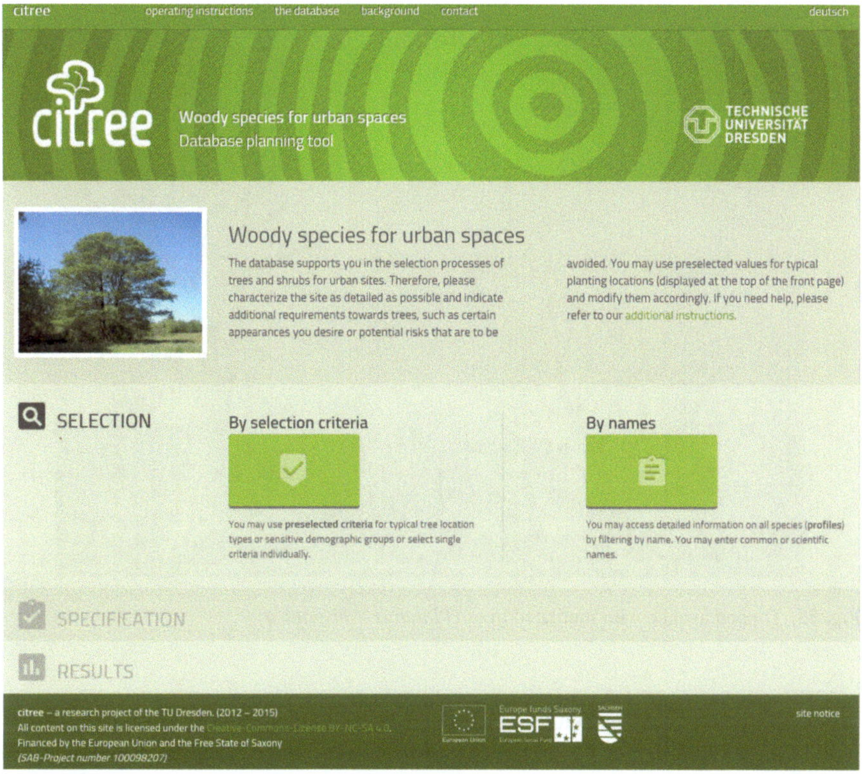


**Fig. 54** Screenshot of Citree Data Base [2025-01-10]

climate change, taking into account site characteristics such as proximity to nature, degree of sealing, root space, water availability and tree characteristics such as lifespan, final size, crown shape, previous use and experiences in the respective city (Fig. 53). In the process, recommendations for action were created with district and partly street-related species lists in two books for Jena and Erfurt, Germany (Knopf et al. 2016, 2020). These recommendations are now being taken into account in the best possible way in planning and planting 'trees for the future'.

Choosing optimal and suitable trees urban areas can minimize the negative influences and increase the positive effects and the aesthetic acceptance by urban residents. Additional challenges for the selection of trees and shrubs are user requirements and growth conditions at urban sites. Therefore, the selection of planted trees in cities has to incorporate these location-specific factors (Gillner et al. 2015a, b, c; Vogt et al. 2017). Based on an extensive literature review, more than 390 tree species and cultivars were investigated to obtain a comprehensive assessment of specific characteristics by integrating specific urban aspects. Within this study, a database Citree was developed for Central Europe (Fig. 54, https://citree.de/?language=en) that allows users to simultaneously consider site characteristics and natural distribution, tree appearance and characteristics, ecosystem

**Fig. 55** Topped avenue with mutilated trees (*Platanus* × *hispanica*)

services, management activities, and the risks and interferences caused by urban woody plants (altogether 65 variables can be chosen and assessed). The developed 'Citree' database is useful for preventing mistakes in planning, which would otherwise result in high ecologic and economic costs. Choosing the right species for the right location will also increase the floristic biodiversity within urban tree plantings and the sustainable uses of urban trees.

Tree topping and mutilation of older trees remain widespread (Fig. 55), despite the well-established knowledge that these practices are destructive to trees and cause long-lasting, often lifelong negative consequences for their health, lifespan and mechanical safety. The causes, specific characteristics, risks and consequences of such practices are analysed, recommendations for their avoidance are provided, and approaches to dealing with topped trees are explained. As part of an increasing appreciation for old trees, more efforts must be made to raise awareness – so that such cruel sights and clear evidence of unprofessional tree maintenance and safety measures become rarer (Roloff 2024c; Roloff and Weiß 2025). With the growing appreciation for trees, there is currently a good chance of progress in this area, provided that those involved pay greater attention and show that they want it.

For the widespread 'special case' of urban trees with improperly strong shortening and trunk topping (Fig. 55), an approach was developed for estimating their further development with the help of new regeneration stages (Roloff 2017d, 2018a). First of all, it becomes clear from the illustration Fig. 56 that severe crown topping and trunk pruning ('stub pruning') are almost always accompanied by a long-term

**Fig. 56** Schematic, simplified visualization of four regeneration stages (RS 0–3, in the leafless winter state) after trunk shortening/topping in older trees; for comparison of reference: an untreated old tree of VS 2 (vitality stage 2). The same old tree is always shown: VS 2 without pruning (top left), with professional shortening according to the technical guidelines and instructions (RS 0, top right) and after improper trunk cuttings with worse and worse prospects for the tree's further development (RS 1–3). State two growing seasons after the measure, green: new shoots of the last 2 years after cutting

high need for maintenance, which causes correspondingly high costs in the long term. Therefore, the seemingly cheapest, improper crown topping is usually a more expensive pruning measure in the medium term. In addition, the goal of crown reduction is often only achieved temporarily, as the possibly positive regeneration through strong re-sprouting close to the top can lead to the return of the original tree height in a timely manner. In principle, this can only be prevented by professional crown shortening with 'supply branches' (growing lateral branches in the cutting area for the supply of water and assimilates to the wound) or even before planting by using smaller tree species, which then do not require such shortening later.

Future functional requirements on road-side trees cannot any longer be met by the ten suitable native tree species only left, according to our investigations. Urban site stress factors (see Sect. 9) and climate change are aggravating the situation. The use of non-native species is therefore necessary, but there are risks to be taken into account concerning their invasive potential, especially on sites near the open countryside or close to nature reserves. However, in urban areas with high sealing levels (Fig. 53), many of these alien species pose little threat. On the contrary, they can have significant advantages due to their high drought and heat tolerance – as long as we know how to manage these tree species, minimizing the risks and optimizing their potential. To achieve that we need to know their bio-ecological features in greater detail (Roloff 2017b, c, 2018b, 2022a).

In some cases, neophytes can cause undesirable ecological and economic effects, exceptionally also health effects, maybe damages after range expansion and formation of dominant populations, and are then referred to as invasive species. For example, they can compete with other plants for habitat and resources and displace them. For Central Europe, the following tree species in particular are classified as potentially invasive: *Acer negundo*, *Ailanthus altissima*, *Fraxinus pennsylvanica*, *Populus × canadensis*, *Prunus serotina*, *Quercus rubra*, *Robinia pseudoacacia*. When assessing their invasiveness, it should be kept in mind that one and the same tree species can be completely unproblematic in one habitat (Fig. 57), but

**Fig. 57** Invasive potential of *Ailanthus altissima*? in this situation of a completely sealed area, a hopeful tree species

**Fig. 58** Adaptation of a park at Meissen (under nature and monument protection) with ten planted 'climate tree species'

problematic in another (Roloff 2017b, 2018b). For example, *Quercus rubra* is an important and suitable street tree species in many cities, which also copes well with the various stress factors at this location, while in nature reserves with a high proportion of rocks, it can displace the native flora. *Pseudotsuga menziesii* is also often classified as invasive by nature conservationists, but this is not comprehensible due to its defensive dispersal behaviour.

As a result of severe drought stress during the summers 2018–2020, many German parks are currently in a very critical condition due to the massive death of numerous older trees – in many parks hundreds of the ancient trees died. Private and municipal park managers in cities and municipalities are very concerned about this situation and hope for support from tree experts (Gillner et al. 2020; Roloff et al. 2021a; Roloff and Rust 2014). Using the example of Siebeneichen Palace Park of the city of Meissen (Eastern Germany near Dresden), the current situation was examined through accompanying scientific research on management and natural regeneration. An innovative concept was developed how the tree population can be optimized in its adaptation potential and thus the parks in their functional fulfilment and made fit for the future through careful conversion and with some new tree species (Fig. 58). The project approach pursues a synthesis of garden monument preservation and nature conservation, visitor experience, road safety aspects, tree maintenance, sustainability and resilience.

**Fig. 59** Short-lived (SL) pioneer tree species *Betula pendula*

# 11 Tree Ageing – Processes, Effects, Methuselah Trees (Mainly 2017–2025)

*In the course of my recent career, the awareness grew stronger and stronger that different tree species develop different strategies to survive as long as possible – about ten tree species make it up to 1,000 years. The characteristics and strategies responsible for this have led to a breakthrough in the understanding of ageing processes in trees during the last years.*

For the topic of tree ageing, it is important and helpful to distinguish between three groups of tree species according to their longevity (Roloff 2018c), which have meanwhile proven and been established for various questions:

- Short-lived (SL) with 80–100 years of life expectancy: e.g. *Ailanthus altissima*, *Alnus glutinosa*, *Betula pendula* (Fig. 59), *B. pubescens*, *Malus domestica*, *Paulownia tomentosa*, *Populus* × *canadensis*, *Prunus avium*, *Pyrus communis*, *Salix* × *rubens*, *Sorbus aucuparia*
- Middle-aged (MA) with up to 300 years of life expectancy: e.g. *Acer platanoides*, *A. pseudoplatanus*, *Aesculus hippocastanum*, *Carpinus betulus*, *Corylus colurna*, *Fagus sylvatica*, *Fraxinus excelsior*, *Gleditsia triacanthos*, *Juglans regia*, *Liquidambar styraciflua*, *Picea abies*, *P. pungens*, *Pinus sylvestris*, *P. nigra*, *Populus nigra*, *Quercus rubra*, *Salix alba*, *Sorbus aria*, *Robinia pseudoacacia*, *Styphnolobium japonicum*

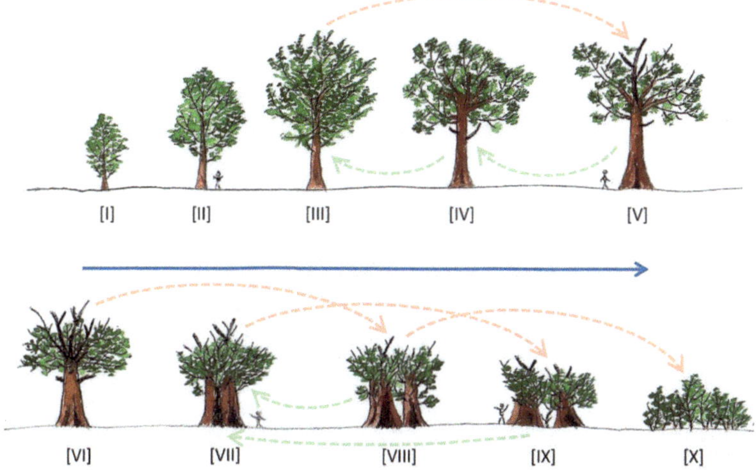


**Fig. 60** Ageing process of long-lived tree species (e.g. *Quercus*, *Tilia*): schematic representation in ten different stages over a life cycle of 500–1,000 years (explanation in text)

- Long-lived (LL) with a life expectancy of over 400 years: e.g. *Castanea sativa*, *Cedrus libani*, *Ginkgo biloba*, *Platanus* × *hispanica*, *Quercus petraea*, *Q. robur*, *Thuja plicata*, *Sequoiadendron giganteum*, *Taxus baccata*, *Tilia cordata*, *T.* × *hollandica*, *T. platyphyllos*, *Ulmus laevis* (in heigher mountains *Abies alba*, *Acer pseudoplatanus*, *Larix decidua*, *Pinus cembra*)

In the third group (LL), the long-lived tree species and so-called 'Methuselah trees' with a life expectancy of 400 up to 1,000 years ideally show ten life stages (I–IX in Fig. 60, Roloff 2018a), which occur during the ageing process (exact age information/ranges are not possible or useful for this, as it depends too much on the tree species, the location and possible damage caused):

(i) Youth phase: crown construction, establishment or planting at the site
(ii) Exploration phase: network of long shoots in the crown top (= VS 0, see Sect. 2.2)
(iii) Degeneration phase: elongated brush structures in the crown top (= VS 1)
(iv) Stagnation phase: short shoot claws or brush structures in the crown top (= VS 2)
(v) Retraction phase: dieback of some main axes in the crown top (= VS 3)
(vi) Retreat phase: Establishment of a secondary crown at medium tree height
(vii) Veteran phase: trunk splitting/openings/rots, compaction of the secondary crown
(viii) Decay phase: decay of the trunk into subtrees
(ix) Tree monument phase: several independent individuals
(x) Rejuvenation phase (re-sprouting from the stump)

**Fig. 61** 'Octopus shape' of a *Salix alba* tree in decay stage 2 (DS 2): some of the branch/stem ends lying on the meadow and rooting

In willow trees (*Salix alba*), four decay stages (DS) can be distinguished if they develop undisturbed (no pruning) during tree ageing (Roloff 2020c):

- DS 0: In tree willows, multi-trunking occurs widely and often due to competition between several branches/stems in adolescence that grow next to each other, or after natural damage in early tree life.
- DS 1: When the lowest side branches and the outermost stems sink, they do not break off immediately, but first lie down on the meadow around the tree, as the branching at the end keeps them at a distance above the ground for a while.
- DS 2 (Fig. 61): The fate of SD 1 affects more and more branches and stems, as rot occurs in the lower trunk areas and spreads upwards until finally (almost) all branches and stems are affected. Some of the branch/stem ends lying on the meadow and rooting ('Octopus shape'), and in addition, lower trunk parts are sprouting again, as more light now falls into these formerly dark crown areas.
- DS 3: This ultimately leads to the 'resurrection' of a new generation of willows from the same tree, which is genetically identical to the previous tree. This creates a multi-stemmed successor tree close to the trunk, or a clone of hundreds of successor trees over a larger area or in groups around the previous mother tree.

This four-stage process can be repeated several times in the tree's life and thus lead to a cloning age of up to 1,000 years. In other tree species (e.g. *Acer pseudoplatanus*, *Castanea sativa*, *Ulmus laevis*), these phases can also occur one

**Fig. 62** 600-year-old tumulus lime tree (*Tilia platyphyllos*) on Bronze Age grave-mound (Evessen, Central Germany)

after the other if and where they can grow free-standing and undisturbed, but not nearly as often and regularly as in willow trees.

Above all, the old age of impressive-looking tree shapes is often the reason for people visiting these trees (Roloff 2020a). And the many individual features of the surroundings, on the trunk or crown as well as their long histories (Fig. 62), also highlight the uniqueness of these tree monuments. These specimens of the long-lived tree species have a high aesthetic, ecological and cultural value. In order to be able to determine this value, two recording sheets for old trees (which should be at least 300 years old) were developed in a comprehensive study (Riedenklau and Roloff 2021). By recording 48 different parameters, the tree habitus and surroundings, root collar, trunk, crown base, crown shape, measures taken as well as the history and special features of the site are to be recorded in detail with a detailed recording sheet. Each of these parameters gets a score, so that the tree reaches an overall score at the end. Optional a quick entry sheet was developed in the study: by concentrating on the 15 most important parameters, the surveyor should receive a comprehensive overview of the tree in a short time in order to be able to decide on a further assessment. The newly developed recording forms for the recording of old trees with detailed explanations can be accessed and downloaded from the institute's website www.tu-dresden.de/forstbotanik (opening page at bottom) in order to test them and use them for application questions.

After further extensive studies on very old trees, various interesting 'strategies' can be identified that are important for achieving an old tree age (Roloff 2024a, b):

- Fungal inhibition by heartwood, 'tricking' pests/damage by current lammas shoots (e.g. *Quercus*) or by evergreen needles (e.g. yew)
- High reaction potential through repair, sprouting (reiterations), cultivation and inner roots: e.g. in *Acer pseudoplatanus*, *Castanea sativa*, *Tilia platyphyllos*; by branch drop in *Quercus*
- Very good compartmentalization: e.g. for *Taxus*, *Quercus*, *Tilia*, *Platanus*
- Embolism tolerance: in conifers and diffuse-porous deciduous trees
- Dominant top with pre-growth to keep up with the light: e.g. in *Larix*, *Sequoiadendron*, *Abies* (but this makes crown retreat more difficult)
- Taproot potential up to an older age for better anchoring and water supply: e.g. *Quercus*, *Ginkgo*, *Larix*
- Slow growth: thus less resource consumption and fewer problems with a big tree height and weight: particularly pronounced in *Taxus*
- High wood density
- Thick bark (none of the long-lived tree species has smooth bark in old age: this would provide too little protection against damage and overheating in the long run)

These characteristics do not all have to be present at the same time, but as many of them as possible and, above all: reiteration potential, compartmentalization and a thick bark.

Many people are aware of the climate effects of old trees – but how can it be quantified, monitored and the positive effects assessed? Tree protection is currently

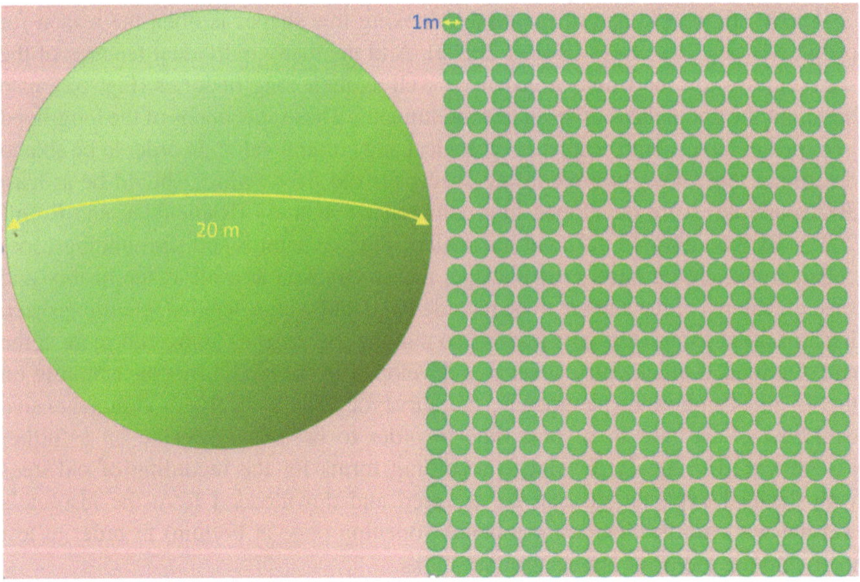


**Fig. 63** The loss/removal of an old tree with a crown diameter of 20 m (left) requires the planting of 400 young trees with a crown diameter of 1 m (right) to replace its climate functions – with planting cost of almost one million Euro's

focusing on functions with regard to climate impacts, especially in cities due to ongoing warming. These effects mainly take place over the entire leaf area of the tree, which can be roughly derived from the crown surface (Roloff 2024b). Thus, a free-standing old tree with an almost spherical crown and a crown diameter of 20 m (e.g. see Fig. 57) has a crown surface of approx. 1,250 $m^2$ in the theoretical ideal case. You would therefore have to plant about 400 young trees with a crown diameter of 1 m to achieve the same effect as from the one crown of the old tree (Fig. 63). This should impressively prove that it is better to first think again about whether the old tree really needs to be removed.

## 12 Long-Lived Tree Species, National Heritage Trees (Mainly 2018–2028)

*The topics of Sect. 11 and the search for and examination of many tree Methuselahs at my home country and studies abroad gave rise to the desire for making an active contribution here in Germany to enable more old trees in this country to survive longer. In addition to hotspots of biodiversity and epigenetics, they are also cultural monuments, can impressively illustrate the importance of adaptation to us researchers, but also to tree interested, and are therefore invaluable. This is how the idea of the National Heritage Trees in 2019 was born, and 3 days after the publication of this idea in a magazine, a foundation was actually found, and in the meantime there are even two foundations that finance everything up to the 100th National Heritage Tree – now 50 trees have already been selected. What will then be possible with 100 trees is a great vision (that still has to mature a bit).*

There is an increasing need to notice, appreciate and protect old trees more. The so-called long-lived tree species (see Sect. 11 above) are particularly important for this, and they therefore understandably represent the highest proportion among tree natural monuments. This makes it clear what responsibility we have for such old trees and that everything must be done to preserve them for posterity. It is impressive to realize that they must be able to tolerate all events, changes and variabilities of location, environment and climate over such a long time of living. Consequently, it can also be assumed that they have a particularly high potential for adaptation (see Sect. 11).

Trees with a high age or at least with a long potential life expectancy are not only natural monuments but also cultural history, as can be seen in more intensive research for all these old trees (Figs. 62, 64 and 65, Roloff 2023b, 2024a, b). The already existing protection category of natural monument is definitely not sufficient for such nationally outstanding individual trees, as it is too local/regional and its implementation and compliance is also noticeably dependent on the available finances and their priorities as well as the motivation of the responsible processors and districts.

**Fig. 64** 900-year-old Methuselah tree *Quercus robur*, supposedly the oldest tree of Germany with a girth of 13 m (Raesfeld, western Germany)

**Fig. 65** The thickest full-wood tree (*Tilia platyphyllos*) in Germany with a girth of 17.5 m and the first National Heritage Tree (Heede, NW-Germany)

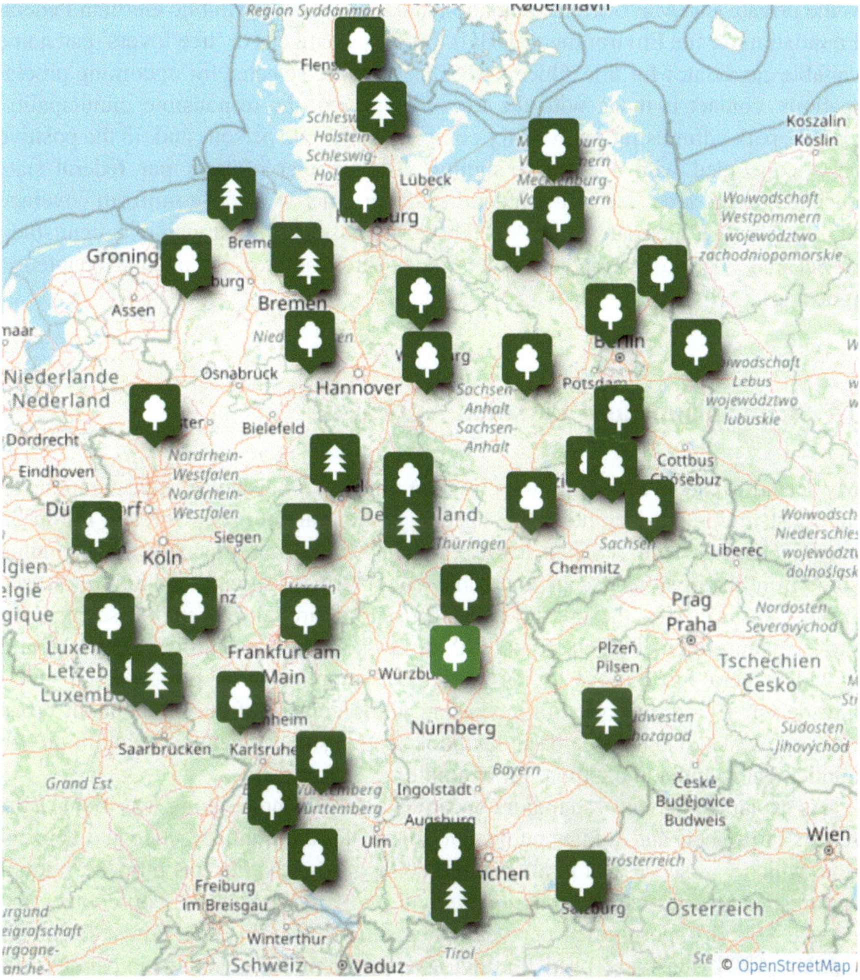


**Fig. 66** Map of the first 50 National Heritage Trees in Germany (as of July 2025, graphic: R. Georgi; see website www.nationalerbe-baeume.de)

As a result, up to 100 trees are now selected as National Heritage Trees in Germany since 2019 (Roloff and DDG 2023; Roloff 2025; https://nationalerbe-baeume.de) – the time horizon for this is a total of 10 years (until 2029), as the selection of suitable trees, the conclusion of an agreement with the tree owners, the preparation and implementation of the proclamation takes a lot of time and a maintenance concept for each of these trees has to be developed, what the planning of measures is then based on. For this purpose, 50 trees with a trunk girth of over 400 cm and preferably more than 400 years of age have already been sought and selected by a responsible board of trustees (chairman is the author) in the German Dendrological Society (DDG). For these candidates, necessary maintenance, security, protection and environmental measures are now funded by two foundations, one is the private Eva Mayr-Stihl Foundation and the other is the public German Federal Foundation for the Environment (DBU). Interested citizens or 'tree lovers' can name suitable candidates for this. Once a tree has been pre-selected for upcoming proclamations, contact is made with the tree owners: e.g. the responsible municipality, church, park administration, forestry company or private persons and, in the positive case, concluded with a contract signing. The number of trees per federal state depends on the area size of the federal states in order to achieve uniform coverage of all regions of Germany (Fig. 66). These Methuselah trees are an excellent study object and communication possibilty for research on tree ageing and adaptation processes.

## 13 Conclusions and Personal Appraisal

My scientific career was initially focused on forestry: it began with investigations on the most important deciduous tree species in Central European forests: European beech (*Fagus sylvatica*). Then it was expanded to 20 more tree species, including some from the open landscape and some from North America like *Acer saccharum* (from Canada). After the interim period in silviculture and vegetation science, the professorship in Goettingen from 1990 onwards marked a significant expansion to include general tree biology topics. With the assumption of the professorship at TU Dresden in 1994, there was an increasing reinforcement of questions about urban forestry, which also led to a basic textbook on this subject, gaining recognition and application in India and China, for example.

As explained in Sect. 1, a major concern in research and teaching was to make the results of projects and studies public and to transfer them into practice and application, which was successful in almost all intended subjects. For this purpose, a lot of lectures, excursions and exercises were held on many workshops, conferences and events: 30–50 opportunities per year. In addition, the Tharandt Forest Botanical Garden was extensively used for this and significantly upgraded, expanded and made more attractive, including a large Indian Summer area with North American forests and a highly sought-after environmental education facility. Since 2007, we organize an annual Urban Tree Conference, the location alternating between Dresden city and Tharandt campus. The success of this orientation and the very great interest from public, politics and practice has motivated us to recommendations for action – on trees in public green spaces as well as in gardens and parks. Therefore, a lot of time was invested in media work and in publications for practice. The initiative 'Tree of the Year' in Germany has also been supported and shaped for a long time.

Basic books on tree identification, tree biology, tree maintenance, tree diagnostics, drought stress, vitality assessment have greatly promoted the general understanding of the special features of trees and have already been published in part in the second, third or fifth edition. In the process, there was an increasing awareness that there are known too few very old trees in Germany. And so the National Heritage Trees Initiative was finally created in 2019 (https://nationalerbe-baeume.de), with

**Fig. 67** National Heritage Tree *Ulmus laevis* shortly after proclamation – an attraction for tree-interested visitors

which a high-ranking new protection category has now been established in Germany. The aim is to search, evaluate, select and proclaim 100 exquisite ancient trees by 2029, and 50 trees of 15 species have already been elevated to this status during the first 5 years since 2019. In the meantime, the initiative has received a lot of attention and significantly increases awareness of the special effects of old trees, in particular their climate impact and importance for biodiversity in urban areas. Thus, this initiative develops as a particularly emotional and motivating finale and as another high-profile result of scientific investigations at the end of my professional development.

I'm really very happy about that, and I would like to take this opportunity to thank the well over 100 companions at my side, without whom much of what I and we have achieved would not have been possible. It's a great honor what I have experienced professionally and what has been able to contribute to the understanding of trees and their influence on us humans, which should not be underestimated (Fig. 67).

## References

AKS Arbeitskreis Standortskartierung (ed) (1980) Forstliche Standortsaufnahme, 4th edn. IHW, Münster, pp 1–243

Bartels H (1993) Gehölzkunde – Einführung in die Dendrologie. Ulmer, Stuttgart, pp 1–336

Böll S, Roloff A, Bauer K, Paeth H, Melzer M (2022) Anpassungsstrategien von Stadtbaumarten an Dürre- und Hitzeperioden. In: Jahrbuch der Baumpflege 2022. Haymarket Media, Braunschweig, pp 125–142
Bolte A (1999) Abschätzung von Trockensubstanz-, Kohlenstoff- und Nährelementvorräten der Waldbodenflora – Verfahren, Anwendung und Schätztafeln. Contrib For Sci 7:1–129 + Attachment
Bonn S, Roloff A (2002) Konkurrenz und Ökophysiologie der wichtigsten Baumarten in Hartholzauenwäldern an der Mittleren Elbe. In: Roloff A, Bonn S (eds) Ergebnisse ökologischer Forschung zur nachhaltigen Bewirtschaftung von Auenwäldern an der Mittleren Elbe. Contrib For Sci 17:7–59
Bolte A, Anders S, Roloff A (2001) Schätztafeln zum oberirdischen Trockensubstanz- und Elementvorrat der Waldbodenflora. Allg Forstzeitschr 56:1080–1082
Bolte A, Anders S, Roloff A (2002) Schätzmodelle zum oberirdischen Vorrat der Waldbodenflora an Trockensubstanz-, Kohlenstoff- und Makronährelementen. Allg Forst Jagdztg 173:57–66
Bolte A, Hilbrig L, Grundmann BM, Kampf F, Brunet J, Roloff A (2010) Climate change impacts on stand structure and competitive interactions in a southern Swedish spruce-beech forest. Eur J For Res 129:261–276
Bolte A, Hilbrig L, Grundmann BM, Roloff A (2014) Understorey dynamics after disturbance accelerate succession from spruce to beech dominated forest – the Siggabody case study. Ann For Sci 71:139–147
Carreiro MM, Song Y-C, Wu J (eds) (2008) Ecology, planning, and management of urban forests. Springer, New York, pp 1–467
Cherubini P, Humbel T, Beeckman H, Gärtner H, Mannes D, Pearson C, Schock W, Tognetti R, Lev-Yadun S (2013) Olive tree-ring problematic dating: a comparative analysis on Santorini (Greece). PLoS One 8:1–5
Dachsel K, Stowasser A, Roloff A (2018) Bestimmung wichtiger Salix-Arten am Naturstandort und ihre Verwendung in der Ingenieurbiologie. ProBaum 3:15–20
Dieler J, Pretzsch H (2013) Morphological plasticity of European beech (*Fagus sylvatica* L.) in pure and mixed-species stands. For Ecol Manag 295:97–108
Dobbertin M (2005) Tree growth as indicator of tree vitality and of tree reaction to environmental stress: a review. Eur J For Res 124:319–333
Dover JW (2015) Green infrastructure – incorporating plants and enhancing biodiversity in buildings and urban environments. Routledge, London/New York, pp 1–337
Dreßler A, Korn S, Scheewe P, Roloff A (2017) Trockenstress-Projekt AdapTree – Anzucht von Stadtbäumen bei unterschiedlichem Bewässerungsmanagement. In: Dujesiefken D (ed) Jahrbuch der Baumpflege 2017. Haymarket Media, Braunschweig, pp 77–93
Dujesiefken D, Roloff A (2020) Die neuen Baumkontrollrichtlinien der FLL – Was hat sich am Inhalt und an der Struktur geändert? In: Jahrbuch der Baumpflege 2020. Haymarket Media, Braunschweig, pp 143–149
Dujesiefken D, Fay N, de Groot J-W, de Berker N (2017) Trees – a lifespan approach. Drukarnia Pasaz, Cracow, pp 1–133
Evert RF, Eichhorn SE (2012) Raven biology of plants, 8th edn. Freeman, New York, pp 1–900
Ferrini F, Konijnendijk CC, Fini A (eds) (2017) Routledge handbook of urban forestry. Routledge, Abingdon/New York, pp 1–547
Fischer A (2003) Forstliche Vegetationskunde, 3rd edn. Ulmer, Stuttgart, pp 1–432
Fitschen J (1977) Gehölzflora, 6th edn. Quelle & Meyer, Heidelberg, pp 1–396
FLL (ed) (2020) Baumkontrollrichtlinien – Richtlinien für Baumkontrollen zur Überprüfung der Verkehrssicherheit. Chairman committee (Regelwerksausschuss RWA) Chairman A. Roloff, 3rd reprint. FLL, Bonn, pp 1–54
Forker M, Roloff A (2023) Ergebnisse der Messungen im Klimawald Köln nach 10 Jahren Wachstum. Contrib For Sci 24:91–104
Forker M, Bouwman M, Roloff A (2020) Eignung heimischer Baumarten für Stadtwälder im Klimawandel? ProBaum 3:10–15

Forman RTT (2014) Urban ecology – science of cities. Cambridge University Press, London, pp 1–462
Gärtner BL (ed) (1995) Plant stems – physiology and functional morphology. Academic Press, San Diego, pp 1–440
Gillner S, Roloff A (2013) Dendrochronologische und physiologische Untersuchungen zur Trockenstress-Empfindlichkeit häufig verwendeter Stadtbaumarten in Dresden. In: Dujesiefken D (ed) Jahrbuch der Baumpflege 2013. Haymarket Media, Braunschweig, pp 246–251
Gillner S, Roloff A (2016) Das Alter von Olivenbäumen: Ein wissenschaftlicher Ansatz zur Altersabschätzung von Olivenbäumen (*Olea europaea* L.) auf Basis einer jahrringanalytischen Methode. Mitt Dtsch Dendrol Ges 101:9–18
Gillner S, Rüger N, Roloff A, Berger U (2013a) Low growth rates predict future mortality of common beech (*Fagus sylvatica* L.). For Ecol Manag 302:372–378
Gillner S, Vogt J, Roloff A (2013b) Climatic response and impacts of drought on oaks at urban and forest sites. Urban For Urban Green 12:293–302
Gillner S, Bräuning A, Roloff A (2014) Dendrochronological analysis of urban trees: climate response and impact of drought on frequently used trees. Trees 28:1079–1083
Gillner S, Korn S, Roloff A (2015a) Leaf-gas exchange of five tree species at urban street sites. Arboricult Urban For 41(3):113–124
Gillner S, Vogt J, Tharang A, Dettmann S, Roloff A (2015b) Role of street trees in mitigating effects of heat and drought at highly sealed urban sites. Landsc Urban Plan 143:33–42
Gillner S, Vogt J, Tharang A, Hofmann M, Roloff A (2015c) Citree-Software zur gezielten Gehölzartenauswahl für urbane Räume. Contrib For Sci 17:130–143
Gillner S, Korn S, Hofmann M, Roloff A (2017) Contrasting strategies for tree species to cope with heat and dry conditions at urban sites. Urban Ecosyst 20:853–865. https://doi.org/10.1007/s11252-016-0636-z
Gillner S, Fenner D, Bräuning A, Rohde M, Roloff A, Seliger A, Schmidt-Wiegand A, Kühn N (2020) Response of growth to climate within oaks of the world heritage site of prussian gardens. Int J For Hortic 6(2):1–15
Grundmann BM, Bolte A, Bonn S, Roloff A (2011) Impact of climatic variation on growth of *Fagus sylvatica* and *Picea abies* in Southern Sweden. Scand J For Res 26:64–71
Grundmann BM, Pietzarka U, Roloff A (2012) Die Weißbeerige Mistel (*Viscum album* L.): Biologie, Ökologie, Verwendung und Befallsrisiken. Mitt Dtsch Dendrol Ges 97:75–90
Hallé F, Oldeman RAA, Tomlinson PB (1978) Tropical tress and forests. Springer, Berlin, pp 1–288
Harris RW, Clark JR, Matheny NP (2004) Arboriculture, 4th edn. Pearson Education, Prentice Hall, Upper Saddle River, pp 1–580
Herbst D, Krabel D, Roloff A (2002) Untersuchungen zur direkten Wasseraufnahme von Blättern. Forst Holz 57:145–147
Hornschuch F, Riek W, Roloff A (2008) Die räumliche Struktur der Feinwurzelsysteme in Reinbeständen von Rot-Buche, Trauben-Eiche und Wald-Kiefer auf grundwasserfernen Sandböden. Archiv Forstw Ldsch.ökol 42:26–40
Hosek L, Roloff A (2015) Relations between cityspace-related and palm-inherent variables and the pruning state of urban *Arecaceae* suggest three reasons for overpruning. Urban For Urban Green 14:975–981
Hosek L, Roloff A (2016) Species site matching: selecting palms (Arecaceae) for urban growing spaces. Urban For Urban Green 20:113–119
Kadereit JW, Körner C, Nick P, Sonnewald U (2021) Strasburger Lehrbuch der Pflanzenwissenschaften, 38th edn. Springer-Spektrum, Berlin, pp 1–1188
Kleinschmit JRG, Roloff A (1998) Ein Beitrag der Taxonomie für die ökologische Praxis – sind Stiel- und Traubeneiche getrennte Arten? Allg Forstzeitschr 53:248–249
Kleinschmit JRG, Kremer A, Roloff A (1995) Comparison of morphological and genetic traits of pedunculate oak (*Q. robur* L.) and sessile oak (*Q. petraea* (Matt.) Liebl.). Silvae Genet 44:256–269

Klugmann K, Roloff A (1999) Ökophysiologische Bedeutung von Zweigabsprüngen (Kladoptosis) unter besonderer Berücksichtigung der Symptomatologie von *Quercus robur* L. Forstwiss Cbl 118:271–286
Knopf D, Roloff A, Boock U, Bachmann J (2016) Bäume in Jena – Stadt- und Straßenbäume im Klimawandel, Stadtbaumkonzept. Schriften Stadtentwicklung 7:1–160
Knopf D, Gstach D, Roloff A (2020) Erfurter Stadtgrün im Klimawandel. Stadt Erfurt, pp 1–154
Konijendijk CC, Nilsson K, Randrup TB, Schipperijn J (eds) (2005) Urban forests and trees. Springer, Berlin, pp 1–520
Kozlowski TT, Kramer PJ, Pallardy SG (1991) The physiological ecology of woody plants. Academic Press, San Diego, pp 1–657
Kramer PJ, Boyer JS (1995) Water relations of plants and soils. Academic Press, San Diego, pp 1–495
Kutschera L, Lichtenegger E (2002) Wurzelatlas mitteleuropäischer Waldbäume und Sträucher. Stocker, Graz, pp 1–604
Larcher W (2003) Physiological plant ecology. Springer, Berlin, pp 1–513
Lerch G (1991) Pflanzenökologie. Akademie-Verlag, Berlin, pp 1–535
Liu M, Pietzarka U, Roloff A (2021a) Trockenstresstoleranz-Bewertung anhand von Trieblängenmessungen. ProBaum 3:2–8
Liu M, Zhang D, Pietzarka U, Roloff A (2021b) Assessing the adaptability of urban tree species to climate change impacts: a case study in Shanghai. Urban For Urban Green 62:27–53. https://doi.org/10.1016/j.ufug.2021.127186
Liu M, Pietzarka U, Kniesel B, Roloff A (2022) Annual shoot lengths of temperate broadleaf species responses to drought. Urban For Urban Green 73:1–15. https://doi.org/10.1016/j.ufug.2022.127592
Lüttge U (2017) Faszination Pflanzen. Springer, Berlin, pp 1–389
Lüttge U, Kluge M, Thiel G (2010) Botanik. Wiley-VCH, Weinheim, pp 1–1236
Lyr H, Fiedler H-J, Tranquillini W (1992) Physiologie und Ökologie der Gehölze. Fischer, Jena, pp 1–620
Matyssek R, Fromm J, Rennenberg H, Roloff A (2010) Biologie der Bäume – Von der Zelle zur globalen Ebene. Ulmer, Stuttgart, pp 1–349
Mayer H (1999) Waldbau auf soziologisch-ökologischer Grundlage, 4th edn. Spektrum, Heidelberg, pp 1–522
McBride JR (2017) The world's urban forests – history, composition, design, function and management. Springer, Cham, pp 1–266
Meyer M, Rust S, Krabel D, Roloff A (2005) Combining physiological and genetic studies on hydraulic architecture, gas exchange and wood anatomy in *Populus tremula*. In: Kobbert M, Freyer K (eds) Results of genetic field trials and laboratory studies and their applications in practical forestry. Forum Genetik und Wald, Teisendorf, pp 337–339
Miller RW, Hauer RJ, Werner LP (2015) Urban forestry, 3rd edn. Waveland Press, Long Grove, pp 1–560
Mitscherlich G (1970) Wald, Wachstum und Umwelt. Sauerländers, Frankfurt, pp 1–419
Nebe W, Roloff A, Vogel M (1998) Untersuchung von Waldökosystemen im Erzgebirge als Grundlage für einen ökologisch begründeten Waldumbau. Contrib For Sci 4:1–235
Newton I (2020) Bird migration. Collins, Glasgow, pp 1–608
Newton I (2023) The migration ecology of birds, 2nd edn. Academic Press, London, pp 1–724
Oldeman RAA (2014) Forests – elements of sylvology. Springer, Berlin, pp 1–648
Pretzsch H (2019) Grundlagen der Waldwachstumsforschung, 2nd edn. Springer Spektrum, Berlin, pp 1–684
Pretzsch H, Seifert S, Huang P (2011) Beitrag des terrestrischen Laserscannings zur Erfassung der Struktur von Baumkronen. Schweiz Ztschr Forstwesen 162(6):186–194
Rappole JH (2022) Bird migration – a new understanding. Hopkins University Press, Baltimore, pp 1–323

Riedenklau A, Roloff A (2021) Lebende Zeitzeugen – Wie sich eine Bewertung unserer Baumveteranen durchführen und verbessern lässt. ProBaum 1:2–7

Roloff A (1983) Morphologie und Phänologie der Rotbuche (*Fagus sylvatica* L.) unter besonderer Berücksichtigung des Verzweigungssystems. Diploma thesis, Institut für Forstbotanik der Universität Göttingen, pp 1–105

Roloff A (1984) Bestimmungsschlüssel für die deutschen Waldgräser. Mitt Dtsch Dendrol Ges 75: 221–228

Roloff A (1986a) Morphologie der Kronenentwicklung von *Fagus sylvatica* L. (Rotbuche) unter besonderer Berücksichtigung möglicherweise neuartiger Veränderungen. Ber Forschungsz Waldökosysteme Göttingen 18:1–177. (PhD thesis)

Roloff A (1986b) Bestimmungsschlüssel für die wichtigsten einheimischen Waldmoose nach vegetativen Merkmalen. Forst Holz 41:64–71

Roloff A (1987) Morphologie der Kronenentwicklung von *Fagus sylvatica* L. (Rotbuche) unter besonderer Berücksichtigung neuartiger Veränderungen: I. Morphogenetischer Zyklus, Anomalien infolge Prolepsis und Blattfall. Flora 179:355–378

Roloff A (1988a) Branching structure in hardwoods related to forest decline. In: Cape JN, Mathy P (eds) Scientific basis of forest decline symptomatology. Commission of the European Communities, Edinburgh, pp 73–91

Roloff A (1988b) Morphologie der Kronenentwicklung von *Fagus sylvatica* L. (Rotbuche) unter besonderer Berücksichtigung neuartiger Veränderungen: II. Strategie der Luftraumeroberung und Veränderungen durch Umwelteinflüsse. Flora 180:297–338

Roloff A (1989a) Entwicklung und Flexibilität der Baumkrone und ihre Bedeutung als Vitalitätsweiser: I. Kronenentwicklung und Vitalitätsbeurteilung am Beispiel der Buche. Schweiz Ztschr Forstw 140:775–789

Roloff A (1989b) Entwicklung und Flexibilität der Baumkrone und ihre Bedeutung als Vitalitätsweiser: II. Kronenentwicklung und Vitalitätsbeurteilung der Eichen, Reiterationen, Kurztrieblebensdauer und Beziehungen zwischen Krone und Wurzel. Schweiz Ztschr Forstw 140:943–963

Roloff A (1989c) Kronenentwicklung und Vitalitätsbeurteilung ausgewählter Baumarten der gemäßigten Breiten. Habilitation thesis, Sauerländer's Verlag, Frankfurt a. M., pp 1–258

Roloff A (1991a) Correlations between crown and root system in European Beech (*Fagus sylvatica* L.). In: Edelib C (ed) L'Arbre – Biologie et Developpement. Nat Monspel A7:652–653

Roloff A (1991b) Crown structure and tree vitality. In: Longhurst JWS (ed) Acid deposition – origins, impacts and abatement strategies. Springer, Berlin, Heidelberg, pp 199–213

Roloff A (1992a) Die Baumkrone als Indikator für Umweltbelastungen. VDI-Berichte 901:349–361

Roloff A (1992b) Mögliche Auswirkungen des Treibhauseffektes auf die Konkurrenzsituation in Waldökosystemen. Forstarchiv 63:4–10

Roloff A (1994) Effects of pollutants and environmental stress on the morphology of deciduous trees and on the ground vegetation. In: Hüttermann A (ed) Effects of acid rain on forest processes. Wiley-Liss, New York, pp 325–352

Roloff A (1998) Der Baum des Jahres 1998 – Biologie und Ökologie der Wild-Birne (*Pyrus communis* L.). In: Dujesiefken D, Kockerbeck P (eds) Jahrbuch der Baumpflege 1998. Haymarket Media, Braunschweig, pp 8–14

Roloff A (1999) Tree vigour and branching pattern. J For Sci 5:206–216

Roloff A (2001) Baumkronen – Verständnis und praktische Bedeutung eines komplexen, Naturphänomens. Ulmer, Stuttgart, pp 1–172

Roloff A (2004) Trees – phenomena of adaptation and optimization. Ecomed, Landshut, pp 1–276

Roloff A (2006) Der Baum des Jahres 2006 – die Schwarz-Pappel (*Populus nigra* L.). Eberswalder Forstl Schriftenr 27:8–14

Roloff A (2010) Bäume – Lexikon der praktischen Baumbiologie. Wiley-VCH, Weinheim, pp 1–207

Roloff A (2013) Bäume in der Stadt – Besonderheiten, Funktion, Nutzen, Arten, Riseiken. Ulmer, Stuttgart, pp 1–254
Roloff A (2014a) Straßen- und Stadtbäume der Zukunft – welche Arten sind geeignet? In: Dujesiefken D (ed) Jahrbuch der Baumpflege 2014. Haymarket Media, Braunschweig, pp 173–183
Roloff A (2014b) Tree vigour assessment by branching pattern. Schriftenr Inst Intern Forst Holzwirtsch, Tharandt 23:36–42
Roloff A (2015) Baum des Jahres 2015 – Der Feld-Ahorn (*Acer campestre* L.): Biologie, Ökologie und Verwendung. In: Dujesiefken D (ed) Jahrbuch der Baumpflege 2015. Haymarket Media, Braunschweig, pp 9–17
Roloff A (ed) (2016a) Urban tree management – for a sustainable development of green cities. Wiley-Blackwell, Chichester, pp 1–274
Roloff A (2016b) Urban trees – importance, benefits, problems. In: Roloff A (ed) Urban tree management – for a sustainable development of green cities. Wiley-VCH, Oxford, pp 1–14
Roloff A (2016c) Urban trees: features and requirements. In: Roloff A (ed) Urban tree management – for a sustainable development of green cities. Wiley-VCH, Oxford, pp 15–19
Roloff A (2016d) Vitality assessment, tree architecture. In: Roloff A (ed) Urban tree management – for a sustainable development of green cities. Wiley-VCH, Oxford, pp 82–94
Roloff A (2016e) Body language of trees, tree diagnostics. In: Roloff A (ed) Urban tree management – for a sustainable development of green cities. Wiley-VCH, Oxford, pp 95–110
Roloff A (2017a) Der Charakter unserer Bäume – ihre Eigenschaften und Besonderheiten. Ulmer, Stuttgart, pp 1–252
Roloff A (2017b) Invasive Baumarten in der Stadt – Risiken und Potenziale. ProBaum 3:2–7
Roloff A (2017c) Invasive Baumarten in der Stadt – Status, Risiken, Potenziale und Management. In: Jahrbuch der Baumpflege 2017. Haymarket Media, Braunschweig, pp 108–124
Roloff A (2017d) Regenerationsstufen: Beurteilung von stark eingekürzten Linden, Einschätzung ihrer weiteren Entwicklung und Handlungsempfehlungen. ProBaum 2:2–5
Roloff A (2018a) Vitalitätsbeurteilung von Bäumen – Aktueller Stand und Weiterentwicklung. Haymarket Media, Braunschweig, pp 1–205
Roloff A (2018b) Invasive Baumarten in der Stadt – Status, Risiken, Potenziale und Management. Forstwiss Beitr Tharandt 20:155–175
Roloff A (2018c) 1000-jährige Baumarten: Eibe, Ginkgo, Ess-Kastanie – Charakter, Besonderheiten und Potenzial. In: Dujesiefken D (ed) Jahrbuch der Baumpflege 2018. Haymarket Media, Braunschweig, pp 117–132
Roloff A (2019a) Geeignet für den Schatten – Schattentoleranz als Kriterium der Baumartenwahl in der Stadt. Dt Baumschule 4:26–29
Roloff A (2019b) Schattentoleranz als Kriterium der Baumartenwahl in der Stadt – Bedeutung, Strategien und Beispiele. In: Dujesiefken D (ed) Jahrbuch der Baumpflege 2019. Haymarket Media, Braunschweig, pp 222–239
Roloff A (ed) (2020a) Die starken Bäume Deutschlands – 111 faszinierende Naturerben und ihre Geschichten. Quelle & Meyer, Wiebelsheim, pp 1–272
Roloff A (2020b) An Trockenstress angepasste Stadtbäume – Einflussfaktoren, Kriterien, Eigenschaften, Arten und wie man sie findet. In: Jahrbuch der Baumpflege 2020. Haymarket Media, Braunschweig, pp 166–179
Roloff A (2020c) Krakenweiden – Zerfall und "Wiederauferstehen". BaumZtg 3:24–28
Roloff A (2021a) Ivenacker Eichen – Uraltbäume mit baumbiologischen Überraschungen. ProBaum 4:2–5
Roloff A (ed) (2021b) Trockenstress bei Bäumen – Ursachen, Strategien, Praxis. Quelle & Meyer, Wiebelsheim, pp 1–288
Roloff A (2021c) Baumveteranen – Die Frage des Alter(n)s. BaumZtg 1:36–39
Roloff A (2021d) Stiel- oder Trauben-Eiche? BaumZtg 1:33–34

Roloff A (2022a) Handbuch Baumdiagnostik – Baumbiologische Grundlagen verstehen, Symptome erkennen und für die Baumkontrolle und -beurteilung nutzen, 2nd edn. Ulmer, Stuttgart, pp 1–205
Roloff A (2022b) Wie erfolgt eine Anpassung an Trockenstress bei Bäumen? Einflussfaktoren, Kriterien, Mechanismen, Bewertung. Jahrb Nassauischer Verein Naturkd 143:35–47
Roloff A (ed) (2023a) Baumpflege – Baumbiologische Grundlagen und ihre Anwendung in der Baumkontrolle, Baumbeurteilung, Baumuntersuchung, Baumschnitt, Naturschutz, Verkehrssicherungspflicht und -maßnahmen, 5th edn. Ulmer, Stuttgart, pp 1–276
Roloff A (2023b) Baumveteranen und Nationalerbe-Bäume – Faszination, Besonderheiten, Risiken und Chancen. In: Jahrbuch der Baumpflege 2023. Haymarket Media, Braunschweig, pp 217–231
Roloff A (2024a) Methusalembäume – Wie und warum können manche Baumarten 1000 Jahre alt werden? In: Jahrbuch der Baumpflege 2024. Haymarket Media, Braunschweig, pp 285–299
Roloff A (2024b) Baumveteranen und zum Stand Nationalerbe-Bäume, Besonderheiten und Wirkungen der Baumalterung. Mitt Dtsch Dendrol Ges 109:11–43
Roloff A (2024c) Kappungen und Verstümmelungen an Stadt- und Straßenbäumen. Forstwiss Beitr Tharandt 25:151–176
Roloff A (2025) Alt wie Methusalem – Eine Reise zu den Jahrhundert-Bäumen Deutschlands: 50 Nationalerbe-Bäume im Porträt. Quelle & Meyer, Wiebelsheim, pp 1–312
Roloff A, Bärtels A (1996) Flora der Gehölze – Bestimmung, Eigenschaften, Verwendung, 1st edn. Ulmer, Stuttgart, pp 1–694
Roloff A, Bärtels A (2018) Flora der Gehölze – Bestimmung, Eigenschaften, Verwendung, 5th edn. Ulmer, Stuttgart, pp 1–967
Roloff A, Bonn S (eds) (2002) Ergebnisse ökologischer Forschung zur nachhaltigen Bewirtschaftung von Auenwäldern an der Mittleren Elbe. Ulmer, Stuttgart, pp 1–227
Roloff A, DDG (2023) Nationalerbe-Bäume: Ziele, Konzeption und Stand der Umsetzung – zum Schutz alter Bäume in Deutschland mit den ersten 30 Kandidaten (Okt. 2023). Eigenverlag Dt. Baum-Institut TU Dresden, Tharandt, pp 1–176
Roloff A, Klugmann K (2000) Twig abscission (cladoptosis) and its ecological significance for decline symptoms in *Quercus robur* L. In: Oszako T, Delatour W (eds) Recent advances in oak health in Europe. Forest Research Institute, Warszawa, pp 147–156
Roloff A, Peik K (eds) (1990) Ökologische Grundlagen naturnahen Waldbaus. Universität Göttingen, Göttingen, pp 1–243
Roloff A, Pietzarka U (1996) Der Forstbotanische Garten Tharandt der TU Dresden. Selbstverlag Atelier Forstgarten, Tharandt, pp 1–205
Roloff A, Pietzarka U (2000) *Betula pendula* (L.) Roth (Sand-Birke). Enzyklop Holzgewächse 21: 1–15
Roloff A, Pietzarka U (2006) Der Forstbotanische Garten Tharandt – ein Gartenführer. Addprint, Possendorf, pp 1–228
Roloff A, Römer H-P (1989) Beziehungen zwischen Krone und Wurzel bei der Rotbuche (*Fagus sylvatica* L.). Allg Forst Jagdztg 160:200–205
Roloff A, Rothe M (2000) Sprühen statt Bewässern? Neue Erkenntnisse zu Reaktionen von Bäumen auf Trockenstress und Konsequenzen für die Praxis. Landschaftsarchitektur 30:44–47
Roloff A, Rust S (2014) The influence of climate change on tree biology and its consequences for the use of urban park trees. In: Prussian Palaces and Gardens Foundation Berlin-Brandenburg (ed) Historic gardens and climate change. Edition Leipzig, Potsdam, pp 70–73
Roloff A, Weiß H (2025) Kappungen an Stadt- und Straßenbäumen: Vermeiden, Erkennen und Regeneration bewerten. In: Jahrbuch der Baumpflege 2025. Haymarket Media, Braunschweig, pp 166–189
Roloff A, Ziegler B, Meng V (1993) Leitfaden durch den Forstbotanischen Garten der Universität Göttingen. Selbstverlag University, Göttingen, pp 1–143

Roloff A, Rothe M, Solger A, Rust S, Schmidt C (2000) Der Baum-Wasserhaushalt – neue Erkenntnisse und Konsequenzen für die Praxis. In: Dujesiefken D, Kockerbeck P (eds) Jahrbuch der Baumpflege 2000. Haymarket Media, Braunschweig, pp 55–65

Roloff A, Rust S, Klugmann K (2001) Twig abscission (cladoptosis) and its morphological-ecophysiological significance in *Quercus robur* L. In: Labrecque M (ed) L'arbre/The tree. IQ, Montreal, pp 222–229

Roloff A, Korn S, Gillner S (2009) The Climate Species-Matrix to select tree species for urban habitats considering climate change. Urban For Urban Green 8:295–308

Roloff A, Grundmann BM, Pietzarka U (2012) Misteln an Bäumen – Fluch oder Segen? In: Dujesiefken D (ed) Jahrbuch der Baumpflege 2012. Haymarket Media, Braunschweig, pp 177–188

Roloff A, Grundmann BM, Pietzarka U (2013) Lebewesen auf Bäumen: Misteln. In: Roloff A (ed) Bäume in der Stadt – Besonderheiten, Funktion, Nutzen, Arten, Risiken. Ulmer, Stuttgart, pp 234–245

Roloff A, Gillner S, Kniesel R, Zhang D (2018) Interesting and new street tree species for European cities. J For Landsc Res 1:1–7

Roloff A, Gebhardt T, Karich M (2021a) Anpassung von Parkanlagen – Erarbeitung einer Konzeption für das zukünftige Parkmanagement. ProBaum 2:2–7

Roloff A, Weisgerber H, Lang UM, Stimm B (eds) (2021b) Enzyklopädie der Holzgewächse (687 species). Wiley-VCH, Weinheim, pp 1–6247

Roloff A, Pietzarka U, Gillner S (2022) Baumartenverwendung im Klimawandel: KlimaArtenMatrix 2021 (KLAM 2.0) und Empfehlungen zu Baumgrößen, -pflanzungen und -umfeld. In: Jahrbuch der Baumpflege 2022. Haymarket Media, Braunschweig, pp 204–223

Rothe M (2005) Reaktionen des Wasserhaushaltes der Fichte (Picea abies) auf extremen Trockenstress. Contrib For Sci 23:1–175

Rothe M, Roloff A (1998) Wasserhaushaltsuntersuchungen an Einzelbäumen der Fichte (*Picea abies* (L.) KARST.) unter dem Einfluss von Wassermangel. In: Nebe W, Roloff A, Vogel M (eds) Untersuchung von Waldökosystemen im Erzgebirge als Grundlage für einen ökologisch begründeten Waldumbau. Contrib For Sci 4:143–149

Rothe M, Krabel D, Roloff A (1999) Responses of Norway spruce (*Picea abies* [L.] Karst.) to water deficiency. Phyton 39:183–190

Rothe M, Vogel M, Roloff A (2002) Charakterisierung des Wasserhaushaltes der Fichte (*Picea abies* (L.) Karst.) unter anhaltendem artifiziellen Wasserstress. Allg Forst Jagdztg 173:29–36

Runkel M, Roloff A (1985) Schadstufen bei der Buche im Infrarot-Farbluftbild. Allg Forstzeitschr 40:789–792

Rust S, Roloff A (2002) Reduced photosynthesis in old oak (*Quercus robur*) the impact of crown and hydraulic architecture. Tree Physiol 22:597–601

Rust S, Roloff A (2004) Acclimation of crown structure to drought in *Quercus robur* L. – intra- and inter-annual variation of abscission and traits of shed twigs. Basic Appl Ecol 5:283–291

Rust S, Roloff A (2009) Competition in mixed stands of beech and spruce: effect of climate and ozone. Eur J For Res 128:85–86

Rust S, Klugmann K, Roloff A (2000) Twig abscission (cladoptosis) as a decline symptom in *Quercus robur* L (European oak)? In: Backhaus G, Balder H, Idczak E (eds) Plant health in urban horticulture. Mitt Biol Bundesanst Land Forstw 370:196–203

Rust S, Solger A, Roloff A (2004) Bottlenecks to water transport in *Quercus robur* L.: the abscission zone and its physiological consequences. Basic Appl Ecol 5:293–299

Rysavy T, Roloff A (1994) Ursachen der Vereschung in Mischbeständen und Möglichkeiten zu ihrer Vermeidung. Forst Holz 49:392–395

Schmidt C, Roloff A (2000) Beziehungen zwischen Wurzelsträngen und Kronenteilen bei Bäumen. Allg Forstztschr 55:786–788

Schmidt C, Roloff A (2001) Beziehungen zwischen Kronenteilen und Wurzelsträngen bei Bäumen. In: Dujesiefken D, Kockerbeck P (eds) Jahrbuch der Baumpflege 2001. Haymarket Media, Braunschweig, pp 264–269

Schmidt C, Roloff A (2003) Stamminjektion zur Bekämpfung der Rosskastanien-Miniermotte. Stadt+Grün 52:47–49

Schmidt C, Roloff A (2004) Bekämpfung der Rosskastanien-Miniermotte (*Cameraria ohridella*). Mitt Biol Bundesanst Land Forstw 396:356–357

Schmidt C, Roloff A (2005) Wirksamkeit systemischer Insektizide gegen die Rosskastanien-Miniermotte (*Cameraria ohridella* Deschka & Dimic). In: Dujesiefken D, Kockerbeck P (eds) Jahrbuch der Baumpflege 2005. Haymarket Media, Braunschweig, pp 216–219

Schumacher J, Leonhard S, Grundmann BM, Roloff A (2006) New alder disease in Spreewald biosphere reserve – causes and incidental factors of an epidemic. Nachrichtenbl Deut Pflanzenschutzd 58:141–147

Schütt P, Lang UM (1996) *Pinus longaeva* (Great Basin bristlecone pine). EnzyklopHolzgewächse 5:1–8

Schweingruber FH (2007) Wood structure and environment. Springer, Berlin/Heidelberg, pp 1–279

Schweingruber FH, Börner A, Schulze E-D (2006) Atlas of woody stems – evolution, structure, and environmental modifications. Springer, Berlin/Heidelberg, pp 1–229

Späth V (1988) Zur Hochwassertoleranz von Auenwaldbaumarten. Natur Landschaft 63:312–315

Stribley GH, Ashmore MR (2002) Quantitative changes in twig growth pattern of young woodland beech (*Fagus sylvatica* L.) in relation to climate and ozone pollution over 10 years. For Ecol Manag 157:191–204

Tyree MT, Zimmermann MH (2002) Xylem structure and the ascent of sap, 2nd edn. Springer, Berlin, pp 1–283

UN/ECE (United Nations Economic Commission for Europe) (1989/2016) Manual Part II – Visual assessment of crown condition. http://icp-forests.net/page/icp-forests-manual. Accessed 15 Feb 2025

VDI (Verein Deutscher Ingenieure) (1993) Messen von Vegetationsschäden am natürlichen Standort – Interpretationsschlüssel für die Auswertung von CIR-Luftbildern zur Kronenzustandserfassung von Nadel- und Laubgehölzen. VDI-Richtlinie 3793(Bl. 2):1–57. (with collaboration of A. Roloff)

Vogel M, Roloff A, Rothe M, Sambale C, Peschke G (1998) Der Fichten-Austrocknungsversuch Oberbärenburg zur Beurteilung von Trockenstreß in den Höheren Berglagen des Erzgebirges. In: Nebe W, Roloff A, Vogel M (eds) Untersuchung von Waldökosystemen im Erzgebirge als Grundlage für einen ökologisch begründeten Waldumbau. Contrib For Sci 4: 238–252

Vogt J, Gillner S, Berger U, Roloff A (2017) Citree: a database supporting tree selection for urban areas in temperate climate. Landsc Urban Plan 157:14–25

Volke V, Knapp S, Roloff A (2019) Survey of *Hymenoscyphus fraxineus* in a central European urban area and exploration of its possible environmental drivers. Urban For Urban Green 40: 165–173

Weiß H, Roloff A, Goede M (2004) Tharandter Baumdiagnose – ein integriertes Verfahren zur Beurteilung von Problembäumen. Allg Forstztschr 59:52–55

Worbes M, Hofmann M, Roloff A (1992) Wuchsdynamik der Baumschicht in einem Seggen-Kalkbuchenwald in Nordwestdeutschland. Dendrochronology 10:91–106

Worbes M, Staschel R, Roloff A, Junk WJ (2003) Tree ring analysis reveals age structure, dynamics and wood production of a natural forest stand in Cameroon. For Ecol Manag 173:105–123

# Transcriptome Dynamics in Plant Reproduction

**Diogo A. Lucas and Jörg D. Becker**

## Contents



Communicated by Francisco M. Cánovas

D. A. Lucas
Instituto de Tecnologia Química e Biológica António Xavier, Universidade Nova de Lisboa, Oeiras, Portugal

Gulbenkian Institute for Molecular Medicine, Oeiras, Portugal
e-mail: diogo.lucas@itqb.unl.pt

J. D. Becker (✉)
Instituto de Tecnologia Química e Biológica António Xavier, Universidade Nova de Lisboa, Oeiras, Portugal
e-mail: jbecker@itqb.unl.pt

 
Progress in Botany (2026) 86: 83–107, https://doi.org/10.1007/124_2025_91,
Published online: 15 June 2025

**Abstract** Sexual reproduction in angiosperms requires tight transcriptomic regulation in a short developmental window to guarantee double fertilization and seed production. In the last 40 years, these processes have been intensively studied, particularly at the gene expression level. Most of the studies focused on the male gametophyte because of its easier accessibility. In this chapter, we focus on insights gathered from transcriptomic studies on gametophyte development, pollen-pistil interactions, and the final steps leading to double fertilization. Since heat stress severely affects plant sexual reproduction, we briefly cover studies of its effects on gene transcription. Finally, we provide an outlook on future technological developments and how these could contribute to resolve open questions of plant sexual reproduction.

**Keywords** Angiosperms, Gametophyte, Heat stress, Sexual reproduction, Transcriptomics

## 1 Introduction

Studies of gene expression and regulation underlying plant sexual reproduction of seed-bearing plants are crucial if we want to address the threats climate change imposes on crop productivity. Both female and male gametophytes comprise only a few cells: the embryo sac and the pollen grains, respectively. Since 1981 there have been 4,614 gene expression studies involving the male gametophyte, while knowledge for the female gametophyte lags behind, with only 458 studies (Fig. 1). This difference is mainly because of a better accessibility of the male gametophyte, especially when combined with efficient methods for its isolation, like cell separation by fluorescence-activated cell sorting (Borges et al. 2012). While studies of the evolution of sexual reproduction in land plants have provided unique insights, for example, that male-specific transcriptomes show a higher conservation than those of female gametes (Julca et al. 2021), we will focus exclusively on studies of angiosperms.

Transcriptional regulation has many layers of control. During plant sexual reproduction, different types of transcriptional regulation occur, several of which cannot be explored in depth in this chapter. Some examples of this regulation include: (1) alternative splicing, a post-transcriptional mechanism that plays a role in male gametophyte development (Loraine et al. 2013; Misra et al. 2023; Klodová et al. 2023), and is an important layer of gene regulation under abiotic stress, like heat stress (Laloum et al. 2017), (2) small RNAs, like microRNAs (miRNAs) and small interfering RNAs (siRNAs) (Le Trionnaire and Twell 2010; Wu and Zheng 2019), (3) transcription factors, which have been identified in both gametophytes (Wang et al. 2010, 2023), or (4) epigenetic regulation, which has been extensively reviewed (Takeda and Paszkowski 2006; Baroux et al. 2011; Begcy and Dresselhaus 2018; Ashapkin et al. 2019).

In this chapter, we summarize the current knowledge of transcriptome dynamics in the male and female gametophyte during plant sexual reproduction. A majority of

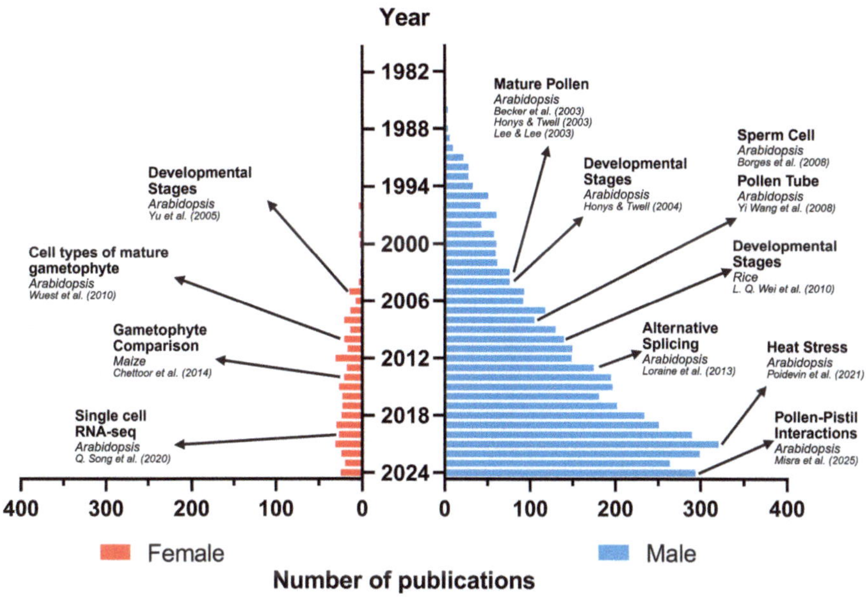


**Fig. 1** A timeline charting the number of transcriptomic studies for both gametophytes, highlighting some of the important advancements in transcriptome studies. There has been an increase in published work related to the terms "transcript*" for both gametophytes. Numbers are from Web of Science™, accessed on 26th February 2025. Male – Terms used: Pollen Transcript*, Male Gametophyte Transcript*, and Pollen Tube Transcript*. Female – Terms used: Female Gametophyte Transcript*

the research done in angiosperms in this area has been in the plant model organism *Arabidopsis thaliana*, although research in crop plants (e.g. *Solanum lycopersicum*, *Zea mays*, or *Oryza sativa*) and other non-crop plants (*Nicotiana tabacum*) has also been performed. We follow the transcriptome changes that occur during gametophyte development (male and female), and during the progamic phase, which comprises pollen germination and pollen tube growth through the female tissues until the male gametes are delivered for fertilization. A brief overview of heat stress effects on each of these stages is also provided.

## 2 The Male Gametophyte

The stamen is the male reproductive organ of flowers and is constituted by an anther and a filament. It is in the anther that the male gametophyte develops. In angiosperms, the development of the male gametophyte is split into two main phases: microsporogenesis and microgametogenesis. During microsporogenesis, diploid microsporocytes (pollen mother cells) undergo meiosis and divide into four haploid microspores, forming a tetrad (Fig. 2). In the next phase, microgametogenesis, the

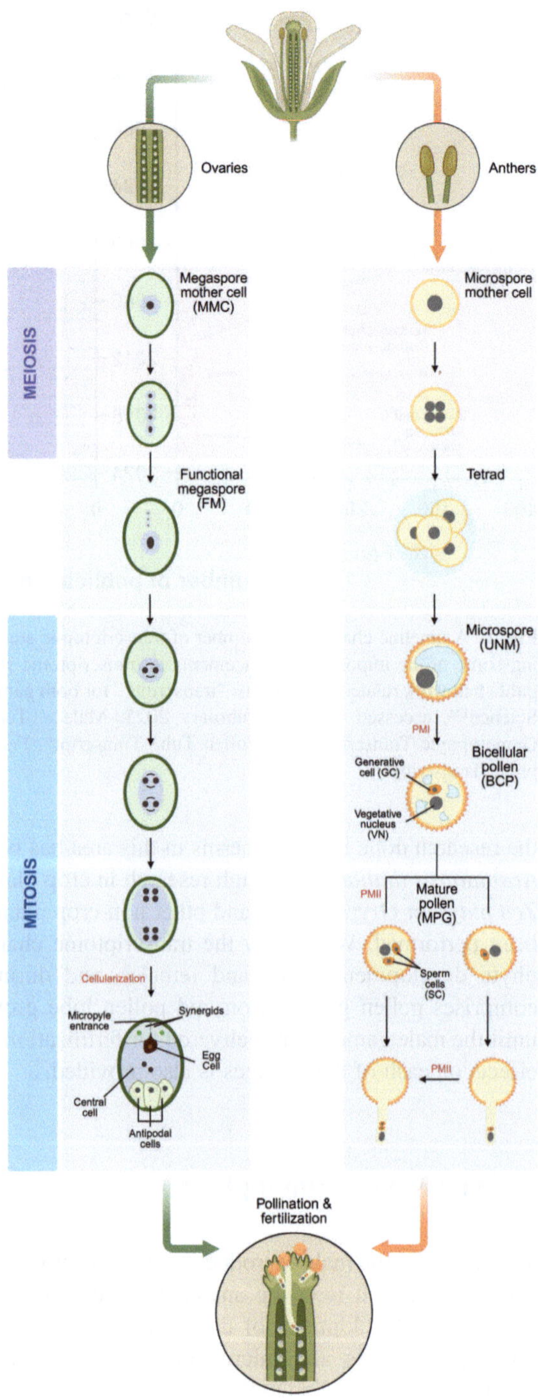


**Fig. 2** Gametophyte development in angiosperms. Female gametophyte development starts with the megaspore mother cell undergoing meiosis, originating four haploid cells. One of the cells originates the functional megaspore, while the other three degenerate. Through three consecutive mitosis events, the functional megaspore will originate the female gametophyte, which is constituted by seven cells: two synergids, a central cell, an egg cell, and three antipodal cells. Male gametophyte development starts with the microspore mother cell undergoing meiosis, originating four haploid microspores. The microspores will develop into pollen grains, by undergoing two mitosis events. Pollen grains can be bicellular or tricellular at anthesis depending on the species. Once gametophytes are developed, pollination, pollen tube growth, and fertilization occur during the progamic phase, culminating in seed production. PMI – Pollen mitosis I. PMII – Pollen mitosis II

microspores will develop into pollen grains, which can then be released for fertilization. The uninucleate microspores (UNMs) will form bicellular pollen (BCP) through pollen mitosis I (PMI) (Fig. 2). The BCP is formed by two cells, a large vegetative cell and a small generative cell (GC). Afterwards, the GC will divide again at pollen mitosis II (PMII), forming tricellular pollen (TCP). In around 30% of angiosperms, PMII occurs before pollen maturation, meaning these shed TCP (e.g. in *Arabidopsis*), while other angiosperms, like tomato, shed BCP undergoing PMII only during pollen tube growth (Brewbaker 1967) (Fig. 2). A programmed dehydration step is required before the pollen grains reach the mature pollen grain (MPG) stage. This dehydration increases the survivability of the MPG upon its release (Firon et al. 2012). When the MPG is released and reaches a compatible female stigma, the progamic phase starts. Here, the pollen rehydrates, and a pollen tube starts growing through the pistil to guide the sperm cells (SCs) into the ovules and achieve double fertilization (Fig. 2). The male gametophyte expresses fewer genes than other plant tissues (Wei et al. 2010; Chettoor et al. 2014; Rutley and Twell 2015; Conze et al. 2017), with the number of expressed genes declining drastically during male gametophyte development, from 15,000 in UNM and BCP, to around 8,000 in TCP, and finally 6,000 in MPG (Klodová et al. 2023).

## 2.1 *Microsporogenesis*

The number of stamen per flower and their type vary across plant families (Åstrand et al. 2021). Albeit this variation, the structure of mature anthers is constant, being comprised of four lobes, with four different cell types, all conserved across flowering plants: epidermis, endothecium, middle layer, and tapetum. These somatic cell types each form a single cell layer, which engulfs the central germinal cells. It is in the germinal cells that microsporogenesis occurs, with these cells first differentiating into archesporial cells, which through mitosis will originate pollen mother cells. These in turn will undergo meiosis, producing a tetrad of four haploid microspores (Nelms and Walbot 2019; Marchant and Walbot 2024) (Fig. 2). In maize, prior to the first meiosis, the transcriptome suffers minor changes, with no sudden shifts in expression. Afterwards, the transcriptional regulation during germinal differentiation is marked by two sharp shifts. After the first meiosis, there is a high expression of transcripts related to the translation machinery (e.g. ribosome biogenesis, translation elongation, and tRNA metabolism) that decreases during germinal differentiation. By contrast, transcripts for core meiotic functions and specific cellular organelles increase their expression during germinal differentiation, highlighting the changes in metabolism and cell biology that occur during the transition from diploid to haploid stages of pollen development (Nelms and Walbot 2019). In cotton, single cell transcriptomics allowed the categorization of cells into different stages of meiosis, and the identification of novel meiotic-related genes expressed during meiosis, providing a more in-depth look into the meiotic process and its regulation (Li et al. 2024b). In rice, the early stages of meiosis are regulated by phased secondary siRNAs that cleave target mRNAs, thus allowing meiosis progression (Zhang et al. 2020b). Research in barley has shown that pre-meiotic anthers are enriched in long non-coding RNAs transcripts that are significantly downregulated upon entry to meiosis. In addition, after meiosis begins, few transcripts show differential expression, pointing to a stable regulatory network during meiosis, after the initial reprogramming (Barakate et al. 2020).

## 2.2 The Start of Microgametogenesis: From Uninucleate Pollen to Bicellular Pollen

The first major event that occurs during microgametogenesis is the polarization of the UNM and its subsequent asymmetric division to BCP during PMI (Fig. 2). The asymmetrical division that produces the vegetative cell and the GC is essential, with microspores that divide symmetrically failing to develop germ cells (Eady et al. 1995). In the UNM of *Arabidopsis*, transcripts associated with mitotic events (e.g. cytokinesis and microtubule-based movement) and the cell cycle are upregulated when compared to BCP, pointing to a preparation of the UNM for PMI (Klodová et al. 2023). An analysis of late UNM revealed an enrichment of transcripts related with the cell cycle, but also with ribosome biogenesis, suggesting that the late microspores are already establishing a transcriptional programme that will be used by the vegetative nucleus (VN) in the BCP (Ichino et al. 2022). The asymmetrical division that occurs at PMI is dependent on microtubules, and their establishment can be impaired by mutants that affect microtubule organization and dynamics. Some examples of these mutants are *gem1*, which affects the function of the MICROTUBULE ORGANIZATION 1/GEMENI POLLEN 1 (MOR1/GEM1) protein (Twell et al. 2002), *sidecar* that disturbs mitotic spindle orientation, affecting the timing of mitotic entry and leading to the formation of an extra vegetative-like cell (Chen and McCormick 1996; Oh et al. 2010), and *two-in-one*, which causes defects in the cytokinesis of PMI (Oh et al. 2005). Remnants of meiosis-related transcripts present from the microsporogenesis phase can still be found in the UNM. Other transcripts related to early pollen tube, cell wall organization, and processes like transmembrane transport and glucan metabolism start increasing in the transition from UNM to BCP (Klodová et al. 2023). After PMI, the BCP is formed. At this stage, the expression of genes related to pollen tube growth and regulation starts and continues to increase throughout pollen development. Although the BCP bears an additional cell, the GC, the transcriptomes of the UNM and BCP are very similar, showing that at this stage, the GC contributes very little to the transcriptome of the BCP (Klodová et al. 2023).

## 2.3 *Major Transcriptomic Changes: Bicellular Pollen to Tricellular Pollen*

The transition from bicellular to tricellular pollen is accompanied by a major transcriptomic switch in *Arabidopsis* (Honys and Twell 2004; Twell et al. 2006; Klodová et al. 2023). While BCP shows an upregulation of transcripts related to translation (e.g. tRNA, ribosome biogenesis, mRNA maturation, protein folding), this enrichment drastically decreases from TCP onwards (Klodová et al. 2023). Contrastingly, from TCP onwards, a notable increase is observed in transcripts related to vesicular transport, pH regulation, and energy metabolism (e.g. ATP synthesis and electron transport chain). Additionally, there is a rise in transcripts related to pollen tube growth and regulation. These transcripts may play a later role in ensuring a rapid germination of the MPG and pollen tube growth.

## 2.4 *Pollen Dehydration and Maturation*

During pollen development, the male gametophyte has to undergo programmed dehydration to reach maturity (Fig. 2). This mechanism increases pollen endurance to the environment, especially during dispersion. Dehydration starts at the BCP to late BCP stage, with the expression of genes that encode LEA proteins, which play a role in desiccation tolerance in seeds (Hoekstra et al. 2001), and other stress-/desiccation-related proteins (Klodová et al. 2023; Sze et al. 2024). The products of these genes probably play a protective role at this stage. The expression of many transporters that mediate nutrient uptake (e.g. $H^+$ and $Ca^{2+}$ pumps, and sugar and amino acid co-transporters), osmotic balance (aquaporins), and cell wall synthesis suggests an active modification of the vegetative cell in the BCP, which will lead to pollen dehydration and differentiation into mature pollen grains (Sze et al. 2024).

The final stage of microgametogenesis is pollen maturation. The vegetative cell, responsible for pollen tube growth and guiding the SC(s) to the female tissue, accumulates a substantial amount of metabolic reserves like carbohydrates, proteins, and lipids, which ensure a rapid pollen tube growth (Pacini et al. 2006). In *Arabidopsis*, from PMI to pollen maturation, mRNAs and protein reserves, in addition to other transcripts related to cell-wall metabolism, signalling, and cytoskeletal dynamics, have been shown to be synthetized and accumulated (Klodová et al. 2023). In *Nicotiana tabacum*, pollen grains accumulate significant reserves of mRNAs that will be later translated to various proteins during pollen tube growth. This accumulation was defined as the pollen sequestrome and is mediated by nontranslating monosomes. The pollen sequestrome is another highly dynamic and important mechanism of translational regulation in pollen grains (Hafidh et al. 2018). During pollen development, the expression of transcripts related to vesicle transport, exocytosis-related proteins, and energy and biosynthetic processes has been steadily declining since the UNM stage, reaching minimum expression at MPG stage.

Transcripts related to pollen, pollen tube germination, and pollen development decrease at this stage, while transcripts related with regulation of pollen tube growth, microgametogenesis, and stress response increase in *Arabidopsis* (Klodová et al. 2023). The diverse transcripts safeguard pollen grain viability during abiotic/biotic stresses and promote rapid pollen tube germination and growth.

Another layer of protection of the MPG from environmental stress is its thick wall. The wall is composed by two layers, the inner layer, intine, and the outer layer, exine (Shi et al. 2015). The intine is formed during the late stage of pollen development, while exine is believed to start being formed as early as at the microsporogenesis stage, and completely formed by the BCP stage (Shi et al. 2015). In rice, three GDSL esterase/lipase proteins (GELPs) were found to be required for exine development, making them also essential for finalizing pollen development (Zhang et al. 2020a). A more detailed look into pollen wall development can be found in the review by Zhang et al. (2024).

## *2.5 Preparing for Release: Anther Dehydration and Dehiscence*

One of the final processes in male gametophyte development is anther dehydration, which happens after the pollen grains are fully developed. The locular fluid of the anthers is removed, enabling pollen dispersal, and the anther walls are dehydrated, facilitating anther dehiscence (small ruptures in the anther) and pollen release (Åstrand et al. 2021). Anther dehydration is caused by water transport in vascular bundles and evaporation through the stomata. Stomatal differentiation in the epidermis of anthers is controlled by INDUCER OF CBF EXPRESSION 1 (ICE1). This gene affects anther dehiscence, and pollen viability and germination, while also positively regulating processes related to water movement (e.g. ion transporters, hydrolases, and dehydration-associated genes) in the anther (Wei et al. 2018). Hormones play an important role in anther dehiscence. Auxin (Cecchetti et al. 2008, 2013; Song et al. 2018; Xu et al. 2019), ethylene (Rieu et al. 2003; Wang and Kumar 2007), jasmonic acid (Sanders et al. 2000; Ishiguro et al. 2001), and cytokinins (Jung et al. 2008) were all reported to be involved in this process. As such, an enrichment of genes, involved in hormone signalling and metabolism for these hormones, can be found during this phase (Hong et al. 2022).

## *2.6 Transcriptomic Dynamics in and Between Sperm Cell and the Vegetative Nucleus*

When analysing pollen developmental stages BCP to MPG and growing pollen tubes, a mixture of GC/SC, VN, and cytoplasm transcriptomes is analysed. In order to understand the dynamics of each cell type, cell/nuclei isolation is necessary. As such, techniques have been developed in order to achieve this isolation, allowing transcriptomic studies of these cells (Tian and Russell 1997; Di Martino et al. 2004; Lu et al. 2015; Santos et al. 2017). The VN and the SC play different functions during plant sexual reproduction. The VN is responsible for pollen tube growth and transport of the SCs, which are responsible for double fertilization of the female gametophyte. The VN presents a more complex transcriptome when compared with the SCs in MPG of *Arabidopsis* (Misra et al. 2023). During early pollen development of *Arabidopsis*, the VN expresses genes related with mRNA processing and ribosome biogenesis that are gradually downregulated during the rest of the development. In the middle stages of pollen development (from BCP to TCP), transcripts related with Golgi vesicle transport and endocytosis are being upregulated. In the final stages, transcripts associated with cell tip growth and pollen tube development were identified as being upregulated (Ichino et al. 2022). The expression pattern is very similar to what happens in the development of the whole pollen grain, which is expected, since it has been shown that the VN transcriptome strongly correlated with that of whole pollen grains (Borg et al. 2021). Expression in the VN initially builds the machinery required for transcription and translation, which will be used in later stages of development to produce proteins required for membrane expansion and pollen tube growth. After PMI, the GC is formed. This cell will originate the SC after PMII. The GC shows an enrichment in transcripts related to cell cycle and chromosome organization, similar to what can be observed for SCs (Ichino et al. 2022; Misra et al. 2023). When comparing both cell types in MPG, the VN is enriched in transcripts related to pollen germination and tube growth, and vesicle-mediated transport, while the SCs present an enrichment of transcripts related to cell-cycle regulation, DNA metabolic process, and chromatin organization (Anderson et al. 2013; Misra et al. 2023).

## 3 The Female Gametophyte

In angiosperms, the development of the female gametophyte, like of its male counterpart, is divided into two main phases: megasporogenesis and megagametogenesis (Yadegari and Drews 2004; Hater et al. 2020). Megasporogenesis starts with the differentiation of a single subepidermal cell located in the ovule primordium into the archesporial cell (AC). The AC will elongate and develop into the megaspore mother cell (MMC). The diploid MMC originates a tetrad of haploid megaspores through meiosis (Webb and Gunning 1990) (Fig. 2). During megagametogenesis, one of the megaspores, the functional megaspore (FM), will develop into a mature female gametophyte by undergoing three sequential rounds of mitotic division (Fig. 2). The female gametophyte is constituted by eight nuclei, which are incorporated by seven cells: three antipodal cells, the egg cell, two synergids, and the central cell, which has two haploid nuclei (Webb and Gunning 1994) (Fig. 2). Due to the small number of cells and their high inaccessibility, molecular and genomic studies have been historically difficult to perform on the female gametophyte, particularly when compared to the male.

## 3.1 Megasporogenesis

In angiosperms, the most common pattern of megasporogenesis is the monosporic (*Polygonum* type), being found in *A. thaliana*, *Zea mays*, *Oryza sativa*, and other crops (Christensen et al. 1997). In the first stage of megasporogenesis, the AC and the MMC are enriched by transcripts related to plant hormone signal transduction and plant-pathogen interactions (Zhao et al. 2020a). In particular, transcripts related to the phytohormone auxin, which plays a fundamental role in ovule initiation (Robert et al. 2015), were found preferentially expressed at the AC and MMC stages of rice (Zhao et al. 2020a). Other enriched transcripts identified in the MMC are related to biogenesis, metabolism, structure, transporter activity, and meiosis, suggesting that to promote the progression of the megasporogenesis, the MMC is metabolically very active (Zhao et al. 2014; Pinto et al. 2024). The gene profile of both male and female meiocytes in *Arabidopsis* is highly similar (96.9% of the female genes were also found expressed in the male), which makes sense considering that both meiocytes undergo meiosis, which is a highly conserved process (Zhao et al. 2014). Genes that promote MMC development have been identified for various angiosperms. In *Arabidopsis*, genes like NOZZLE/SPOROCYTELESS (NZZ/SPL) (Yang et al. 1999; Schiefthaler et al. 1999) and WINDHOSE1 and 2 (WH1/WH2) that are indirectly activated by WUSCHEL (WUS) promote MMC development (Lieber et al. 2011). Genes in other species of angiosperms have also been identified, like MULTIPLE SPOROCYTE1 (MSP1) (Nonomura et al. 2003) and MEIOSIS ARRESTED AT LEPTOTENE1 (MEL1) (Nonomura et al. 2007) for rice, or ARGONAUTE104 (AGO104) (Singh et al. 2011) in maize.

## 3.2 Megagametogenesis

After the functional megaspore is selected, it is subject to a series of developmental processes that will end with the formation of the mature female gametophyte. Megagametogenesis starts with the functional megaspore undergoing three mitotic divisions without cellularization, originating eight nuclei (Christensen et al. 1997) (Fig. 2). During this stage, transcripts related to mitosis and vacuole biogenesis are particularly important for the progression of the development (Erbasol Serbes et al. 2019). Similar to what happens during megasporogenesis, the phytohormone auxin plays a role during early megagametogenesis, with targeted disruption of auxin transporters like PIN-FORMED1 (PIN1) or AUX1 resulting in mitotic arrest (Pagnussat et al. 2009; Ceccato et al. 2013; Panoli et al. 2015).

During megagametogenesis, there is also substantial nuclear relocation. After the first mitotic division, two nuclei will be positioned at opposing ends, separated by the developing vacuole. After two more mitotic divisions, a total of eight nuclei are generated, and one from either pole will migrate into the centre, to originate the central cell. The final position of the nuclei is followed by cellularization, compartmentalizing the female gametophyte into a total of seven cells (Fig. 2). As such, during this nuclear relocation, transcripts related to cytoskeleton elements are key, with mutations that create perturbations of the cytoskeleton, like in genes for γ-tubulin (e.g. *TUBG1*/*TUBG2* and *GCP2*), creating disruption in mitosis, nuclear positioning, and cellularization in *Arabidopsis* (Pastuglia et al. 2006; Nakamura and Hashimoto 2009).

The female gametophyte is composed of four different cell types: the synergids, antipodal cells, the egg cell, and the central cell (Fig. 2). These cells are closely related in terms of their cell lineage, but play different roles during sexual reproduction (Yadegari and Drews 2004). Cell differentiation in the female gametophyte has been said to be determined by the position of the nuclei and cell before differentiation (Lin 1978; Guo et al. 2004). Additional molecular information is, of course, necessary for cell specification in the female gametophyte. The synergid cells' development is regulated by *MYB98* in *Arabidopsis*. This gene is specifically expressed in the synergids and is essential for synergid cell differentiation, formation of the filiform apparatus, and for pollen tube guidance (Kasahara et al. 2005; Punwani et al. 2007). Egg cell specification is still poorly understood; however, RWP-RK domain-containing transcription factors (RKDs) have been reported as being involved in this specification (Koszegi et al. 2011). Central cell differentiation is regulated by transcription factors, AGAMOUS LIKE 80 (AGL80) and AGL61 in *Arabidopsis*. Their absence results in defects in central cell maturation and in a failure to initiate seed endosperm (Portereiko et al. 2006; Bemer et al. 2008; Steffen et al. 2008).

### *3.3 Cell Type Transcriptomic Differences in the Female Gametophyte*

The transcriptome of male and female gametophytes share numerous characteristics that are very distinct from the ones of sporophytic tissues (Wuest et al. 2010). In flowering plants, double fertilization occurs, meaning that two female gametes are fertilized by two sperms cell that were carried by the pollen tube. The egg cell forms the embryo of the seed after fertilization, and, as such, it is very enriched in the expression of transcripts related to fertilization, DNA methylation, small regulatory RNA, and embryo development (Song et al. 2020; Susaki et al. 2021). The central cell is also fertilized and is the progenitor of the endosperm of the future seed. The central cell of *Arabidopsis* was found to be enriched in transcripts related to photosynthesis and response to cytokinins (Song et al. 2020; Susaki et al. 2021), all consistent with its biological role. The central cell also plays a critical role in pollen tube guidance through the expression of the CENTRAL CELL GUIDANCE (CCG) gene (Chen et al. 2007). The synergid cells have an enrichment of transcripts related to small GTPase-mediated signal transduction, protein secretion, and cell wall synthesis, all consistent with the biological role of these cells in guiding the pollen tube and fertilization (Song et al. 2020; Susaki et al. 2021). All three cell types have a high level of transcripts that encode for cysteine-rich peptides (CRPs) (Song et al. 2020), which regulate plant growth and help in the guidance of the pollen tube and gamete recognition during double fertilization (Marshall et al. 2011).

## 4 The Pollen Tube Journey: The Progamic Phase

After both gametophytes finish their development, the progamic phase begins, lasting from pollen release to fertilization (Fig. 2). When a dehydrated pollen grain reaches the female stigmatic papillae, it will rehydrate and germinate a pollen tube. The pollen tube will transport the SCs to the female gametophyte, in order to achieve double fertilization. Differentiation between self and nonself pollen is achieved through membrane receptors in the stigmatic papillae, which detect pollen coat ligand molecules and subsequently trigger pollen acceptance or rejection. Cross pollination is required in the majority of angiosperms, but 10–15% of flowering plants can self-pollinate, like *Arabidopsis* or tomato (Wright et al. 2013). After pollen is accepted, pollen performance is a fundamental aspect for fertilization, as the pollen will compete with other pollen for germination, penetration of the transmitting tract and, later, for female gametes (Williams and Reese 2019). MPG will use their accumulated reserves and female-deposited ones to fuel this race and ensure double fertilization. The speed at which the pollen tube grows depends on the stored nutrition reserves, and is different between species that shed bicellular or tricellular pollen (Williams and Reese 2019).

Due to the complexity of isolating pollen tubes growing inside the female tissue, techniques have been developed to emulate this interaction, allowing the study of the transcriptome of pollen tubes during their journey for double fertilization. While in vitro pollen germination allows the study of pollen tubes without female tissue interaction using pollen germination media that simulates the conditions of the style-stigma (Rodriguez-Enriquez et al. 2013), semi-in vivo pollen germination is done directly at the stigmatic surface and some parts of the transmitting tract, emulating the way a pollen tube would grow in natural conditions (Dickinson et al. 2018).

### *4.1 Pollen Tube Germination and Growth*

A compatible MPG that reaches the papillae will hydrate and initiate pollen tube germination (Fig. 2). Germination is accompanied by a general and unspecified downregulation of gene expression, with a high number of genes being downregulated in the pollen tube, when compared to MPGs (Conze et al. 2017). The activity of the transcription machinery is reduced during pollen tube germination, since the mRNAs required for it (e.g. transcripts related to cell-wall metabolism, signalling, and cytoskeleton dynamics) accumulate during pollen maturation (Wei et al. 2010; Rutley and Twell 2015; Conze et al. 2017). Pollen germination is accompanied by a rapid influx of calcium at the emergence site. Transcripts encoding for kinases, particularly ones related to calcium signalling, display high expression and accumulation in MPG and in germinated pollen grains. This aligns with the important role calcium signalling and ion gradients in general play in pollen germination and growth (Holdaway-Clarke et al. 1997; Iwano et al. 2004; Malhó et al. 2006; Cárdenas et al. 2008; Michard et al. 2008). Aquaporin regulation is a key layer in preventing the initiated pollen tube from rupturing due to hypoosmotic shock (Pérez Di Giorgio et al. 2016). Phytohormones, like auxin and ethylene, also play a role during pollen growth, with an increase in transcripts related to these being observed (Wei et al. 2010). Once pollen has germinated, the race to be the first to reach the female gametophyte starts. The pollen tube still needs to activate specific important genes required for its elongation, with an enrichment of transcripts related to pollen tube growth and fertilization (pollen tube reception, synergid death, and fertilization) during this stage (Wang et al. 2008; Conze et al. 2017).

The interaction between pollen tubes and the pistil cells is also very relevant, but unravelling the transcriptome of each one during pollination is technically complex, as the pollen tubes are growing inside the pistil tissue. To emulate this, pollen tubes can be grown in vitro and semi-in vivo and compared, allowing to understand the effects of pollen-pistil interaction (PPI). In *Arabidopsis*, after PPI, pollen tubes undergo transcriptional change, which occurs in both SC and VN. Some of these changes are triggered exclusively after PPI, particularly in the SCs (Misra et al. 2025). The SCs of growing pollen tubes were shown to upregulate genes involved in proteins transport and transcription regulation, some of which have potential functions in gamete interaction and fusion. SCs were shown to have specific alterations in the transcriptome after PPI, reducing the expression of genes related to pollen tube growth and cell-to-cell signalling, functions usually associated with the VN (Misra et al. 2025). The VN of growing pollen tubes upregulate genes associated with pollen tube growth, vesicle-mediated transport, and DNA replication, in addition to several transmembrane receptors and transcription factors (Misra et al. 2025). Pistils also go through transcriptomic changes after PPI, although the majority of the pistil-expressed genes do not change (Boavida et al. 2011; Leydon et al. 2017). Nevertheless, there is upregulation of genes related to the secretion of glycoproteins and cell wall localized proteins. Genes related to response to chitin and auxin were also enriched after pollination, while a decrease in the expression of genes related to cytochrome p450 and oxidative stress could also be observed (Leydon et al. 2017).

The journey of the pollen tube towards the female gametophyte is divided in two phases, the preovular and the ovular guidance. During preovular guidance, the pollen tube is in constant communication with maternal tissues of the stigma, style, and transmitting tract, until it reaches the proximities of the ovule. The preovular guidance through the transmitting tract is not unidirectional and the female gametophyte is said to be unnecessary for pollen tube growth from stigma to style (Hulskamp et al. 1995; Sogo and Tobe 2005), suggesting that the pollen tubes are kept on track by the tissue barriers and other signals, and not from direct signals from the female gametophyte. Some of the signals that the female gametophyte can release include GABA, in *Arabidopsis* (Palanivelu et al. 2003), or TRANSMITTING TRACT-SPECIFIC (TTS) glycoproteins in *Nicotiana tabacum* (Cheung et al. 1995), both stimulating pollen tube growth in vitro. The pollen tube journey through the transmitting tract cells is accompanied by programmed cell death and the production of a rich extracellular matrix, facilitating movement, with mutations in genes that encode for these processes reducing fertilization efficiency (Crawford et al. 2007; Crawford and Yanofsky 2011).

The second phase of the pollen tube journey starts when it reaches the proximities of the ovule, with the female gametophytic cells taking over the communication, redirecting the pollen tube from the transmitting tract to the micropyle entry of the ovule. Ovular guidance is subdivided into two phases, the funicular guidance and the micropylar guidance. Funicular guidance is the least studied of the two, with micropylar guidance being deeply dissected, which, due to its complexity, will only be briefly discussed (Mizuta and Higashiyama 2018; Kim et al. 2021). Defects in funicular guidance have been found in double mutants in two mitogen-activated protein kinases, *mpk3* and *mpk6*, with pollen tubes growing normally, but impairing during funicular guidance (Guan et al. 2014). Micropylar guidance is a very complex process with many different layers of expression (Dresselhaus and Márton 2009). Different cells of the female gametophyte play different roles and express and secrete different proteins. Synergids express MYB98 (see Sect. 2.2) (Kasahara et al. 2005; Punwani et al. 2007), which regulates the expression of cysteine-rich polypeptides (CRPs), with two in particular, LURE1 and LURE3, functioning as pollen tube attractants derived from the synergids in *Torenia fournieri* (Okuda et al. 2009). The central cell in *Arabidopsis* and the egg cell in maize also play a role in micropylar guidance. *Arabidopsis* mutants for CENTRAL CELL GUIDANCE (CCG) (see Sect. 2.3) fail to attract pollen tubes (Chen et al. 2007). In maize, EGG APPARATUS1 encodes a peptide that plays a critical role in micropylar pollen tube guidance, with its downregulation leading to reduced fertility (Márton et al. 2005).

The final step of the pollen tube journey occurs once it is inside the micropyle, with the first contact with the receptive synergid stopping the growth of the pollen tube. These are the last steps before double fertilization and require constant communication between the pollen tube and the female gametophyte. Upon entry into the synergid, the pollen tube will burst, releasing the SCs, and the receptive synergid cell will undergo programmed cell death, facilitating the double fertilization event. Subsequently, if fertilization is successful, the non-receptive synergid degenerates, guaranteeing the distribution of pollen tubes to other receptive ovules. However if fertilization fails (e.g. defective SCs are delivered), the non-receptive synergid becomes receptive, granting a second chance for the ovule to be fertilized (Beale et al. 2012).

## 4.2 Double Fertilization

After pollen tube rupture in the vicinity of the female gametes (egg and central cell), the SCs need to reposition themselves in order to adhere and fuse with their female counterpart. If two sperms cells adhere to the same gamete, they are able to communicate and readjust their position, ensuring double fertilization (Huang et al. 2015). The contribution of maternal and paternal genomes to the transcriptome of early plant embryos is controversial. Different sources show different ratios, either equal contribution from both parents (Nodine and Bartel 2012), or maternally dominated (Autran et al. 2011). The expression patterns of different parent-of-origin genes vary with developmental stages and cell lineages of the embryo (Weijers et al. 2001; Zhao et al. 2020b).

Paternal delivery of specific transcripts appears to be an important process to ensure fertilization and embryo development. SHORT SUSPENSOR (SSP) transcripts are produced in SCs of *Arabidopsis*, but are not translated, being delivered to the zygote and the endosperm, where they transiently accumulate and regulate the first asymmetric division of the zygote (Bayer et al. 2009). The paternal delivery of the core cell cycle component CYCD7;1 to the central cell, upon fertilization, is essential to degrade RETINOBLASTOMA RELATED1 (RBR1) protein, which arrests the central cell in the S phase. This degradation allows S phase progression and the formation of a functional endosperm (Simonini et al. 2024). Many miRNAs accumulate in the MPG, particularly in the SC (Borges et al. 2011). One example, miR159, is enriched in the SC and transmitted to the central cell during fertilization, promoting endosperm nuclear division by repressing MYB33 and MYB65 (Zhao et al. 2018). Two other paternal expressed genes, EIN3-dependent Ethylene-response 1 (TREE1) and its homologue DUO1-ACTIVATED ZINC FINGER3 (DAZ3), are exclusively expressed in SCs, and are sperm-carried and delivered to the egg cell after fertilization. Mutation of these genes causes no defects in the development of the male gamete, nor in fertilization, but will impact the embryo and seedling by disrupting differentiation of the root stem cell niche (Cheng et al. 2024).

# 5 Effect of Heat Stress During Plant Sexual Reproduction

With global warming, heat waves are increasingly more common. Increased temperatures often have a detrimental effect on plant sexual reproduction, affecting fertility, and thus crop yield. Understanding the effects of heat stress on plant sexual reproduction and how plants respond to it is then key to mitigating its effects. Heat stress affects male and female gametophyte development differently, with the former being more sensitive to it (Zinn et al. 2010; Hedhly 2011). Heat stress affects synchronous development of the pollen grains, which is essential for fertilization, being particularly sensitive during meiosis and PMI (Müller and Rieu 2016; Rieu et al. 2017). For example, in maize, heat stress application at earlier stages of pollen development (tetrad stage) affects starch, lipid, and energy biosynthesis-related genes, but genes related to developmental processes remained unaltered (Begcy et al. 2019). Heat stress at the UNM stage accelerated development and impaired pollen germination, while application at BCP stage impaired SC development and their transport into the pollen tube after germination. SC gene expression is also affected, with a change in expression of genes related to transcription, RNA processing and translation, DNA replication, and cell cycle (Li et al. 2024a). Heat shock proteins (HSPs) are well-established proteins involved in heat stress response (Qu et al. 2013). In maize and cotton, heat stress application at the tetrad stage caused upregulation of HSPs and these were still detectable in the MPG (Begcy et al. 2019; Masoomi-Aladizgeh et al. 2021). Post-transcriptional regulation also plays a role in the heat stress response of pollen. For example, during pollen development of tomato under heat stress, miRNAs that are predicted to target heat stress transcription factors and HSP were found to be downregulated, while their target transcripts were upregulated (Keller et al. 2020), pointing to a possible role of these miRNAs in pollen thermotolerance. A more detailed overview of the heat stress effect on pollen development can be found in Chaturvedi et al. (2021).

The female gametophyte, as previously said, is more resilient to heat stress than its male counterpart. There is a lack of transcriptomic studies on the developing female gametophyte under heat stress, although there is data providing evidence of the negative effect heat stress has on female gametophyte development (Saini and Aspinall 1982; Saini et al. 1983; Gross and Kigel 1994; Peet et al. 1998; Whittle et al. 2009; Shi et al. 2022). In *Arabidopsis*, the cells of the female gametophyte withstand heat stress differently, with the central cell being more resilient than the egg and antipodal cells, by upregulation of various HSPs (Ambastha and Leshem 2020). HSPs were also found upregulated in the ovules of maize under heat stress (Jagtap et al. 2023). Different species show different degrees of sensitivity to heat stress in the pistil, with some cases showing higher sensitivity than pollen grains. The impact heat stress has on the pistil has been thoroughly reviewed by Wang et al. (2021).

The progamic phase is also susceptible to heat stress, affecting pollen fitness and reducing the pistil's ability to support pollen germination and pollen tube growth due to a reduction in pollen adhesion (Lohani et al. 2020). In *Brassica napus*, a short-term heat stress affected both pollen and the pistil in similar ways – by altering the expression of genes related to functions like protein processing, regulation of transcription, PPI, or cellular organization, transport, and metabolism. In the pistil, heat stress repressed transcripts that encode transcription factors (Lohani et al. 2020). *Arabidopsis* pollen grains germinated in vitro under heat stress show upregulation of transcripts related to heat stress response, protein folding, and endoplasmic reticulum. There was also an increase in the expression of HSPs, and a downregulation of genes related to transport (Poidevin et al. 2021). In tomato, pollen tubes grown under heat stress showed a reduction in growth, which lead to a reduction of seed production. Heat stress during pollen tube growth increased the expression of transcripts related to temperature stimulus responses, to oxidative, osmotic, and salt stress, and to protein and chaperon-mediated protein folding. Pollen tubes of thermotolerant plants showed an enhanced reactive oxygen species homeostasis and pollen tube callose synthesis and deposition, suggesting a role of these transcripts in thermotolerance of tomato (Ouonkap et al. 2024).

# 6 Technological Trends and Future Directions

It is impossible to talk about future directions in the field of transcriptomics without mentioning current and future technological advancements. In the past, whole-genome transcriptomic studies on plant sexual reproduction were based on microarrays and serial analysis of gene expression (SAGE) (Honys and Twell 2003; Lee and Lee 2003; Becker et al. 2003; Tung et al. 2005; Pina et al. 2005; Yu et al. 2005; Johnston et al. 2007). With the emergence of RNA sequencing (RNA-seq), more detailed whole-genome transcriptomic studies were performed, allowing the study of other layers of gene regulation, like alternative splicing (Loraine et al. 2013; Misra et al. 2023; Klodová et al. 2023). RNA-seq can also be combined with techniques like laser capture microdissection, allowing to reach and analyse tissues of difficult access (e.g. developing embryo and ovule) (Casson et al. 2005; Sakai et al. 2018). More recent technologies like single-cell/single-nucleus RNA-seq analysis have been adopted in the field (Song et al. 2020; Hou et al. 2021; Ichino et al. 2022; Li et al. 2024b). Long-read sequencing is another recent technology (Logsdon et al. 2020) that has been scarcely used with plant reproductive tissues. This technology allows improved transcriptome assembly, making it a useful tool to study alternative splicing by enabling improved detection of novel splicing events (Sun et al. 2023; Wu et al. 2023). A wider use of these technologies would allow a deeper understanding of cell-type-specific transcriptome dynamics during sexual reproduction. Equally important is an easy visualization of published transcriptomic data. Tools have been developed to ensure this, like the Evorepro database (https://evorepro.sbs.ntu.edu.sg), a platform that enables the visualization of gene expression and co-expression networks of various species of plants (including angiosperms), with a focus on reproductive tissues and different stages of their development (Julca et al. 2021). Another database is the eFP-Seq Browser (https://bar.utoronto.ca/eFP-Seq_Browser/), focusing on *Arabidopsis* data. It was recently updated to include data from pollen development and progamic phase (Fig. 3) (Sullivan et al. 2019; Klodová et al. 2023; Misra et al. 2025). Updating, maintaining, and improving tools like this are key to improve accessibility and maximize the value of these datasets for research on plant sexual reproduction. Transcriptome dynamics are only a snapshot of what happens inside plant reproductive cells. To obtain a wider picture, proteomic and metabolomic studies are also important, as they would allow better correlation between transcriptome, proteome, and metabolome. This, of course, has some obstacles, mainly related to the difficulty in obtaining large quantities of material for each specific stage of plant sexual reproduction, especially at the single cell/nucleus level. With the rapid development of low input proteomics and metabolomics, we are moving closer to the single

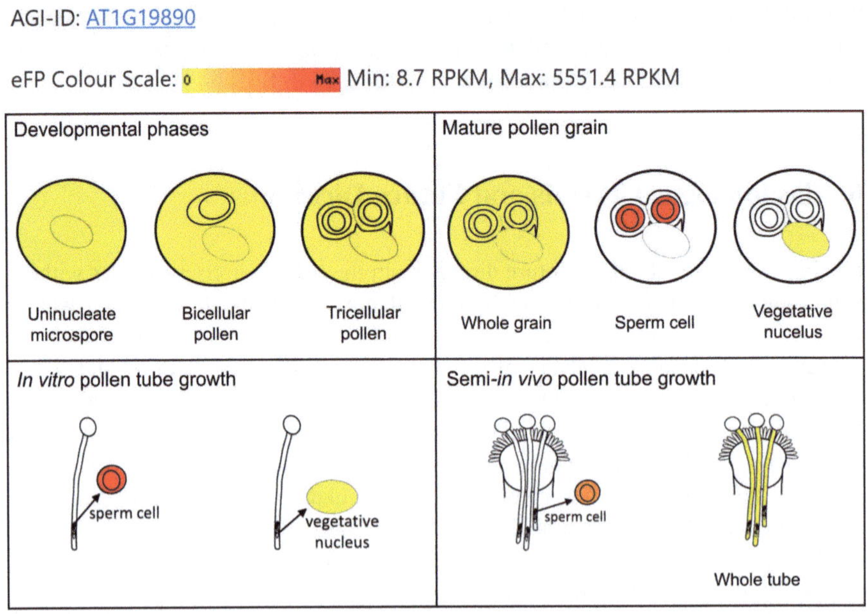


**Fig. 3** Example of the visualization of gene expression in reproductive tissues based on the eFP-Seq Browser and the integrated pollen dataset. AT1G19890 is a male-gamete-specific histone H3 that is highly expressed in the sperm cells of pollen grains and tubes. Cell types with higher RPKMs are represented in red. RPKM – Reads per kilobase per million reads mapped

cell/nucleus resolution required to achieve this correlation. These new technological advancements will allow to address persisting knowledge gaps, like whether isomorphic SCs differ at transcriptome or proteome level and if this may lead to preferential fertilization of the egg or central cell. In addition, this would allow to address the extent of parental contribution of each gamete to the early embryo, and the cell-type-specific transcriptional changes elicited in male and female tissues and gametes by combinations of abiotic stresses.

**Acknowledgements** The authors are supported by Fundação para a Ciência e a Tecnologia (FCT) through fellowships and grants UI/BD/154551/2022 (DOI:10.54499/UI/BD/154551/2022), PTDC/ASP-PLA/2007/2020 and CEECIND/03345/2018, GREEN-IT "Bioresources for Sustainability" R&D Unit base (DOI:10.54499/UIDB/04551/2020) and programmatic (DOI: 10.54499/UIDP/04551/2020) funding, and LS4FUTURE Associated Laboratory (DOI: 10.54499/LA/P/0087/2020).

# References


Ambastha V, Leshem Y (2020) Differential cell persistence is observed in the Arabidopsis female gametophyte during heat stress. Plant Reprod 33:111–116

Anderson SN, Johnson CS, Jones DS et al (2013) Transcriptomes of isolated Oryza sativa gametes characterized by deep sequencing: evidence for distinct sex-dependent chromatin and epigenetic states before fertilization. Plant J 76:729–741

Ashapkin VV, Kutueva LI, Aleksandrushkina NI, Vanyushin BF (2019) Epigenetic regulation of plant gametophyte development. Int J Mol Sci 20:3051

Åstrand J, Knight C, Robson J et al (2021) Evolution and diversity of the angiosperm anther: trends in function and development. Plant Reprod 34:307–319

Autran D, Baroux C, Raissig MT et al (2011) Maternal epigenetic pathways control parental contributions to Arabidopsis early embryogenesis. Cell 145:707–719

Barakate A, Orr J, Schreiber M et al (2020) Barley anther and meiocyte transcriptome dynamics in meiotic prophase I. Front Plant Sci 11:619404

Baroux C, Raissig MT, Grossniklaus U (2011) Epigenetic regulation and reprogramming during gamete formation in plants. Curr Opin Genet Dev 21:124–133

Bayer M, Nawy T, Giglione C et al (2009) Paternal control of embryonic patterning in Arabidopsis thaliana. Science 323:1485–1488

Beale KM, Leydon AR, Johnson MA (2012) Gamete fusion is required to block multiple pollen tubes from entering an Arabidopsis ovule. Curr Biol 22:1090–1094

Becker JD, Boavida LC, Carneiro J et al (2003) Transcriptional profiling of Arabidopsis tissues reveals the unique characteristics of the pollen transcriptome. Plant Physiol 133:713–725

Begcy K, Dresselhaus T (2018) Epigenetic responses to abiotic stresses during reproductive development in cereals. Plant Reprod 31:343–355

Begcy K, Nosenko T, Zhou L-Z et al (2019) Male sterility in maize after transient heat stress during the tetrad stage of pollen development. Plant Physiol 181:683–700

Bemer M, Wolters-Arts M, Grossniklaus U, Angenent GC (2008) The MADS domain protein DIANA acts together with AGAMOUS-LIKE80 to specify the central cell in Arabidopsis ovules. Plant Cell 20:2088–2101

Boavida LC, Borges F, Becker JD, Feijó JA (2011) Whole genome analysis of gene expression reveals coordinated activation of signaling and metabolic pathways during pollen-pistil interactions in Arabidopsis. Plant Physiol 155:2066–2080

Borg M, Papareddy RK, Dombey R et al (2021) Epigenetic reprogramming rewires transcription during the alternation of generations in Arabidopsis. elife 10:e61894.

Borges F, Gomes G, Gardner R, et al (2008) Comparative transcriptomics of Arabidopsis sperm cells. Plant Physiol 148:1168–1181

Borges F, Pereira PA, Slotkin RK et al (2011) MicroRNA activity in the Arabidopsis male germline. J Exp Bot 62:1611–1620

Borges F, Gardner R, Lopes T et al (2012) FACS-based purification of Arabidopsis microspores, sperm cells and vegetative nuclei. Plant Methods 8:44

Brewbaker JL (1967) The distribution and phylogenetic significance of binucleate and trinucleate pollen grains in the angiosperms. Am J Bot 54:1069–1083

Cárdenas L, Lovy-Wheeler A, Kunkel JG, Hepler PK (2008) Pollen tube growth oscillations and intracellular calcium levels are reversibly modulated by actin polymerization. Plant Physiol 146: 1611–1621

Casson S, Spencer M, Walker K, Lindsey K (2005) Laser capture microdissection for the analysis of gene expression during embryogenesis of Arabidopsis: transcriptome of the Arabidopsis embryo. Plant J 42:111–123

Ceccato L, Masiero S, Sinha Roy D et al (2013) Maternal control of PIN1 is required for female gametophyte development in Arabidopsis. PLoS One 8:e66148

Cecchetti V, Altamura MM, Falasca G et al (2008) Auxin regulates Arabidopsis anther dehiscence, pollen maturation, and filament elongation. Plant Cell 20:1760–1774

Cecchetti V, Altamura MM, Brunetti P et al (2013) Auxin controls Arabidopsis anther dehiscence by regulating endothecium lignification and jasmonic acid biosynthesis. Plant J 74:411–422

Chaturvedi P, Wiese AJ, Ghatak A et al (2021) Heat stress response mechanisms in pollen development. New Phytol 231:571–585

Chen YC, McCormick S (1996) Sidecar pollen, an Arabidopsis thaliana male gametophytic mutant with aberrant cell divisions during pollen development. Development 122:3243–3253

Chen Y-H, Li H-J, Shi D-Q et al (2007) The central cell plays a critical role in pollen tube guidance in Arabidopsis. Plant Cell 19:3563–3577

Cheng T, Liu Z, Li H et al (2024) Sperm-origin paternal effects on root stem cell niche differentiation. Nature 634:220–227

Chettoor AM, Givan SA, Cole RA et al (2014) Discovery of novel transcripts and gametophytic functions via RNA-seq analysis of maize gametophytic transcriptomes. Genome Biol 15:414

Cheung AY, Wang H, Wu HM (1995) A floral transmitting tissue-specific glycoprotein attracts pollen tubes and stimulates their growth. Cell 82:383–393

Christensen CA, King EJ, Jordan JR, Drews GN (1997) Megagametogenesis in Arabidopsis wild type and the Gf mutant. Sex Plant Reprod 10:49–64

Conze LL, Berlin S, Le Bail A, Kost B (2017) Transcriptome profiling of tobacco (Nicotiana tabacum) pollen and pollen tubes. BMC Genomics 18:581

Crawford BCW, Yanofsky MF (2011) HALF FILLED promotes reproductive tract development and fertilization efficiency in Arabidopsis thaliana. Development 138:2999–3009

Crawford BCW, Ditta G, Yanofsky MF (2007) The NTT gene is required for transmitting-tract development in carpels of Arabidopsis thaliana. Curr Biol 17:1101–1108

Di Martino D, Giuffrè G, Staiti N et al (2004) Single sperm cell isolation by laser microdissection. Forensic Sci Int 146(Suppl):S151–S153

Dickinson H, Rodriguez-Enriquez J, Grant-Downton R (2018) Pollen germination and pollen tube growth of Arabidopsis thaliana: in vitro and semi in vivo methods. Bio Protoc 8:e2977

Dresselhaus T, Márton ML (2009) Micropylar pollen tube guidance and burst: adapted from defense mechanisms? Curr Opin Plant Biol 12:773–780

Eady C, Lindsey K, Twell D (1995) The significance of microspore division and division symmetry for vegetative cell-specific transcription and generative cell differentiation. Plant Cell 7:65–74

Erbasol Serbes I, Palovaara J, Groß-Hardt R (2019) Development and function of the flowering plant female gametophyte. Curr Top Dev Biol 131:401–434

Firon N, Nepi M, Pacini E (2012) Water status and associated processes mark critical stages in pollen development and functioning. Ann Bot 109:1201–1214

Gross Y, Kigel J (1994) Differential sensitivity to high temperature of stages in the reproductive development of common bean (Phaseolus vulgaris L.). Field Crop Res 36:201–212

Guan Y, Lu J, Xu J et al (2014) Two mitogen-activated protein kinases, MPK3 and MPK6, are required for funicular guidance of pollen tubes in Arabidopsis. Plant Physiol 165:528–533

Guo F, Huang B-Q, Han Y, Zee S-Y (2004) Fertilization in maize indeterminate gametophyte1 mutant. Protoplasma 223:111–120

Hafidh S, Potěšil D, Müller K et al (2018) Dynamics of the pollen sequestrome defined by subcellular coupled omics. Plant Physiol 178:258–282

Hater F, Nakel T, Groß-Hardt R (2020) Reproductive multitasking: the female gametophyte. Annu Rev Plant Biol 71:517–546

Hedhly A (2011) Sensitivity of flowering plant gametophytes to temperature fluctuations. Environ Exp Bot 74:9–16

Hoekstra FA, Golovina EA, Buitink J (2001) Mechanisms of plant desiccation tolerance. Trends Plant Sci 6:431–438

Holdaway-Clarke TL, Feijo JA, Hackett GR et al (1997) Pollen tube growth and the intracellular cytosolic calcium gradient oscillate in phase while extracellular calcium influx is delayed. Plant Cell 9:1999–2010

Hong W-J, Lee SK, Kim S-H et al (2022) Comparative transcriptome analysis of pollen and anther wall reveals novel insights into the regulatory mechanisms underlying anther wall development and its dehiscence in rice. Plant Cell Rep 41:1229–1242
Honys D, Twell D (2003) Comparative analysis of the Arabidopsis pollen transcriptome. Plant Physiol 132:640–652
Honys D, Twell D (2004) Transcriptome analysis of haploid male gametophyte development in Arabidopsis. Genome Biol 5:R85
Hou Z, Liu Y, Zhang M et al (2021) High-throughput single-cell transcriptomics reveals the female germline differentiation trajectory in Arabidopsis thaliana. Commun Biol 4:1149
Huang J, Ju Y, Wang X et al (2015) A one-step rectification of sperm cell targeting ensures the success of double fertilization: sperm cell rectification for double fertilization. J Integr Plant Biol 57:496–503
Hulskamp M, Schneitz K, Pruitt RE (1995) Genetic evidence for a long-range activity that directs pollen tube guidance in Arabidopsis. Plant Cell 7:57–64
Ichino L, Picard CL, Yun J et al (2022) Single-nucleus RNA-seq reveals that MBD5, MBD6, and SILENZIO maintain silencing in the vegetative cell of developing pollen. Cell Rep 41:111699
Ishiguro S, Kawai-Oda A, Ueda J et al (2001) The DEFECTIVE IN ANTHER DEHISCIENCE gene encodes a novel phospholipase A1 catalyzing the initial step of jasmonic acid biosynthesis, which synchronizes pollen maturation, anther dehiscence, and flower opening in Arabidopsis. Plant Cell 13:2191–2209
Iwano M, Shiba H, Miwa T et al (2004) Ca2+ dynamics in a pollen grain and papilla cell during pollination of Arabidopsis. Plant Physiol 136:3562–3571
Jagtap AB, Yadav IS, Vikal Y et al (2023) Transcriptional dynamics of maize leaves, pollens and ovules to gain insights into heat stress-related responses. Front Plant Sci 14:1117136
Johnston AJ, Meier P, Gheyselinck J et al (2007) Genetic subtraction profiling identifies genes essential for Arabidopsis reproduction and reveals interaction between the female gametophyte and the maternal sporophyte. Genome Biol 8:R204
Julca I, Ferrari C, Flores-Tornero M et al (2021) Comparative transcriptomic analysis reveals conserved programmes underpinning organogenesis and reproduction in land plants. Nat Plants 7:1143–1159
Jung KW, Oh S-I, Kim YY et al (2008) Arabidopsis histidine-containing phosphotransfer factor 4 (AHP4) negatively regulates secondary wall thickening of the anther endothecium during flowering. Mol Cells 25:294–300
Kasahara RD, Portereiko MF, Sandaklie-Nikolova L et al (2005) MYB98 is required for pollen tube guidance and synergid cell differentiation in Arabidopsis. Plant Cell 17:2981–2992
Keller M, Schleiff E, Simm S (2020) miRNAs involved in transcriptome remodeling during pollen development and heat stress response in Solanum lycopersicum. Sci Rep 10:10694
Kim M-J, Jeon BW, Oh E et al (2021) Peptide signaling during plant reproduction. Trends Plant Sci 26:822–835
Klodová B, Potěšil D, Steinbachová L et al (2023) Regulatory dynamics of gene expression in the developing male gametophyte of Arabidopsis. Plant Reprod 36:213–241
Koszegi D, Johnston AJ, Rutten T et al (2011) Members of the RKD transcription factor family induce an egg cell-like gene expression program: RKD controls the egg cell transcriptional program. Plant J 67:280–291
Laloum T, Martín G, Duque P (2017) Alternative splicing control of abiotic stress responses. Trends Plant Sci 23:140–150
Le Trionnaire G, Twell D (2010) Small RNAs in angiosperm gametophytes: from epigenetics to gamete development. Genes Dev 24:1081–1085
Lee J-Y, Lee D-H (2003) Use of serial analysis of gene expression technology to reveal changes in gene expression in Arabidopsis pollen undergoing cold stress. Plant Physiol 132:517–529
Leydon AR, Weinreb C, Venable E et al (2017) The molecular dialog between flowering plant reproductive partners defined by SNP-informed RNA-sequencing. Plant Cell 29:984–1006

Li X, Bruckmann A, Dresselhaus T, Begcy K (2024a) Heat stress at the bicellular stage inhibits sperm cell development and transport into pollen tubes. Plant Physiol 195:2111–2128

Li Y, Ma H, Wu Y et al (2024b) Single-cell transcriptome atlas and regulatory dynamics in developing cotton anthers. Adv Sci (Weinh) 11:e2304017

Lieber D, Lora J, Schrempp S et al (2011) Arabidopsis WIH1 and WIH2 genes act in the transition from somatic to reproductive cell fate. Curr Biol 21:1009–1017

Lin B-Y (1978) Structural modifications of the female gametophyte associated with the indeterminate gametophyte (IG) mutant in maize. Can J Genet Cytol 20:249–257

Logsdon GA, Vollger MR, Eichler EE (2020) Long-read human genome sequencing and its applications. Nat Rev Genet 21:597–614

Lohani N, Singh MB, Bhalla PL (2020) RNA-seq highlights molecular events associated with impaired pollen-pistil interactions following short-term heat stress in Brassica napus. Front Plant Sci 11:622748

Loraine AE, McCormick S, Estrada A et al (2013) RNA-seq of Arabidopsis pollen uncovers novel transcription and alternative splicing. Plant Physiol 162:1092–1109

Lu Y, Wei L, Wang T (2015) Methods to isolate a large amount of generative cells, sperm cells and vegetative nuclei from tomato pollen for "omics" analysis. Front Plant Sci 6:391

Malhó R, Liu Q, Monteiro D et al (2006) Signalling pathways in pollen germination and tube growth. Protoplasma 228:21–30

Marchant DB, Walbot V (2024) The establishment of the anther somatic niche with single-cell sequencing. Dev Biol 518:37–47

Marshall E, Costa LM, Gutierrez-Marcos J (2011) Cysteine-rich peptides (CRPs) mediate diverse aspects of cell-cell communication in plant reproduction and development. J Exp Bot 62:1677–1686

Márton ML, Cordts S, Broadhvest J, Dresselhaus T (2005) Micropylar pollen tube guidance by egg apparatus 1 of maize. Science 307:573–576

Masoomi-Aladizgeh F, Najeeb U, Hamzelou S et al (2021) Pollen development in cotton (Gossypium hirsutum) is highly sensitive to heat exposure during the tetrad stage. Plant Cell Environ 44:2150–2166

Michard E, Dias P, Feijó JA (2008) Tobacco pollen tubes as cellular models for ion dynamics: improved spatial and temporal resolution of extracellular flux and free cytosolic concentration of calcium and protons using pHluorin and YC3.1 CaMeleon. Sex Plant Reprod 21:169–181

Misra CS, Sousa AGG, Barros PM et al (2023) Cell-type-specific alternative splicing in the Arabidopsis germline. Plant Physiol 192:85–101

Misra CS, Sousa A, Khan H et al (2025) Transcriptome dynamics in the Arabidopsis male germline during pollen-pistil interactions. Plant J 121:e70095

Mizuta Y, Higashiyama T (2018) Chemical signaling for pollen tube guidance at a glance. J Cell Sci 131:jcs208447

Müller F, Rieu I (2016) Acclimation to high temperature during pollen development. Plant Reprod 29:107–118

Nakamura M, Hashimoto T (2009) A mutation in the Arabidopsis gamma-tubulin-containing complex causes helical growth and abnormal microtubule branching. J Cell Sci 122:2208–2217

Nelms B, Walbot V (2019) Defining the developmental program leading to meiosis in maize. Science 364:52–56

Nodine MD, Bartel DP (2012) Maternal and paternal genomes contribute equally to the transcriptome of early plant embryos. Nature 482:94–97

Nonomura K-I, Miyoshi K, Eiguchi M et al (2003) The MSP1 gene is necessary to restrict the number of cells entering into male and female sporogenesis and to initiate anther wall formation in rice. Plant Cell 15:1728–1739

Nonomura K-I, Morohoshi A, Nakano M et al (2007) A germ cell specific gene of the ARGONAUTE family is essential for the progression of premeiotic mitosis and meiosis during sporogenesis in rice. Plant Cell 19:2583–2594

Oh SA, Johnson A, Smertenko A et al (2005) A divergent cellular role for the FUSED kinase family in the plant-specific cytokinetic phragmoplast. Curr Biol 15:2107–2111
Oh SA, Park KS, Twell D, Park SK (2010) The SIDECAR POLLEN gene encodes a microspore-specific LOB/AS2 domain protein required for the correct timing and orientation of asymmetric cell division: SCP, LBD27/ASL29, for microspore division asymmetry. Plant J 64:839–850
Okuda S, Tsutsui H, Shiina K et al (2009) Defensin-like polypeptide LUREs are pollen tube attractants secreted from synergid cells. Nature 458:357–361
Ouonkap SVY, Palaniappan M, Pryze K et al (2024) Enhanced pollen tube performance at high temperature contributes to thermotolerant fruit and seed production in tomato. Curr Biol 34: 5319
Pacini E, Guarnieri M, Nepi M (2006) Pollen carbohydrates and water content during development, presentation, and dispersal: a short review. Protoplasma 228:73–77
Pagnussat GC, Alandete-Saez M, Bowman JL, Sundaresan V (2009) Auxin-dependent patterning and gamete specification in the Arabidopsis female gametophyte. Science 324:1684–1689
Palanivelu R, Brass L, Edlund AF, Preuss D (2003) Pollen tube growth and guidance is regulated by POP2, an Arabidopsis gene that controls GABA levels. Cell 114:47–59
Panoli A, Martin MV, Alandete-Saez M et al (2015) Auxin import and local auxin biosynthesis are required for mitotic divisions, cell expansion and cell specification during female gametophyte development in Arabidopsis thaliana. PLoS One 10:e0126164
Pastuglia M, Azimzadeh J, Goussot M et al (2006) Gamma-tubulin is essential for microtubule organization and development in Arabidopsis. Plant Cell 18:1412–1425
Peet MM, Sato S, Gardner RG (1998) Comparing heat stress effects on male-fertile and male-sterile tomatoes. Plant Cell Environ 21:225–231
Pérez Di Giorgio JA, Barberini ML, Amodeo G, Muschietti JP (2016) Pollen aquaporins: what are they there for? Plant Signal Behav 11:e1217375
Pina C, Pinto F, Feijó JA, Becker JD (2005) Gene family analysis of the Arabidopsis pollen transcriptome reveals biological implications for cell growth, division control, and gene expression regulation. Plant Physiol 138:744–756
Pinto SC, Leong WH, Tan H et al (2024) Germline β-1,3-glucan deposits are required for female gametogenesis in Arabidopsis thaliana. Nat Commun 15:5875
Poidevin L, Forment J, Unal D, Ferrando A (2021) Transcriptome and translatome changes in germinated pollen under heat stress uncover roles of transporter genes involved in pollen tube growth. Plant Cell Environ 44:2167–2184
Portereiko MF, Lloyd A, Steffen JG et al (2006) AGL80 is required for central cell and endosperm development in Arabidopsis. Plant Cell 18:1862–1872
Punwani JA, Rabiger DS, Drews GN (2007) MYB98 positively regulates a battery of synergid-expressed genes encoding filiform apparatus localized proteins. Plant Cell 19:2557–2568
Qu A-L, Ding Y-F, Jiang Q, Zhu C (2013) Molecular mechanisms of the plant heat stress response. Biochem Biophys Res Commun 432:203–207
Rieu I, Wolters-Arts M, Derksen J et al (2003) Ethylene regulates the timing of anther dehiscence in tobacco. Planta 217:131–137
Rieu I, Twell D, Firon N (2017) Pollen development at high temperature: from acclimation to collapse. Plant Physiol 173:1967–1976
Robert HS, Crhak Khaitova L, Mroue S, Benková E (2015) The importance of localized auxin production for morphogenesis of reproductive organs and embryos in Arabidopsis. J Exp Bot 66:5029–5042
Rodriguez-Enriquez MJ, Mehdi S, Dickinson HG, Grant-Downton RT (2013) A novel method for efficient in vitro germination and tube growth of Arabidopsis thaliana pollen. New Phytol 197: 668–679
Rutley N, Twell D (2015) A decade of pollen transcriptomics. Plant Reprod 28:73–89
Saini HS, Aspinall D (1982) Abnormal sporogenesis in wheat (Triticum aestivum L.) induced by short periods of high temperature. Ann Bot 49:835–846

Saini HS, Sedgley M, Aspinall D (1983) Effect of heat stress during floral development on pollen tube growth and ovary anatomy in wheat (Triticum aestivum L.). Funct Plant Biol 10:137

Sakai K, Taconnat L, Borrega N et al (2018) Combining laser-assisted microdissection (LAM) and RNA-seq allows to perform a comprehensive transcriptomic analysis of epidermal cells of Arabidopsis embryo. Plant Methods 14:10

Sanders PM, Lee PY, Biesgen C et al (2000) The Arabidopsis DELAYED DEHISCENCE1 gene encodes an enzyme in the jasmonic acid synthesis pathway. Plant Cell 12:1041–1061

Santos MR, Bispo C, Becker JD (2017) Isolation of Arabidopsis pollen, sperm cells, and vegetative nuclei by fluorescence-activated cell sorting (FACS). Methods Mol Biol 1669:193–210

Schiefthaler U, Balasubramanian S, Sieber P et al (1999) Molecular analysis of NOZZLE, a gene involved in pattern formation and early sporogenesis during sex organ development in Arabidopsis thaliana. Proc Natl Acad Sci USA 96:11664–11669

Shi J, Cui M, Yang L et al (2015) Genetic and biochemical mechanisms of pollen wall development. Trends Plant Sci 20:741–753

Shi W, Yang J, Kumar R et al (2022) Heat stress during gametogenesis irreversibly damages female reproductive organ in rice. Rice (N Y) 15:32

Simonini S, Bencivenga S, Grossniklaus U (2024) A paternal signal induces endosperm proliferation upon fertilization in Arabidopsis. Science 383:646–653

Singh M, Goel S, Meeley RB et al (2011) Production of viable gametes without meiosis in maize deficient for an ARGONAUTE protein. Plant Cell 23:443–458

Sogo A, Tobe H (2005) Intermittent pollen-tube growth in pistils of alders (Alnus). Proc Natl Acad Sci USA 102:8770–8775

Song S, Chen Y, Liu L et al (2018) OsFTIP7 determines auxin-mediated anther dehiscence in rice. Nat Plants 4:495–504

Song Q, Ando A, Jiang N et al (2020) Single-cell RNA-seq analysis reveals ploidy-dependent and cell-specific transcriptome changes in Arabidopsis female gametophytes. Genome Biol 21:178

Steffen JG, Kang I-H, Portereiko MF et al (2008) AGL61 interacts with AGL80 and is required for central cell development in Arabidopsis. Plant Physiol 148:259–268

Sullivan A, Purohit PK, Freese NH et al (2019) An "eFP-Seq Browser" for visualizing and exploring RNA sequencing data. Plant J 100:641–654

Sun Q, Han Y, He J et al (2023) Long-read sequencing reveals the landscape of aberrant alternative splicing and novel therapeutic target in colorectal cancer. Genome Med 15:76

Susaki D, Suzuki T, Maruyama D et al (2021) Dynamics of the cell fate specifications during female gametophyte development in Arabidopsis. PLoS Biol 19:e3001123

Sze H, Klodová B, Ward JM et al (2024) A wave of specific transcript and protein accumulation accompanies pollen dehydration. Plant Physiol 195:1775–1795

Takeda S, Paszkowski J (2006) DNA methylation and epigenetic inheritance during plant gametogenesis. Chromosoma 115:27–35

Tian HQ, Russell SD (1997) Micromanipulation of male and female gametes of Nicotiana tabacum: I. Isolation of gametes. Plant Cell Rep 16:555–560

Tung C-W, Dwyer KG, Nasrallah ME, Nasrallah JB (2005) Genome-wide identification of genes expressed in Arabidopsis pistils specifically along the path of pollen tube growth. Plant Physiol 138:977–989

Twell D, Park SK, Hawkins TJ et al (2002) MOR1/GEM1 has an essential role in the plant-specific cytokinetic phragmoplast. Nat Cell Biol 4:711–714

Twell D, Oh S-A, Honys D (2006) Pollen development, a genetic and transcriptomic view. In: The pollen tube. Springer-Verlag, Berlin/Heidelberg, pp 15–45

Wang Y, Kumar PP (2007) Characterization of two ethylene receptors PhERS1 and PhETR2 from petunia: PhETR2 regulates timing of anther dehiscence. J Exp Bot 58:533–544

Wang Y, Zhang W-Z, Song L-F et al (2008) Transcriptome analyses show changes in gene expression to accompany pollen germination and tube growth in Arabidopsis. Plant Physiol 148:1201–1211

Wang D, Zhang C, Hearn DJ et al (2010) Identification of transcription-factor genes expressed in the Arabidopsis female gametophyte. BMC Plant Biol 10:110

Wang Y, Impa SM, Sunkar R, Jagadish SVK (2021) The neglected other half – role of the pistil in plant heat stress responses. Plant Cell Environ 44:2200–2210

Wang Y, Zhou H, He Y et al (2023) MYB transcription factors and their roles in the male reproductive development of flowering plants. Plant Sci 335:111811

Webb M, Gunning BS (1990) Embryo sac development in Arabidopsis thaliana: I. Megasporogenesis, including the microtubular cytoskeleton. Sex Plant Reprod 3:244–256

Webb M, Gunning BS (1994) Embryo sac development in Arabidopsis thaliana II. The cytoskeleton during megagametogenesis. Sex Plant Reprod 7:153–163

Wei LQ, Xu WY, Deng ZY et al (2010) Genome-scale analysis and comparison of gene expression profiles in developing and germinated pollen in Oryza sativa. BMC Genomics 11:338

Wei D, Liu M, Chen H et al (2018) INDUCER OF CBF EXPRESSION 1 is a male fertility regulator impacting anther dehydration in Arabidopsis. PLoS Genet 14:e1007695

Weijers D, Geldner N, Offringa R, Jürgens G (2001) Seed development: early paternal gene activity in Arabidopsis: seed development (communication arising). Nature 414:709–710

Whittle CA, Otto SP, Johnston MO, Krochko JE (2009) Adaptive epigenetic memory of ancestral temperature regime in *Arabidopsis thaliana*. Botany 87:650–657

Williams JH, Reese JB (2019) Evolution of development of pollen performance. Curr Top Dev Biol 131:299–336

Wright SI, Kalisz S, Slotte T (2013) Evolutionary consequences of self-fertilization in plants. Proc Biol Sci 280:20130133

Wu W, Zheng B (2019) Intercellular delivery of small RNAs in plant gametes. New Phytol 224:86–90

Wu H, Lu Y, Duan Z et al (2023) Nanopore long-read RNA sequencing reveals functional alternative splicing variants in human vascular smooth muscle cells. Commun Biol 6:1104

Wuest SE, Vijverberg K, Schmidt A et al (2010) Arabidopsis female gametophyte gene expression map reveals similarities between plant and animal gametes. Curr Biol 20:506–512

Xu X-F, Wang B, Feng Y-F et al (2019) AUXIN RESPONSE FACTOR17 directly regulates MYB108 for anther dehiscence. Plant Physiol 181:645–655

Yadegari R, Drews GN (2004) Female gametophyte development. Plant Cell 16(Suppl):S133–S141

Yang WC, Ye D, Xu J, Sundaresan V (1999) The SPOROCYTELESS gene of Arabidopsis is required for initiation of sporogenesis and encodes a novel nuclear protein. Genes Dev 13:2108–2117

Yu H-J, Hogan P, Sundaresan V (2005) Analysis of the female gametophyte transcriptome of Arabidopsis by comparative expression profiling. Plant Physiol 139:1853–1869

Zhang H, Wang M, Li Y et al (2020a) GDSL esterase/lipases OsGELP34 and OsGELP110/OsGELP115 are essential for rice pollen development. J Integr Plant Biol 62:1574–1593

Zhang Y-C, Lei M-Q, Zhou Y-F et al (2020b) Reproductive phasiRNAs regulate reprogramming of gene expression and meiotic progression in rice. Nat Commun 11:6031

Zhang Z, Sun M, Xiong T et al (2024) Development and genetic regulation of pollen intine in Arabidopsis and rice. Gene 893:147936

Zhao L, He J, Cai H et al (2014) Comparative expression profiling reveals gene functions in female meiosis and gametophyte development in Arabidopsis. Plant J 80:615–628

Zhao Y, Wang S, Wu W et al (2018) Clearance of maternal barriers by paternal miR159 to initiate endosperm nuclear division in Arabidopsis. Nat Commun 9:5011

Zhao H, Guo M, Yan M et al (2020a) Comparative expression profiling reveals genes involved in megasporogenesis. Plant Physiol 182:2006–2024

Zhao P, Zhou X, Zheng Y et al (2020b) Equal parental contribution to the transcriptome is not equal control of embryogenesis. Nat Plants 6:1354–1364

Zinn KE, Tunc-Ozdemir M, Harper JF (2010) Temperature stress and plant sexual reproduction: uncovering the weakest links. J Exp Bot 61:1959–1968

# Biotechnological Production of Plant Bioactive Compounds as a Strategy to Obtain Natural Plant Biostimulants

**José Manuel Martí-Guillén, Sara Esperanza Martínez-Lorente, María Ángeles Pedreño, Lorena Almagro, and Ana Belén Sabater-Jara**

## Contents



**Abstract** Climate change and a growing world population challenge modern agriculture to maintain sustainable production over time on increasingly scarce arable land. The use of plant biostimulants emerges as a safe and environmentally friendly alternative to improve crop production and yield under these unfavorable conditions. Among all plant biostimulants, higher plant-derived biostimulants (hPDBs) are of

---

Communicated by Francisco M. Cánovas

Authors José Manuel Martí-Guillén and Sara Esperanza Martínez-Lorente have contributed equally to this work.

**Supplementary Information** The online version contains supplementary material available at https://doi.org/10.1007/124_2025_90.

---

J. M. Martí-Guillén, S. E. Martínez-Lorente, M. Á. Pedreño, L. Almagro, and A. B. Sabater-Jara (✉)
Department of Plant Biology, Faculty of Biology, University of Murcia, Campus of Espinardo, Murcia, Spain
e-mail: josemanuel.martig@um.es; sesperanza.martinez@um.es; mpedreno@um.es; lorena.almagro@um.es; anabelen.sabater@um.es

 
Progress in Botany (2026) 86: 109–129, https://doi.org/10.1007/124_2025_90,
Published online: 27 June 2025

special interest, due to all the advantages they have in terms of sustainability, production costs, safety, and mainly due to the diversity of bioactive compounds that compose them. In this work, a novel type of plant biostimulant, based on yellow carrot cell cultures, was developed through biotechnological tools. Using 25 mM Me-β-cyclodextrins (Me-β-CD) and 100 μM methyljasmonate (MJ) as elicitors, the production of bioactive compounds with a described biostimulant effect, mainly phytosterols and phenolic compounds, was increased in cell suspensions. A compositionally stable product, yellow carrot biostimulant (YCB), was prepared, and the evaluation of its biostimulant effect was corroborated by applying it as seed priming on tomato seeds germinated under salinity conditions, resulting in the improvement of germination index (+13%), $H_2O_2$ (−13%), MDA (−50%), and proline (+12%) contents, compared with untreated seeds, when the dose of 1% (v/v) YCB was applied. Therefore, we have developed a novel potential biostimulant product based on yellow carrot cell cultures, enriched in phenolic compounds and phytosterols, which has demonstrated its effect on tomato seeds subjected to salt stress.

**Keywords** Bioactive compounds, Biotechnological production, Climate change, Higher plant-derived biostimulants, Salt stress, Tomato seeds, Yellow carrot biostimulant, Yellow carrot cell cultures

## 1 Introduction

The main challenge facing modern agriculture is to maintain a sustainable production over time, capable of feeding a growing world population, in the adverse conditions resulting in climate change that are increasingly present in agricultural lands (United Nations 2019). Additionally, strict regulations limit the use of agrochemicals to improve crop yield, due to their prolonged abuse over the years and the negative impact they have on the environment (European Commission 2019, 2020).

The adverse conditions produced by climate change are perceived by plants as abiotic stresses, among which the most notable ones are the exposure to extreme temperatures, prolonged water deficits, or soils salinized or contaminated with potentially hazardous elements. In areas dedicated to the production of agricultural crops of global importance, such as tomato in Region of Murcia and throughout southeastern Spain, one of the main problems for productivity is irrigation with poor low quality irrigation water, resulting in soil salinization (Acosta et al. 2011). As a result, an imbalance in plant homeostasis occurs, which negatively affects its development and, therefore, crop yield (Ullah et al. 2021; Martí-Guillén et al. 2022).

On the road to a more sustainable agriculture, in a scenario where the use of agrochemicals is limited and the arable land is increasingly smaller due to these adverse phenomena, the use of plant biostimulants emerges as a safer and more sustainable agricultural system, respectful of the environment and which enables the improvement of crop yields, even under high edaphoclimatic and anthropogenic pressure (Del Buono 2021).

Plant biostimulants, by definition, are products whose function is to stimulate the plant physiological processes (nutrient use efficiency, tolerance to abiotic stress, etc.) regardless of their nutrient content (European Parliament 2019). Plant biostimulants can be applied using three different methods: soil application, foliar spray, and seed priming (Martínez-Lorente et al. 2024). The latter is of particular interest, since the germination process determines the vigor and performance of the developed seedlings. A defective germination process, developed when germination is initiated in marginal environments, limits crop yield. Seed priming is a technique that involves a controlled hydration of the seed with biostimulants, which improves the germination process, including the germination rate and vigor of the developed seedlings, even under adverse conditions (Ofoe et al. 2022).

Among all those classified as plant biostimulants, higher plant-derived biostimulants (hPDBs) are of special interest, due to their advantages in sustainability, production costs, and human and environmental safety. Additionally, the diversity of bioactive compounds that compose them, ranging from peptides to specialized metabolites, makes them potentially capable of inducing multiple physiological responses in plants, which can lead to an increased crop yield (Ahmad et al. 2022; Martínez-Lorente et al. 2024). However, maintaining homogeneity in the composition of this type of biostimulants during their production is often complex, as it depends, to a large extent, on the origin of the plant raw materials used, doses and treatments, or plant stress conditions to which they have been subjected during their cultivation, as well as the processing techniques used to obtain the final biostimulant product (Zulfiqar et al. 2020). This lack of homogeneity is problematic, since differences in the content of bioactive compounds contained in the biostimulant can lead to the effect of the latter not being consistent between different production periods.

For this reason, in vitro plant cell culture represents an alternative to the use of plant raw materials, minimizing the heterogeneity of the resulting final product due to its independence from geographical and environmental factors, and ensuring the continuous production of bioactive compounds, especially when they are subjected to elicitation processes (Espinosa-Leal et al. 2018; Marchev et al. 2020; Chandran et al. 2020). These processes consist of the application of inducing agents on the plant cell culture to trigger the synthesis and accumulation of bioactive compounds, mainly specialized metabolites and plant defense proteins (Ramirez-Estrada et al. 2016).

Our research group has previously demonstrated the ability to increase the multitude of bioactive compounds by applying inducing agents, such as methyljasmonate (MJ), coronatine, and different modified cyclodextrins (CDs), in plant cell cultures. Thus, the use of coronatine and MJ increased the production of phenolic compounds in grapevine (Almagro et al. 2022) and broccoli (Sánchez-Pujante et al. 2020) cell cultures, respectively. Furthermore, the use of CD increased the production of triterpenes in tomato (Sabater-Jara et al. 2022), phenylpropanoids in *Silybum marianum* (Corchete et al. 2022), and phytosterols and phenolic compounds in green (Miras-Moreno et al. 2016a) and orange (Sabater-Jara and Pedreño 2013) carrot cell cultures.

In particular, carrot cell cultures allow the production of phenolic compounds and phytosterols through the application of inducing agents, specifically CD and MJ. Both phytosterols and phenolic compounds have been described as bioactive metabolites in multiple processes, both in human health and in plant physiology (Rahman et al. 2021; Liu et al. 2024). In the latter, a multitude of phenolic compounds have been described with a biostimulant effect on plants, improving growth parameters, yield, and resistance to stress conditions (Kisiriko et al. 2021; Kumar et al. 2023a). Similarly, this positive effect has also been described for phytosterols and their derivatives (Valitova et al. 2024).

Thus, this methodology represents a promising biotechnological tool to produce homogeneous products with a potential biostimulant effect, enriched in a huge diversity of bioactive compounds, especially phytosterols and phenolic compounds. The objective of this work, therefore, is the preparation of a product enriched in bioactive compounds using the techniques of elicitation in a yellow carrot cell culture, and the evaluation of its potential biostimulant effect on tomato seeds subjected to salt stress.

# 2 Materials and Methods

## 2.1 *Plant Cell Cultures*

### 2.1.1 Plant Cell Material and Growth Conditions

Yellow carrot (*Daucus carota* L.) calli from root explants were previously established in our laboratory (Supplementary Fig. 1a). Calli were maintained on petri dishes with solid culture medium composed of 4.3 g $L^{-1}$ Murashige and Skoog basal medium (Murashige and Skoog 1962), 30 g $L^{-1}$ sucrose, 0.25 g $L^{-1}$ casein hydrolysate, supplemented with Morel (Morel and Wetmore 1951), B1 and B3 vitamins, 1 mg $L^{-1}$ of 2,4-dichlorophenoxyacetic acid as a growth regulator, and with 6 g $L^{-1}$ plant agar (Supplementary Fig. 1b). The pH was adjusted to 6.0 prior autoclaving. Calli were maintained under a photoperiod of 16 h light/8 h dark (85 μmol $m^{-2}$ $s^{-1}$), 23°C temperature, and were subcultured every month.

### 2.1.2 Determination of Yellow Carrot Cell Culture Growth Curve (Time Course)

Yellow carrot cell suspensions were started by inoculating 40 g of friable calli into 250 mL flasks, containing a total volume of 100 mL of the same medium described above without plant agar (Supplementary Fig. 1c). Suspensions were maintained in an orbital shaker at 105 rpm and the same photoperiod and temperature conditions described above. The cell suspensions were subcultured every 9–11 days by 1:1 (v/v) dilution with fresh culture medium. After at least four subcultures, cell suspensions were vacuum-filtered at the time of the next subculture; 2 or 4 g of the fresh weight (FW) of the obtained cells were used to inoculate 100 mL flasks, containing a volume of 20 mL of liquid culture medium, implying a cell density of 100 g $L^{-1}$ or 200 g $L^{-1}$, respectively.

Temporary samples of the different cell suspensions were taken up to 14 days. Cell volume after sedimentation (CVS), cell FW, and °Brix and pH of the filtered extracellular medium were determined to stablish a time course of yellow carrot cell growth. °Brix were determined with a portable Brix refractometer. Cell growth kinetic parameters were defined as nutrient depletion time ($T_a$), maximum cell volume after sedimentation [$CVS_{max}$ (%)], and exponential growth rate ($V_{exp}$).

### 2.1.3 Elicitation of Cell Suspensions

Cell suspensions were also started from vacuum-filtered cells with at least four subcultures. For elicitation treatment, 4 g of yellow carrot cells were inoculated into 100 mL flasks containing liquid culture medium as a control, and liquid culture medium supplemented with 10, 25, or 50 mM methyl-β-CD (Me-β-CD), or 10, 25, or 50 mM hydroxypropyl-β-CD (HP-β-CD), in a total volume of 20 mL. Growth conditions were identical to those described above. Three temporary biological samples were taken from the different suspensions up to 9 days. Cells were separated from the media by vacuum filtration, and both were immediately frozen at −20°C until use.

For the evaluation of the effect of MJ, similarly to the previous methodology, 4 g of yellow carrot cells were inoculated into 100 mL flasks containing liquid culture medium (C), culture medium supplemented with 25 mM Me-β-CD (CD), 100 μM MJ (MJ), or the combination of both elicitors (25 mM Me–βCD + 100 μM MJ, CDMJ), in a total volume of 20 mL. Cell suspensions were removed after 7 days of growth. Cells were separated from the media by vacuum filtration, and both were immediately stored at −20°C until use.

### 2.1.4 Metabolite Extraction

The extracellular media were subjected to a phase partitioning extraction with 1:1 (v/v) ethyl acetate. The volumes were homogenized, and the organic phase was recovered after phase definition. The process was repeated in the aqueous phase and both volumes of organic phase were mixed. Subsequently, dry residue was obtained after evaporation at 40°C in a rotary vacuum evaporator. The dry residue was resuspended in 1 mL of methanol, and the obtained extract was filtered through 0.22 μm nylon filters for subsequent analysis.

For the extraction of metabolites from the cells, the methodology described in Burns et al. (2003) was followed with some modifications. Previously lyophilized cells were homogenized in methanol (2:3 w/v). Next, 750 μL of 50 mM Tris-HCl pH 7.5 containing 1 mM NaCl were added, followed by sonication for 15 min. A phase partitioning extraction was performed by adding 3 mL of chloroform, recovering the organic phase once defined. The extraction was repeated by adding 3 mL of chloroform to the aqueous phase. Both volumes of organic phase were mixed and distilled in a rotary evaporator under vacuum at 40°C. The samples were resuspended in 1 mL of methanol and filtered with 0.22 μm nylon filters for later analysis.

#### 2.1.5 Identification and Quantification of Phytosterols

The analysis of metabolites in yellow carrot extracts was performed on a gas chromatograph (Agilent Technologies 6890 Network GS System), equipped with a mass selective detector (Agilent Technologies 5973), using a capillary column of 30 m × 0.25 mm (Agilent 19091 S-433HP-5MS). For the analysis of metabolites, the GC oven temperature was programmed from 60 to 310°C and helium was used as carrier gas, with a constant flow rate of 0.1 mL $min^{-1}$. 1 μL of the extracts was injected. The mass range was recorded from m/z 50 to 800, with an ionization energy of 70 eV (Sabater-Jara and Pedreño 2013).

The identification of the detected compounds was performed using Chemstation software based on their mass spectra, compared to published mass data and retention times from NIST11 spectral library. The concentration of the main phytosterols identified (β-sitosterol, campesterol, fucosterol, and stigmasterol) was determined based on a standard curve prepared with standards acquired from Sigma-Aldrich (Spain).

#### 2.1.6 Determination of Total Phenolic Content

Total phenolic content (TPC) in cells and extracellular media was determined by a Folin-Ciocalteu assay (Almagro et al. 2022). Briefly, 50 μL of the extracts obtained above were incubated with 120 μL of 1:16 (v/v) Folin-Ciocalteu reagent for 3 min. Then, 30 μL of 20% (w/v) $Na_2CO_3$ was added, and the reaction was incubated at room temperature for 2 h in darkness. The TPC was calculated by measuring the absorbance at 765 nm, based on a gallic acid standard curve.

### *2.2 Evaluation of Biostimulant Effect in Tomato Seeds*

#### 2.2.1 Production of Yellow Carrot Biostimulant

Seven-day-old cell suspensions elicited with CDMJ were homogenized for 5 min with a blender. For the microbiological control of the product, a broad-spectrum preservative based on a combination of sodium benzoate and potassium sorbate was added. The resulting final product, yellow carrot biostimulant (YCB), was used for seed treatment.

### 2.2.2 Plant Material and Seed Priming

Commercial tomato seeds (*Solanum lycopersicum* L. cv. Gades) purchased from Intersemillas S.A. were used. For seed priming, the seeds were soaked in different doses of the biostimulant product: 1% (v/v) YCB (YCB1), 0.1% (v/v) YCB (YCB0.1), and 0.01% (v/v) YCB (YCB0.01). Priming was carried out for 2 h in orbital shaking at 200 rpm, in darkness and at room temperature. Additionally, priming with distilled water (hydropriming, HP) was carried out as a hydration control of the process. Subsequently, the seeds were kept in an oven with ventilation at room temperature until the original moisture content was recovered. Seeds without undergoing any priming treatment (no priming, NP) were used as a control.

### 2.2.3 Germination Conditions and Parameters

Seeds were germinated in petri dishes with a sterile filter paper disk, containing 5 mL of autoclaved distilled water as control conditions, or 5 mL of 100 mM NaCl to induce salt stress, a total of 25 seeds per plate, at 25°C and darkness. Germination Index (GI) was determined using the mathematical formulas described by Maguire (1962). Seedlings were sampled 8 days after the start of germination, immediately frozen in liquid nitrogen and stored at −80°C until use.

### 2.2.4 Determination of $H_2O_2$ Content

For the determination of hydrogen peroxide ($H_2O_2$) concentration, the method described by Junglee et al. (2014) with some modifications was used. Previously grounded plant material was homogenized in 0.1% (w/v) trichloroacetic acid (TCA) (4:1 w/v). The mixture was centrifuged at 14,600× *g* for 15 min and the supernatant was recovered. The reaction was triggered by adding 50 μL of 50 mM potassium phosphate buffer pH 7.0, 100 μL of 0.1% (w/v) TCA and 250 μL of 1 M KI to 100 μL of extract and incubated at room temperature for 1 h in the dark. The absorbance was measured at 390 nm, and the $H_2O_2$ concentration was determined using a $H_2O_2$ standard curve.

### 2.2.5 Determination of Lipid Peroxidation

Lipid peroxidation was estimated by quantifying malondialdehyde (MDA) content, based on the reaction with thiobarbituric acid (TBA) described by Heath and Packer (1968), with some modifications. The same extract obtained for the determination of

$H_2O_2$ was used for the determination of MDA. 150 μL of this extract was added to 450 μL of 20% (w/v) TCA containing 0.5% (w/v) TBA and incubated for 25 min at 95°C. Tubes with 20% (w/v) TCA without TBA were used as a negative control. Measurements were made at 440, 532, and 600 nm based on Hodges et al. (1999), after which MDA content was calculated based on Landi (2017).

#### 2.2.6 Determination of Proline Content

Quantification of proline concentration was carried out based on the protocol of Bates et al. (1973) with modifications. Briefly, previously grounded plant material was homogenized in 3% (w/v) sulfosalicylic acid (2:1 w/v), and samples were centrifuged for 15 min at 14,600× *g* to recover the supernatant. The reaction was developed in tubes containing extract, glacial acetic acid, and acidic ninhydrin 1:1:1 (v/v/v) and incubated at 100°C for 1 h. Then, the reaction was stopped on ice and the chromophore was extracted by adding 1 mL of toluene. Proline content was determined by measuring absorbance at 520 nm. Different concentrations of L-proline were used as a standard curve.

### *2.3 Statistical Analysis*

All data analysis was developed using SPSS software package (SPSS Inc., Chicago, USA) version 28.0. One-way and two-way analyses of variance (ANOVA) were performed, followed by Duncan's test, and $p$ values $< 0.05$ were considered as statistically significant.

## 3 Results and Discussion

### *3.1 Characterization of Yellow Carrot Cell Growth*

Yellow carrot (*Daucus carota* L.) cell growth at two cell densities (100 and 200 g FW $L^{-1}$) was characterized by measuring CVS (%) and pH and °Brix of the culture media for 14 days (Fig. 1). As observed, beginning from 22.9% to 45% CVS at 100 (Fig. 1a) and 200 (Fig. 1b) g FW $L^{-1}$, respectively, cell growth exhibited a typical S-shaped growth curve. Thus, when an initial cell density of 100 g FW $L^{-1}$ was used (Fig. 1a), CVS increased up to 94.4%, and then was stabilized upon reaching the stationary phase of growth, distinguishing a short latency period or "lag" phase from the beginning of the curve (from day 0 to day 2) until exponential growth begins; the "log" phase that comprises from the beginning of the exponential growth phase until a decrease in growth is observed (from day 3 to day 11); and the "stationary" phase that covered from day 11 to the end of the culture (day 14).

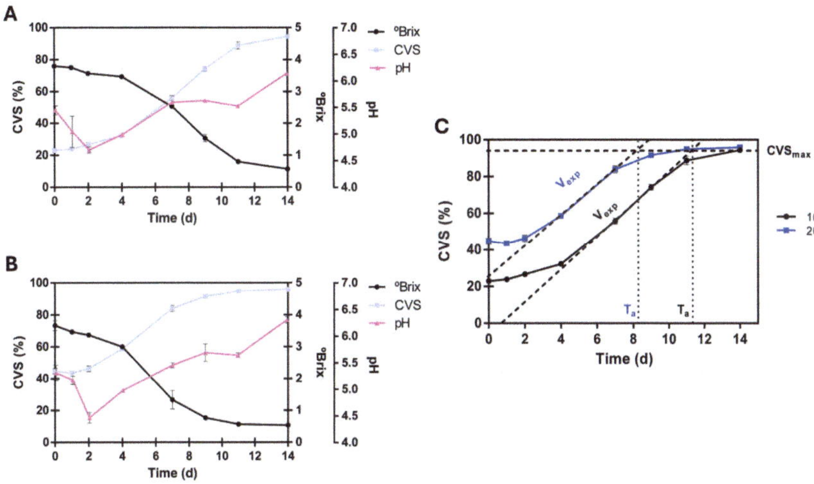


**Fig. 1** Time course of yellow carrot cell growth measured as cell volume after sedimentation (CVS) in %. Also, time course of °Brix and pH for 14 days. The initial cell density was 100 g FW $L^{-1}$ (**a**) or 200 g FW $L^{-1}$ (**b**). (**c**) Characterization of cell growth parameters of yellow carrot cell cultures. Cell volume after sedimentation (CVS (%)); nutrient depletion time ($T_a$); maximum cell volume after sedimentation ($CVS_{max}$ (%)); exponential growth rate ($V_{exp}$). The initial cell density was 100 or 200 g FW $L^{-1}$. Points show means ± SD ($n = 3$)

However, at 200 g FW $L^{-1}$ of cell density (Fig. 1b), cell growth quickly reached the exponential phase, with a lag phase shortened (1 day), reaching the stationary phase of growth on day 8 of growth with an increase of 96% CVS to the end of the culture.

In addition, CVS measurement at both cell densities was accompanied by the measurement of pH and °Brix of the culture medium in which the cells were growing (Fig. 1a, b). In relation to the °Brix, a variation inversely proportional to the growth curve was observed in both cases, which is related to the concentration of soluble solids in the medium. Therefore, when the cells reach the exponential or "log" phase (the phase of greatest activity of primary metabolism), they consume a greater amount of nutrients, decreasing their concentration in the culture medium. During "lag" and "stationary" phases, the variations in this parameter are minimum due to the lower activity of primary metabolism, reaching 0.5 °Brix at the end of the culture. On the other hand, the results shown in Fig. 1a, b indicated that the initial pH of the culture medium was around 5.5 and this value decreased by approximately 1 unit until day 2, and from that day on a progressive increase was observed until day 7, remaining stable until the end of the culture. This variation in the pH of the extracellular medium was mainly due to the availability of nitrogen sources and the energy status of the cells. In vitro plant cells cultured tend to use the nitrogen source whose incorporation involves a lower energy cost. However, the incorporation of nitrogen also depends on the ratio of ammonium and nitrate concentrations in the culture medium (McDonald and Jackman 1989; Loulakakis and Roubelakis-Angelakis 2001) and even on the availability of molecules capable of reducing nitrates to ammonium in the cell, which is related to their metabolism and carbon sources.

As regards growth kinetic parameters (Fig. 1c), in both cases, the maximum CVS ($CVS_{max}$) reached was approximately 95%, although at different times. Thus, while starting with 200 g FW $L^{-1}$, the $CVS_{max}$ was reached after 10 days of growth; with a cell density of 100 g FW $L^{-1}$, the $CVS_{max}$ was reached at the end of the culture (14 days), being the $V_{exp}$ of 9.19% $day^{-1}$ and 8.43% $day^{-1}$ for 100 and 200 g FW $L^{-1}$, respectively. Based on these results, the $T_a$ was reached at about 11 and 8 days using an initial cell density of 100 and 200 g FW $L^{-1}$, respectively (Fig. 1c), determining the need for subculture at that time in each case. Therefore, the starting cell density of 200 g FW $L^{-1}$ was selected for subsequent trials, since it represents a sufficient cell density which could make the scaling process viable using conventional bioreactors, enabling industrial production economically profitable.

## 3.2 Effect of Different Cyclodextrins on Total Phytosterol Content in Yellow Carrot Cell Cultures

CDs have been previously described as elicitors capable of inducing the production of metabolites of interest in cell cultures. In particular, in previous studies, Sabater-Jara and Pedreño (2013) described the increase in phytosterol content when applying different types of CDs (Me-β-CD and HP-β-CD) in orange carrot cell suspensions. Therefore, the effect of applying 10, 25, and 50 mM of both types of CDs on phytosterol production in yellow carrot cell cultures was evaluated (Fig. 2).

Thus, after elicitation with Me-β-CD and HP-β-CD at 10, 25, and 50 mM, the extraction and quantification of the phytosterols was performed outside the cells for 9 days. According to the two-way ANOVA analysis, total phytosterol content was significantly affected by the treatment and time of elicitation, all differences being significant with a $p$ value $< 0.05$. As shown in Fig. 2, the total phytosterol content of the culture medium increased up to 7 days of treatment, whereas there were no statistically significant differences between 7 and 9 days of elicitation. The maximal concentration of phytosterols on the seventh day was reached when the cell cultures were treated with 25 and 50 mM Me-β-CD, production being around 46-fold higher than in control treatment (81.65 ± 4.55 mg $L^{-1}$ vs. 1.77 ± 0.5 mg $L^{-1}$), not finding significant differences between both Me-β-CD concentrations. It is also important to note that phytosterol levels were always higher in the presence of Me-β-CD than in the presence of HP-β-CD (52.35 ± 0.46 mg $L^{-1}$) for the same concentration at the seventh day of treatment (Fig. 2). Besides, the results showed that the phytosterols (which represented 92% of the identified compounds) found in the highest proportion in yellow carrot cell cultures treated with 25 mM Me-β-CD for 7 days was β-sitosterol (27.91%), followed by fucosterol (19.67%), campesterol (12.20%), and stigmasterol (9.65%). Other precursors and derivatives of phytosterols (other PS), such as cycloartenol and elasterol, were identified (Supplementary Fig. 2). Other

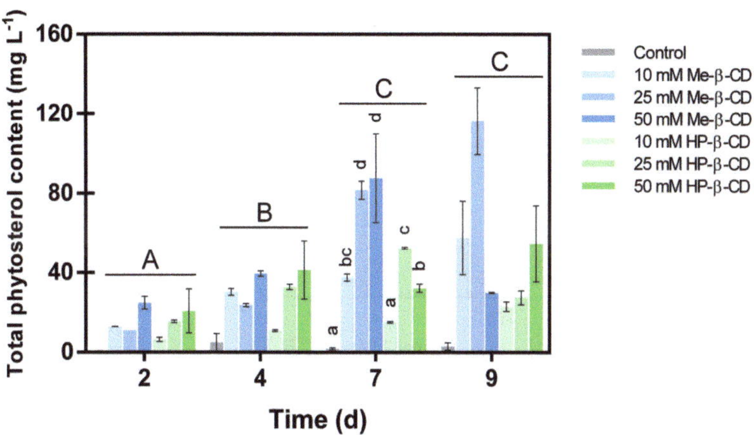


**Fig. 2** Time course of extracellular phytosterol levels in yellow carrot cell cultures elicited with Me-β-CD (10, 25, and 50 mM) and HP-β-CD (10, 25, and 50 mM). Columns represent means ± SD expressed as mg $L^{-1}$ ($n = 3$). Bars with different uppercase letters within each time denote statistically significant differences according to Duncan's test ($p < 0.05$). Bars with different lowercase letters denote significant differences ($p < 0.05$) between the treatments at 7 days of elicitation according to Duncan's test ($p < 0.05$)

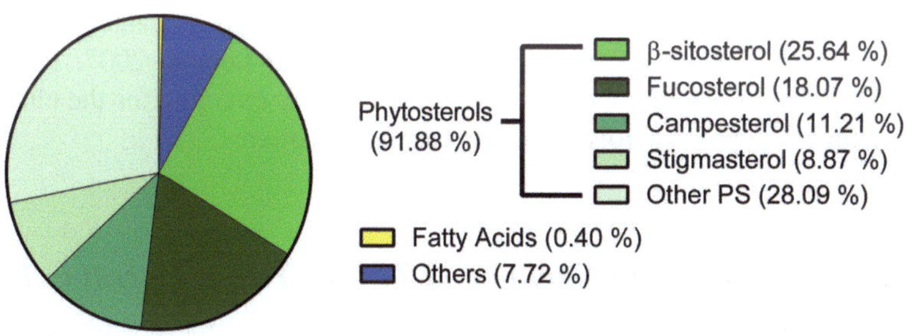


**Fig. 3** Total content of extracellular bioactive compounds in yellow carrot cell cultures elicited with 25 mM Me-β-CD for 7 days

than phytosterol, fatty acids (0.40%) and other bioactive compounds, such as hydrocarbon derivatives, were present (Fig. 3). Based on these results, the treatment with 25 mM Me-β-CD for 7 days was selected as the most efficient, since it maximized the production of phytosterols, minimizing the input of Me-β-CD and the time needed to achieve it.

However, when the intracellular content of phytosterols was analyzed, it was observed that treatment with 25 mM Me-β-CD caused a 40% decrease compared to the control treatment (Fig. 4), with no significant differences being found between each of the days analyzed. In contrast, the extracellular content of phytosterols increased exponentially with the days of treatment, reaching the maximum at 7 days after CD-elicitation (81.65 ± 4.55 mg $L^{-1}$, Fig. 4). This greater accumulation of phytosterols in the extracellular medium, to the detriment of intracellular

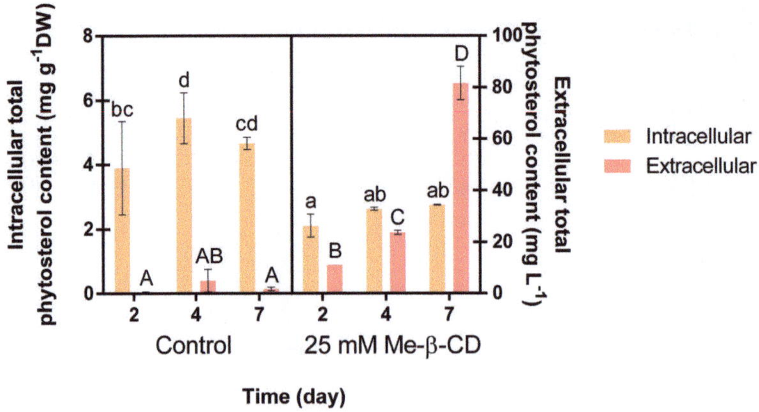


**Fig. 4** Effect of 25 mM Me-β-CD on the intracellular (mg $g^{-1}$ DW) and extracellular (mg $L^{-1}$) phytosterol content in yellow carrot cell cultures. Columns represent means ± SD ($n = 3$). Bars with different lowercase letters denote significant differences ($p < 0.05$) in intracellular phytosterol between the treatments throughout 7 days of elicitation, according to Duncan's test ($p < 0.05$). Bars with different uppercase letters denote significant differences ($p < 0.05$) in extracellular phytosterol between the treatments throughout 7 days of elicitation according to Duncan's test ($p < 0.05$)

phytosterol content, is due to the ability of CDs to bind lipids, extracting them from cell membranes and solubilizing them (Zidovetzki and Levitan 2007; Piel et al. 2007). Therefore, treatment with 25 mM Me-β-CD was effective for the elicitation and enrichment in phytosterols of yellow carrot cell cultures.

## 3.3 *Effect of Different Elicitors on TPC in Yellow Carrot Cell Cultures*

Given the results obtained on the increase of phytosterols with the application of 25 mM Me-β-CD, and with the aim of increasing the content of other types of metabolites in yellow carrot suspensions, the possibility of adding a second elicitor was discussed. The application of MJ in cell cultures has been previously described as an effective mechanism to improve the production of specialized metabolites, especially a wide variety of phenolic compounds. Specifically, combined applications of CD and MJ have been described with a synergistic effect in the production of this type of compounds in cell cultures of grapevine (Belchí-Navarro et al. 2012) and tomato (Briceño et al. 2012; Sabater-Jara et al. 2022). Therefore, the effect of the application of both elicitors in the production of phenolic compounds was evaluated individually or in combination.

TPCs were determined using the Folin-Ciocalteu method in yellow carrot cell cultures elicited with 25 mM Me-β-CD and/or 100 μM MJ for 7 days. As can be observed in Fig. 5, a marked enhancement in the intracellular TPC (0.77 mg $g^{-1}$

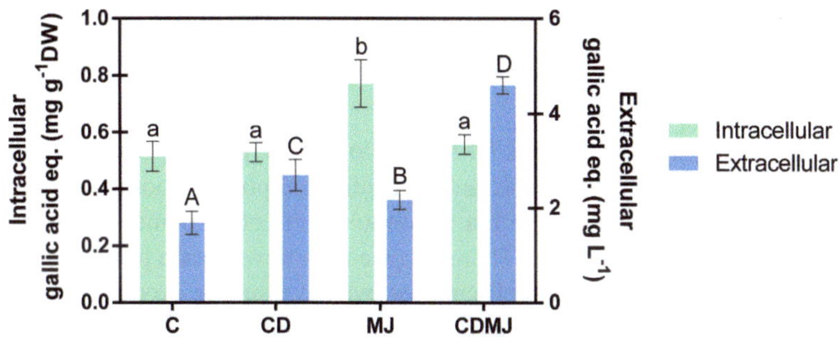


**Fig. 5** Effect of 25 mM Me-β-CD and/or 100 μM MJ on the intracellular (mg $g^{-1}$ DW) and extracellular (mg $L^{-1}$) total phenolic content (expressed as gallic acid equivalents) in yellow carrot cell cultures. Columns represent means ± SD ($n = 3$). Bars with different lowercase letters denote significant differences ($p < 0.05$) in intracellular phenolic content between the treatments according to Duncan's test ($p < 0.05$). Bars with different uppercase letters denote significant differences ($p < 0.05$) in extracellular total phenolic content between the treatments according to Duncan's test ($p < 0.05$)

DW) was determined when yellow carrot cell cultures were elicited only with MJ, whereas in CD-treated cells (CD and CDMJ), the TPC was similar to control cells (0.5 mg $g^{-1}$ DW, Fig. 5). Furthermore, all treatments used in this experiment increased the extracellular TPC, but the largest increase was found in the combined treatment (CDMJ), being 2.73, 1.70, and 2.11 times higher than in the control, CD- or MJ-treated cells, respectively (Fig. 5).

Therefore, the application of MJ allows to increase TPC in yellow carrot cell cultures, especially in combination with Me-β-CD. Therefore, it is an advantage for the enrichment of the diversity of compounds in the cell suspension, despite the decrease in the speed of cell growth, and the associated decrease in phytosterol production (up to 20%) (Pauwels et al. 2008; Sabater-Jara 2013). However, the enrichment of the cell culture in other types of bioactive compounds, such as phenolic compounds, allows us to obtain a much more complete bioproduct.

Both phytosterols and phenolic compounds have been shown to take part in multiple physiological processes, having also a biostimulant effect on plant crops, improving growth and yield parameters, as well as strengthening plants' mechanisms to resist abiotic stress (Kisiriko et al. 2021; Valitova et al. 2024). Therefore, the results demonstrated that the combined treatment of 25 mM Me-β-CD and 100 μM MJ is a tool to enrich yellow carrot cell cultures not only with phytosterols, but also in phenolic compounds, which maximizes its potential as a biostimulant.

## 3.4 Evaluation of YCB as Biostimulant on Tomato Seeds in Germination Speed and Stress Markers

The result after the application of CDMJ in yellow carrot cell suspensions resulted in a product enriched in phytosterols and phenolic compounds, both types of compounds having been described with a biostimulant effect on plants (Miras-Moreno et al. 2016b; Kumar et al. 2023a). Therefore, we have developed a production system based on yellow carrot elicited cell cultures, whose resulting product, YCB, has a potential biostimulant effect on plants. To evaluate this potential effect, a study on tomato seeds primed with different doses of YCB (0.01%, 0.1%, and 1% v/v) and germinated in sterile water (control conditions) or in 100 mM NaCl (salt stress) (Fig. 6) was carried out as salt stress has been described to inhibit seed germination by reducing seed's ability to absorb water and disrupting seed metabolism (de la Reguera et al. 2020).

Germination speed was determined by measuring germination index (GI), a parameter that represents the number of seeds that germinate per day from the

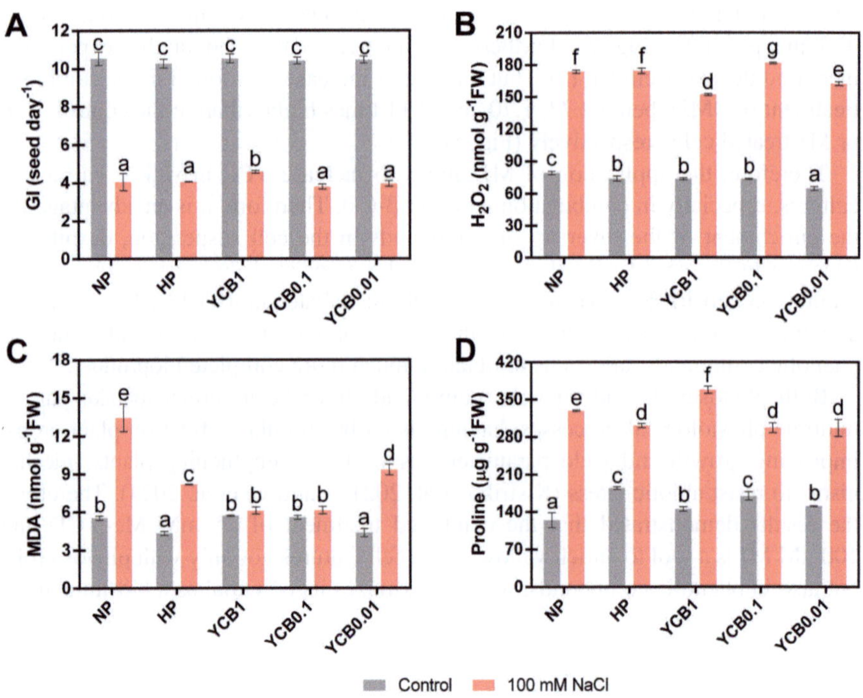


**Fig. 6** Effect of seed priming treatments in germination index (GI) (**a**) and $H_2O_2$ (**b**), MDA (**c**) and proline (**d**) content of tomato seedlings germinated under control or salinity conditions (100 mM NaCl). Columns represent mean ± SD ($n = 4$). Bars with different letters show significant differences ($p < 0.05$) between the treatments according to Duncan's test ($p < 0.05$). NP: no priming; HP: hydropriming; YCB: yellow carrot biostimulant at 1%, 0.1%, and 0.01%

beginning of the germination process. As shown in Fig. 6a, under control conditions, no significant differences were observed in the GI of primed seeds with respect to NP. Seeds subjected to salt stress, however, were affected in this parameter, showing a severe decrease in it. On the other hand, under these conditions, treatment with YCB1 increased the GI value significantly with respect to the other treatments, as well as with respect to NP (+13%) (Fig. 6a).

The first response of plants to abiotic stress is triggering the overproduction of reactive oxygen species (ROS), such as $H_2O_2$. These are signaling molecules, whose overproduction disrupts redox homeostasis, leading to an oxidative stress and causing damage to the cell by causing lipid peroxidation, impaired enzyme activity, and DNA damage (Martí-Guillén et al. 2022; Martínez-Lorente et al. 2022; Zhou et al. 2022). To evaluate the biostimulant effect of YCB on this response, typical molecular markers of oxidative stress were determined: $H_2O_2$ and lipid peroxidation.

Regarding total $H_2O_2$ content (Fig. 6b), all seed priming treatments decreased $H_2O_2$ concentration with respect to NP under control conditions, the decrease being greater in the treatment with YCB0.01 (−18%). Under salt stress conditions, its content was doubled with respect to control conditions, with no significant differences being observed between NP and HP. However, when the biostimulant was applied, both YCB1 and YCB0.01 decreased $H_2O_2$ content with respect to NP, this decrease being greater with the application of YCB1 (−13%) (Fig. 6b), which correlates with the higher GI (Fig. 6a).

On the other hand, lipid peroxidation was determined by measuring total MDA content in the samples (Fig. 6c). The results showed that the priming treatment decreased MDA content in HP and YCB0.01 with respect to NP under control conditions. In seedlings germinated under salt stress, MDA content increased significantly. However, under salinity conditions, it stands out that any seed treatment decreased MDA content with respect to NP, especially YCB1 and YCB0.1 doses, where its content reverted to the one in control conditions (Fig. 6c).

Proline has been described as an osmoprotectant molecule able to scavenge and modulate ROS detoxification pathways, constituting one of the main non-enzymatic defense systems against oxidative stress (Anwar Hossain et al. 2014; Zulfiqar and Ashraf 2023). Regarding proline content, the results showed that under control conditions, seed priming increased its content in all treatments with respect to NP. As shown in Fig. 6d, in the presence of salt stress, the proline content increased generally. On the other hand, under salinity conditions, all seed priming treatments decreased the proline content with respect to NP. However, the response found in seeds subjected to treatment with YCB1 is especially noteworthy, as it showed an increase of +12%.

Therefore, seed priming treatment with YCB has shown an effect alleviating salt stress, especially 1% (v/v) dose, which had a positive effect on all the parameters measured (Fig. 6). Several studies have shown that biostimulants applied as seed priming are a sustainable tool to increase crop productivity. Specifically, they have been proved to have a positive effect by increasing the germination rate under salt stress conditions, detoxifying ROS, and increasing osmoprotectant content, while maintaining the integrity of cell membranes (Gupta et al. 2022).

In multiple studies, these effects of plant-derived biostimulants on germination processes and oxidative stress have been associated with the composition of the biostimulant, especially its metabolite content. Specialized plant metabolites have been described as agents capable of stimulating plant development and mitigating the negative effects of abiotic stress (Aremu et al. 2015; Ertani et al. 2016; Ben Mrid et al. 2021). Specifically, phenolic compounds can act either by directly detoxifying ROS or by promoting plant defense mechanisms (Kumar et al. 2023a). hPDBs rich in phenolic compounds have been shown to be a powerful tool in stress situations, promoting germination and seedling development; and reinforcing both enzymatic and non-enzymatic antioxidant defense, mainly by increasing proline content, resulting in a reduction in ROS and lipid peroxidation levels (García-Sánchez et al. 2012; Hayat et al. 2018; Naz et al. 2022; Kumar et al. 2023b; Toscano Adamo et al. 2024; Albaladejo-Marico et al. 2024). Similar effects have been demonstrated for phytosterols. Phytosterols have been described as essential compounds for the correct development of the germination process (Suo et al. 2021) and stress response (Aboobucker and Suza 2019) with potential to modulate antioxidant response. Specifically, stigmasterol, campesterol, and β-sitosterol are able to alleviate oxidative stress, acting as radical scavengers or promoting the activity of antioxidant enzymes, decreasing lipid peroxidation (Sharma et al. 2021; Shen et al. 2024). In addition, it has been shown that treatment with fucosterol and stigmasterol promotes the growth of seedlings and plants subjected to abiotic stress, decreasing the levels of $H_2O_2$ and MDA, as well as increasing proline content (Hanafy and Sadak 2023; Yu et al. 2024).

Thus, our results indicate that YCB has a biostimulant effect on tomato seeds, especially those germinated under salinity conditions, by modulating oxidative stress response, mainly due to its composition in phenolic compounds and phytosterols. However, YCB constitutes a biostimulant with a complex composition, and other bioactive compounds could be participating in the biostimulant effect observed.

## 4 Conclusions and Future Perspectives

Our work has demonstrated that the elicitation of yellow carrot cell cultures with 25 mM Me-β-CD and 100 μM MJ for 7 days is an effective tool for increasing the production of phytosterols and phenolic compounds, constituting a stable system for producing a biostimulant product. The evaluation of the biostimulant effect of the product was corroborated by applying it as seed priming on tomato seeds, resulting in the improvement of germination parameters and markers of oxidative damage and cellular defense measured when the dose of 1% (v/v) YCB was applied. Therefore, we have developed a novel potential biostimulant product based on yellow carrot cell cultures, enriched in phenolic compounds and phytosterols, which has demonstrated its effect on tomato seeds subjected to salt stress.

Additionally, we consider that our YCB could develop similar effects in other types of seeds subjected to salt stress, or even to other abiotic stresses. Further research is needed to elucidate whether the application of YCB, or even other cell culture-based biostimulants developed with the same biotechnological tools, provide seeds with the capacity to improve the germination process in marginal environments.

**Acknowledgments** This study forms part of the Agroalnext Programme and was supported by MCIN with funding from European Union NextGenerationEU (PRTR-C17.I1) and by Comunidad Autónoma de la Región de Murcia-Fundación Séneca and by Fundación Séneca-Agencia de Ciencia y Tecnología from Murcia (project 22016/PI/22), via “Ayudas a proyectos para el desarrollo de investigación científica y técnica por grupos competitivos”, included in “Programa Regional de Fomento de la Investigación Científica y Técnica (Plan de Actuación 2022)”. S.E. M.-L. received grants from Ministerio de Ciencia, Innovación y Universidades of Spain FPU Grants (Grant No. FPU21/01593), and J.M.M.-G. received grants from Universidad de Murcia PhD contracts (Registry number 109144/2022).

**CRediT Authorship Contribution Statement** S.E.M.-L. and J.M.M.-G. conducted experiments. S.E.M.-L., J.M.M.-G., and A.B.S.-J. conceived and designed research. S.E.M.-L., J.M.M.-G., A.B. S.-J., L.A., and M.A.P. analyzed data. S.E.M.-L., J.M.M.-G., L.A., and M.A.P. wrote the manuscript. M.A.P. and A.B.S.-J. developed funding acquisition. All authors read and approved the manuscript.

**Competing Interests** The authors declare that they have no known competing financial interests or personal relationships that could have appeared to influence the work reported in this paper.

## References

Aboobucker SI, Suza WP (2019) Why do plants convert sitosterol to stigmasterol? Front Plant Sci 10:354. https://doi.org/10.3389/fpls.2019.00354

Acosta JA, Faz A, Jansen B et al (2011) Assessment of salinity status in intensively cultivated soils under semiarid climate, Murcia, SE Spain. J Arid Environ 75:1056–1066. https://doi.org/10.1016/j.jaridenv.2011.05.006

Ahmad A, Blasco B, Martos V (2022) Combating salinity through natural plant extracts based biostimulants: a review. Front Plant Sci 13:862034. https://doi.org/10.3389/fpls.2022.862034

Albaladejo-Marico L, Carvajal M, Yepes-Molina L (2024) Involvement of glucosinolates and phenolics in the promotion of broccoli seedling growth through the modulation of primary and secondary metabolism. Plant Sci 347:112205. https://doi.org/10.1016/j.plantsci.2024.112205

Almagro L, Calderón AA, Pedreño MA, Ferrer MA (2022) Differential response of phenol metabolism associated with antioxidative network in elicited grapevine suspension cultured cells under saline conditions. Antioxidants 11:388. https://doi.org/10.3390/antiox11020388

Anwar Hossain M, Hoque MA, Burritt DJ, Fujita M (2014) Chapter 16 – Proline protects plants against abiotic oxidative stress: biochemical and molecular mechanisms. In: Ahmad P (ed) Oxidative damage to plants. Academic Press, San Diego, pp 477–522

Aremu AO, Masondo NA, Rengasamy KRR et al (2015) Physiological role of phenolic biostimulants isolated from brown seaweed *Ecklonia maxima* on plant growth and development. Planta 241:1313–1324. https://doi.org/10.1007/s00425-015-2256-x

Bates LS, Waldren RP, Teare ID (1973) Rapid determination of free proline for water-stress studies. Plant Soil 39:205–207. https://doi.org/10.1007/BF00018060

Belchí-Navarro S, Almagro L, Lijavetzky D et al (2012) Enhanced extracellular production of *trans*-resveratrol in *Vitis vinifera* suspension cultured cells by using cyclodextrins and methyljasmonate. Plant Cell Rep 31:81–89. https://doi.org/10.1007/s00299-011-1141-8

Ben Mrid R, Benmrid B, Hafsa J et al (2021) Secondary metabolites as biostimulant and bioprotectant agents: a review. Sci Total Environ 777:146204. https://doi.org/10.1016/j.scitotenv.2021.146204

Briceño Z, Almagro L, Sabater-Jara AB et al (2012) Enhancement of phytosterols, taraxasterol and induction of extracellular pathogenesis-related proteins in cell cultures of *Solanum lycopersicum* cv Micro-Tom elicited with cyclodextrins and methyl jasmonate. J Plant Physiol 169:1050–1058. https://doi.org/10.1016/j.jplph.2012.03.008

Burns J, Fraser PD, Bramley PM (2003) Identification and quantification of carotenoids, tocopherols and chlorophylls in commonly consumed fruits and vegetables. Phytochemistry 62:939–947. https://doi.org/10.1016/S0031-9422(02)00710-0

Chandran H, Meena M, Barupal T, Sharma K (2020) Plant tissue culture as a perpetual source for production of industrially important bioactive compounds. Biotechnol Rep 26:e00450. https://doi.org/10.1016/j.btre.2020.e00450

Corchete P, Almagro L, Gabaldón JA et al (2022) Phenylpropanoids in *Silybum marianum* cultures treated with cyclodextrins coated with magnetic nanoparticles. Appl Microbiol Biotechnol 106: 2393–2401. https://doi.org/10.1007/s00253-022-11886-2

de la Reguera E, Veatch J, Gedan K, Tully KL (2020) The effects of saltwater intrusion on germination success of standard and alternative crops. Environ Exp Bot 180:104254. https://doi.org/10.1016/j.envexpbot.2020.104254

Del Buono D (2021) Can biostimulants be used to mitigate the effect of anthropogenic climate change on agriculture? It is time to respond. Sci Total Environ 751:141763. https://doi.org/10.1016/j.scitotenv.2020.141763

Ertani A, Pizzeghello D, Francioso O et al (2016) Biological activity of vegetal extracts containing phenols on plant metabolism. Molecules 21:205. https://doi.org/10.3390/molecules21020205

Espinosa-Leal CA, Puente-Garza CA, García-Lara S (2018) In vitro plant tissue culture: means for production of biological active compounds. Planta 248:1–18. https://doi.org/10.1007/s00425-018-2910-1

European Commission (2019) Communication from the Commission to the European Parliament, The European Council, The Council, The European Economic and Social Committee and The Committee of the Regions The European Green Deal

European Commission (2020) Communication from the Commission to the European Parliament, The Council, The European Economic and Social Committee and The Committee of the Regions A Farm to Fork Strategy for a fair, healthy and environmentally-friendly food system

European Parliament (2019) Regulation (EU) 2019/1009 of the European Parliament and of the Council of 5 June 2019 laying down rules on the making available on the market of EU fertilising products and amending Regulations (EC) No 1069/2009 and (EC) No 1107/2009 and repealing Regulation (EC) No 2003/2003 (Text with EEA relevance)

García-Sánchez M, Garrido I, Casimiro IJ et al (2012) Defence response of tomato seedlings to oxidative stress induced by phenolic compounds from dry olive mill residue. Chemosphere 89: 708–716. https://doi.org/10.1016/j.chemosphere.2012.06.026

Gupta S, Doležal K, Kulkarni MG et al (2022) Role of non-microbial biostimulants in regulation of seed germination and seedling establishment. Plant Growth Regul 97:271–313. https://doi.org/10.1007/s10725-021-00794-6

Hanafy RS, Sadak MS (2023) Foliar spray of stigmasterol regulates physiological processes and antioxidant mechanisms to improve yield and quality of sunflower under drought stress. J Soil Sci Plant Nutr 23:2433–2450. https://doi.org/10.1007/s42729-023-01197-4

Hayat S, Ahmad H, Ali M et al (2018) Aqueous garlic extract stimulates growth and antioxidant enzymes activity of tomato (*Solanum lycopersicum*). Sci Hortic 240:139–146. https://doi.org/10.1016/j.scienta.2018.06.011

Heath RL, Packer L (1968) Photoperoxidation in isolated chloroplasts: I. Kinetics and stoichiometry of fatty acid peroxidation. Arch Biochem Biophys 125:189–198. https://doi.org/10.1016/0003-9861(68)90654-1

Hodges DM, DeLong JM, Forney CF, Prange RK (1999) Improving the thiobarbituric acid-reactive-substances assay for estimating lipid peroxidation in plant tissues containing anthocyanin and other interfering compounds. Planta 207:604–611. https://doi.org/10.1007/s004250050524

Junglee S, Urban L, Sallanon H, Lopez-Lauri F (2014) Optimized assay for hydrogen peroxide determination in plant tissue using potassium iodide. Am J Anal Chem 5:730–736. https://doi.org/10.4236/ajac.2014.511081

Kisiriko M, Anastasiadi M, Terry LA et al (2021) Phenolics from medicinal and aromatic plants: characterisation and potential as biostimulants and bioprotectants. Molecules 26:6343. https://doi.org/10.3390/molecules26216343

Kumar K, Debnath P, Singh S, Kumar N (2023a) An overview of plant phenolics and their involvement in abiotic stress tolerance. Stresses 3:570–585. https://doi.org/10.3390/stresses3030040

Kumar S, Shukla V, Tripathi YN et al (2023b) Biochemical changes, antioxidative profile, and efficacy of the bio-stimulant in plant defense response against Sclerotinia sclerotiorum in common bean (*Phasaeolus vulgaris* L.). Heliyon 10:e23030. https://doi.org/10.1016/j.heliyon.2023.e23030

Landi M (2017) Commentary to: "Improving the thiobarbituric acid-reactive-substances assay for estimating lipid peroxidation in plant tissues containing anthocyanin and other interfering compounds" by Hodges et al., Planta (1999) 207:604–611. Planta 245:1067–1067. https://doi.org/10.1007/s00425-017-2699-3

Liu B, Chen K, Chen X et al (2024) Health outcomes associated with phytosterols: an umbrella review of systematic reviews and meta-analyses of randomized controlled trials. Phytomedicine 122:155151. https://doi.org/10.1016/j.phymed.2023.155151

Loulakakis KA, Roubelakis-Angelakis KA (2001) Nitrogen assimilation in grapevine. In: Roubelakis-Angelakis KA (ed) Molecular biology & biotechnology of the grapevine. Springer, Dordrecht, pp 59–85

Maguire JD (1962) Speed of germination—aid in selection and evaluation for seedling emergence and vigor. Crop Sci 2:176–177. https://doi.org/10.2135/cropsci1962.0011183X000200020033x

Marchev AS, Yordanova ZP, Georgiev MI (2020) Green (cell) factories for advanced production of plant secondary metabolites. Crit Rev Biotechnol 40:443–458. https://doi.org/10.1080/07388551.2020.1731414

Martí-Guillén JM, Pardo-Hernández M, Martínez-Lorente SE et al (2022) Redox post-translational modifications and their interplay in plant abiotic stress tolerance. Front Plant Sci 13:1027730. https://doi.org/10.3389/fpls.2022.1027730

Martínez-Lorente SE, Pardo-Hernández M, Martí-Guillén JM et al (2022) Interaction between melatonin and NO: action mechanisms, main targets, and putative roles of the emerging molecule NOmela. Int J Mol Sci 23:6646. https://doi.org/10.3390/ijms23126646

Martínez-Lorente SE, Martí-Guillén JM, Pedreño MÁ et al (2024) Higher plant-derived biostimulants: mechanisms of action and their role in mitigating plant abiotic stress. Antioxidants 13:318. https://doi.org/10.3390/antiox13030318

McDonald KA, Jackman AP (1989) Bioreactor studies of growth and nutrient utilization in alfalfa suspension cultures. Plant Cell Rep 8:455–458. https://doi.org/10.1007/BF00269047

Miras-Moreno B, Almagro L, Pedreño MA, Sabater-Jara AB (2016a) Enhanced accumulation of phytosterols and phenolic compounds in cyclodextrin-elicited cell suspension culture of *Daucus carota*. Plant Sci 250:154–164. https://doi.org/10.1016/j.plantsci.2016.06.008

Miras-Moreno B, Sabater-Jara AB, Pedreño MA, Almagro L (2016b) Bioactivity of phytosterols and their production in plant in vitro cultures. J Agric Food Chem 64:7049–7058. https://doi.org/10.1021/acs.jafc.6b02345

Morel G, Wetmore RH (1951) Fern callus tissue culture. Am J Bot 38:141–143. https://doi.org/10.1002/j.1537-2197.1951.tb14804.x

Murashige T, Skoog F (1962) A revised medium for rapid growth and bio assays with tobacco tissue cultures. Physiol Plant 15:473–497. https://doi.org/10.1111/j.1399-3054.1962.tb08052.x

Naz H, Aisha Akram N, Ashraf M et al (2022) Leaf extract of neem (*Azadirachta indica*) alleviates adverse effects of drought in quinoa (*Chenopodium quinoa* Willd.) plants through alterations in biochemical attributes and antioxidants. Saudi J Biol Sci 29:1367–1374. https://doi.org/10.1016/j.sjbs.2022.01.038

Ofoe R, Gunupuru LR, Wang-Pruski G et al (2022) Seed priming with pyroligneous acid mitigates aluminum stress, and promotes tomato seed germination and seedling growth. Plant Stress 4:100083. https://doi.org/10.1016/j.stress.2022.100083

Pauwels L, Morreel K, De Witte E et al (2008) Mapping methyl jasmonate-mediated transcriptional reprogramming of metabolism and cell cycle progression in cultured Arabidopsis cells. Proc Natl Acad Sci USA 105:1380–1385. https://doi.org/10.1073/pnas.0711203105

Piel G, Piette M, Barillaro V et al (2007) Study of the relationship between lipid binding properties of cyclodextrins and their effect on the integrity of liposomes. Int J Pharm 338:35–42. https://doi.org/10.1016/j.ijpharm.2007.01.015

Rahman MM, Rahaman MS, Islam MR et al (2021) Role of phenolic compounds in human disease: current knowledge and future prospects. Molecules 27:233. https://doi.org/10.3390/molecules27010233

Ramirez-Estrada K, Vidal-Limon H, Hidalgo D et al (2016) Elicitation, an effective strategy for the biotechnological production of bioactive high-added value compounds in plant cell factories. Molecules 21:182. https://doi.org/10.3390/molecules21020182

Sabater-Jara AB (2013) Estudio de las respuestas de defensa inducidas por elicitores en cultivos de células vegetales. Ph.D. thesis, University of Murcia. Available at https://digitum.um.es/digitum/handle/10201/35414. Accessed 3 Dec 2024

Sabater-Jara AB, Pedreño MA (2013) Use of β-cyclodextrins to enhance phytosterol production in cell suspension cultures of carrot (*Daucus carota* L.). Plant Cell Tissue Organ Cult 114:249–258. https://doi.org/10.1007/s11240-013-0320-4

Sabater-Jara AB, Marín-Marín MJ, Almagro L, Pedreño MA (2022) Cyclodextrins increase triterpene production in *Solanum lycopersicum* cell cultures by activating biosynthetic genes. Plants 11:2782. https://doi.org/10.3390/plants11202782

Sánchez-Pujante PJ, Gionfriddo M, Sabater-Jara AB et al (2020) Enhanced bioactive compound production in broccoli cells due to coronatine and methyl jasmonate is linked to antioxidative metabolism. J Plant Physiol 248:153136. https://doi.org/10.1016/j.jplph.2020.153136

Sharma N, Tan MA, An SSA (2021) Phytosterols: potential metabolic modulators in neurodegenerative diseases. Int J Mol Sci 22:12255. https://doi.org/10.3390/ijms222212255

Shen M, Yuan L, Zhang J et al (2024) Phytosterols: physiological functions and potential application. Foods 13:1754. https://doi.org/10.3390/foods13111754

Suo X, Xu F, Tan K et al (2021) Functions of phytosterols in seed development of upland cotton (*Gossypium hirsutum L.*). Ind Crops Prod 170:113802. https://doi.org/10.1016/j.indcrop.2021.113802

Toscano Adamo ML, Yonny ME, Villalba GF, Nazareno MA (2024) Natural biostimulants foliar application as sustainable mitigation strategy of drought stress damage on the melon crop (*Cucumis melo* L.). Sci Hortic 323:112471. https://doi.org/10.1016/j.scienta.2023.112471

Ullah A, Bano A, Khan N (2021) Climate change and salinity effects on crops and chemical communication between plants and plant growth-promoting microorganisms under stress. Front Sustain Food Syst 5:618092. https://doi.org/10.3389/fsufs.2021.618092

United Nations (2019) World population prospects 2019. Department of Economic and Social Affairs, United Nations, New York

Valitova J, Renkova A, Beckett R, Minibayeva F (2024) Stigmasterol: an enigmatic plant stress sterol with versatile functions. Int J Mol Sci 25:8122. https://doi.org/10.3390/ijms25158122
Yu T-F, Hou Z-H, Wang H-L et al (2024) Soybean steroids improve crop abiotic stress tolerance and increase yield. Plant Biotechnol J 22:2333–2347. https://doi.org/10.1111/pbi.14349
Zhou X, Joshi S, Patil S et al (2022) Reactive oxygen, nitrogen, carbonyl and sulfur species and their roles in plant abiotic stress responses and tolerance. J Plant Growth Regul 41:119–142. https://doi.org/10.1007/s00344-020-10294-y
Zidovetzki R, Levitan I (2007) Use of cyclodextrins to manipulate plasma membrane cholesterol content: evidence, misconceptions and control strategies. Biochim Biophys Acta 1768:1311–1324. https://doi.org/10.1016/j.bbamem.2007.03.026
Zulfiqar F, Ashraf M (2023) Proline alleviates abiotic stress induced oxidative stress in plants. J Plant Growth Regul 42:4629–4651. https://doi.org/10.1007/s00344-022-10839-3
Zulfiqar F, Casadesús A, Brockman H, Munné-Bosch S (2020) An overview of plant-based natural biostimulants for sustainable horticulture with a particular focus on moringa leaf extracts. Plant Sci 295:110194. https://doi.org/10.1016/j.plantsci.2019.110194

# Mutualistic Relationships Between Roots and Soil Microbes Facilitate Plant Potassium Acquisition

**Danielle R. Cooney, Benjamin D. Rose, Alena P. Jones, Hannah E. R. Frank, Gitanjali Nanda Kafle, Arjun Kafle, and Kevin Garcia**

## Contents



**Abstract** Soil supports diverse life forms and ecosystems through complex interactions among its components, many of which remain poorly understood. Potassium (K) is a vital macronutrient that plays an essential role in plant growth, enzymatic activities, and stress tolerance, but its bioavailability in soils is limited due to complexation in minerals, such as muscovite, biotite, and feldspars. Therefore, only a small fraction is in exchangeable, plant-accessible form. While K fertilizers are commonly used in agriculture and agroforestry to address K deficiencies, the contributions of soil microbial communities to K solubilization and plant acquisition are increasingly recognized. Soil microbes, including K-solubilizing fungi and

Communicated by Francisco M. Cánovas

D. R. Cooney, B. D. Rose, A. P. Jones, H. E. R. Frank, G. N. Kafle, and K. Garcia (✉)
Department of Crop and Soil Sciences, North Carolina State University, Raleigh, NC, USA
e-mail: danielle.cooney@corteva.com; bdrose@ncsu.edu; apjones5@ncsu.edu; hef43@cornell.edu; kgarcia2@ncsu.edu

A. Kafle
Department of Crop and Soil Sciences, North Carolina State University, Raleigh, NC, USA

Agriculture Research Service, Plant Science Research Unit, USDA, Raleigh, NC, USA
e-mail: akafle@ncsu.edu



Progress in Botany (2026) 86: 131–149, https://doi.org/10.1007/124_2025_88,
Published online: 23 May 2025

bacteria, enhance nutrient availability by decomposing soil components, facilitating K release and plant uptake. Despite early recognition of microbial K solubilization in the late nineteenth century, the role of microbes in supporting plant K acquisition remains underexplored, particularly in natural and agro-ecosystems. This chapter reviews current knowledge on plant K uptake mediated by soil microbes, including endophytic and K-solubilizing fungi, bacteria, and mycorrhizal fungi, to highlight their potential for improving soil K cycling, plant K-use-efficiency, and sustainable agricultural practices.

# 1 Introduction

Soil functions as a fundamental substrate and mediator of diverse environmental conditions, supporting a wide variety of life forms. The maintenance of soil ecosystems relies on a complex network of interactions and processes among its components, many of which remain poorly understood. Of particular interest are the mutualistic relationships between soil microbes, nutrients, and root functions, which are essential for plant growth and survival. For plants to thrive, they require the acquisition of macronutrients, micronutrients, and water from the soil. Among these elements, potassium (K) uptake is of specific significance due to its role in many vital processes, such as enzymatic activities, photosynthesis, tolerance to environmental stressors, and many other factors contributing to plant growth and development. Therefore, K deficiency hinders overall plant physiology, leading to poor growth, lesser yields (White and Karley 2010), and disturbances of the immune system that can increase susceptibility to diseases and pest attack (Troufflard et al. 2010; Armengaud et al. 2010).

In soils, K bioavailability is often low, with a concentration in the soil matrix of the earth's crust ranging from 0.5% to 2.5%, and generally complexed with minerals (Tisdale et al. 1985). About 90–98% of the total K is plant-unavailable and bound in mineral complexes such as muscovite ($K[(Si_3Al)Al_2O_{10}(OH)_2]$), biotite ($K[(Si_3Al)(Mg,Fe)_3O_{10}(OH)_2]$), illite, and orthoclase (KAlSi3O8) (Zörb et al. 2014). K is a monovalent cation, which mostly occurs as a structural component in 2:1 clays in the soil matrix, and originates from various sources like micas, feldspars, and micaceous minerals in the clay fraction (Pratt 1965). The exchangeable form of K is either water soluble or loosely held in association with the negatively charged surface of clay particles. Thus, soils with higher sand (and lower clay) contents are more susceptible to losses of K when environmental conditions are conducive to water leaching (Alfaro et al. 2004; Goulding et al. 2021). When soil testing reveals that plant-available K in the soil is limited, a common solution is the addition of fertilizers, often applied in the form of salts. The most common fertilizer choice is K chloride (KCl), but other forms of K salts such as nitrate, carbonate, bromide, sulfate, and cyanide can be utilized (Mikkelsen and Roberts 2021). Other inputs of K into the soil solution can originate from the atmosphere, irrigation, runoff and erosion, and plant biomass residues (Mikkelsen and Roberts 2021). The production of row-crop agriculture can intensify the demands of plant-available K in the soil. A plant's ability to acquire K from the soil solution is dependent upon the soil type, properties, and available K concentration. The concentration drives the selection and uptake of ions via either high-affinity or low-affinity transport mechanisms, spread among multiple large protein families (Wang and Wu 2017).

In addition to the molecular mechanisms allowing root nutrient acquisition when resources are physically close and readily available, root interaction with surrounding microbes is necessary to ensure a continuous uptake of soil resources, including K ions. Therefore, plant roots have developed intricate relationships with various soil microbes to enhance and maintain nutrient acquisition. Although the role of soil bacteria in solubilizing K from rock structures was first reported more than a century ago (Muentz 1890) and multiple species of bacteria and fungi have been described with K-solubilizing properties (Sarikhani et al. 2018; Haro and Benito 2019), the actual involvement of these microbes on plant root K acquisition in natural and agroecosystems is largely understudied and probably underestimated.

Therefore, in this chapter, we aim to evaluate the current state of knowledge relative to plant K uptake facilitated and/or directly mediated by various soil microbes, including arbuscular and ectomycorrhizal fungi, as well as endophytic fungi and bacteria.

## 2 Plant Potassium Uptake Facilitated by Filamentous Fungi

### *2.1 Arbuscular Mycorrhizal Fungi*

The dynamic cycle of nutrient formation, transformation, and movement in the soils is a complex process driven by root-microbe, soil-microbe, and microbe-microbe interactions. Among these microbial organisms, many filamentous fungi can solubilize soil K through the secretion of organic acids or anions, leading to a breakdown in the structure of the minerals (Haro and Benito 2019). Other fungi can improve plant K acquisition through the direct transfer of ions from the soil to colonized roots, such as arbuscular mycorrhizal (AM) fungi (Garcia et al. 2016).

AM fungi have been present for more than 400 million years and helped plants to colonize land (Rich et al. 2021). These fungi are ubiquitous in the soil and can form mutualistic symbiosis with approximately 80% of terrestrial plant species (Lanfranco et al. 2016). AM fungi are obligate symbionts belonging to the subphylum Glomeromycotina (Spatafora et al. 2016), and they need to colonize a host plant to complete their life cycle. AM fungi rely on sugars and lipids from the plant, and in return the fungi provide the plant with water and nutrients obtained from the soil (Smith and Read 2008). AM fungi develop a diversity of structures in the soil and within plant roots, including arbuscules for plant-fungus nutrient exchanges, sometimes vesicles for resource storage, external/internal hyphae, and asexual spores (Smith and Read 2008). Interacting with these fungal symbionts provides the host plant with multiple benefits, including improved water and nutrient acquisition, abiotic stress tolerance, and protection from pathogenic microbes (Chen et al. 2018; Begum et al. 2019; Tang et al. 2023). Furthermore, hyphae of AM fungi connected to roots facilitate the transfer of plant-originating carbon to the rhizosphere communities (Kaiser et al. 2015). They also influence the community-based decomposition of soil organic matter and thus the availability of many essential nutrients associated with organic matter breakdown (Frey 2019). Due to the large prevalence of AM fungi in native and agro-ecosystems, the study of this symbiosis is highly relevant to improve our agriculture practices. Indeed, agricultural crop productivity is linked to essential macro- and micro-nutrient availability in the soil solution for plant or fungal acquisition, so understanding AM symbiosis establishment and function could help in reducing dependency on chemical fertilizers. AM-related research on plant nutrition improvement has mainly focused on the macronutrients phosphorus (P) (e.g., George et al. 1995; Smith et al. 2011; Plassard et al. 2019; Thioub et al. 2019; Adeyemi et al. 2021), nitrogen (N) (e.g. George et al. 1995; Fellbaum et al. 2012; Ngwene et al. 2013; Pérez-Tienda et al. 2014; Kafle et al. 2019b), as well as some micro-nutrients (e.g. Liu et al. 2000; Watts-Williams et al. 2013, 2017; Watts-Williams and Cavagnaro 2018; Ruytinx et al. 2020). Although limited data is available on the ability of AM fungi to take up K from the soil and transfer it to colonized root, it has been shown with multiple model and crop species that interacting with AM fungi generally improves plant K accumulation (Scheloske et al. 2004; Pallon et al. 2007; Estrada et al. 2013; Garcia and Zimmermann 2014; Garcia and Ané 2017; Ye et al. 2017; Haro and Benito 2019; Liu et al. 2019). K ions also play a crucial role in alleviating abiotic stress responses in plants, particularly high sodium (Na) accumulation (Tester and Davenport 2003; Benito et al. 2014), and it has been demonstrated that AM fungi can play a pivotal role in improving these responses (Campanelli et al. 2013; Sarwat et al. 2016; Santander et al. 2019). For example, Santander et al. (2021) demonstrated in a study on lettuce that additions of AM inoculants significantly impacted the K/Na ratio in plant roots.

Tracking K transport in AM symbiosis has been challenging for a long time due to the difficulty of using K isotopes. Recently, the use of rubidium (Rb), and to a lesser extent cesium (Cs), as proxies for K has shown promising results to achieve this goal. Although Rb ions have a larger radius than K ones, they are transported through K transporters and channels, since no specific Rb transport protein has been identified so far in any organism. Therefore, by placing Rb in a compartment only available for the AM fungus but not for the roots, detection of Rb in plant shoots would indicate that the K transport pathway exists in AM symbiosis. Using this approach, it has been recently demonstrated in the model legume *Medicago truncatula* that shoot Rb concentrations were significantly higher in AM plants compared to non-mycorrhizal plants under limiting and sufficient K conditions (Kafle et al. 2022). Positive correlations between shoot Rb concentration, shoot K concentration, and AM colonization were observed, which strongly suggests that Rb was being transported to the plant from the fungus, and there is a symbiotic K transport pathway. Another recent study investigated the use of Cs as a proxy for K transport in AM symbiosis (Kafle and Garcia 2022). Plants were inoculated with AM

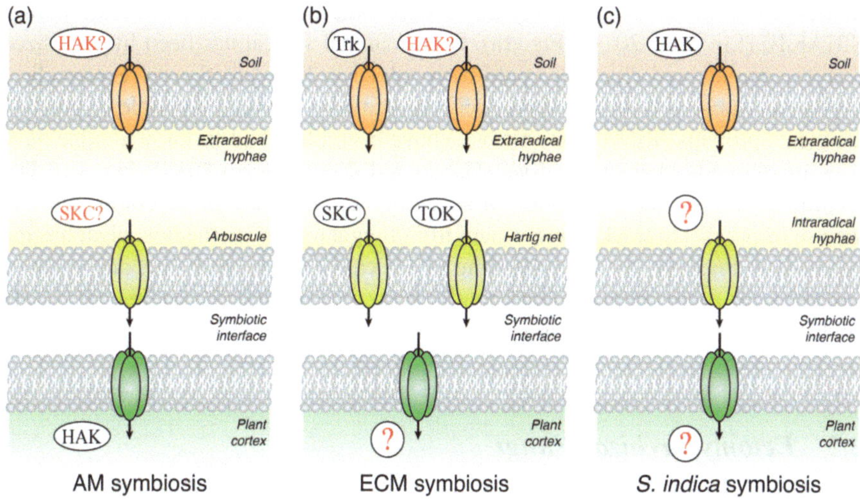


**Fig. 1** Current knowledge of the molecular mechanisms governing potassium transport from the soil to plant roots colonized by arbuscular mycorrhizal, ectomycorrhizal, and *Serendipita indica* fungi. (**a**) In arbuscular mycorrhizal (AM) symbiosis, no K fungal transport protein has been functionally characterized so far. Expression data showed that putative members from the KT/KUP/HAK transporter (HAK) and *Shaker*-like K channel (SKC) families might be involved in K acquisition from the soil and transfer from the arbuscules to the symbiotic interface, respectively. One plant HAK transporter has been functionally characterize in tomato roots colonized by AM fungi (SlHAK10) and participates in K acquisition from the symbiotic interface. (**b**) In ectomycorrhizal (ECM) symbiosis, transporters are involved in K acquisition from the soil by extraradical hyphae, with one member of the Trk/Ktr/HKT family being functionally characterized in the fungus *Hebeloma cylindrosporum* (HcTrk1), while HAK members have been identified but not functionally characterized yet. Members from the Tandem-pore Outward-rectifying K (TOK) and SKC channel families have been described as participating in K transport from the Hartig net to the symbiotic interface (HcTOK2.2 and HcSKC). No plant transporter responsible for K uptake from the symbiotic interface has been identified so far in ECM symbiosis. (**c**) In root-*Serendipita indica* interactions, only one fungal HAK (SiHAK1) transporter has been described to participate in K acquisition from the soil. However, the transfer of K from this fungus to colonized roots is strongly debated and no fungal or plant K transport proteins have been identified at the symbiotic interface thus far

fungi and placed in K-deprivation or K-sufficient conditions. Cs chloride was added in various concentrations into an adjacent compartment only accessible to the fungus. No significant difference in shoot Cs concentration between non-mycorrhizal and AM plants was observed under K-deprived conditions. With sufficient K supply, a significant difference of shoot Cs concentration was only observed between non-mycorrhizal and AM plants at the highest Cs condition (Kafle and Garcia 2022). Although it was concluded that Cs could also be used as a proxy for K transport, it appeared that the use of Rb was more conclusive than Cs. Although multiple families of K transport proteins have been identified and/or found differentially regulated upon colonization (Garcia and Zimmermann 2014; Garcia and Ané 2017; Keller-Pearson et al. 2023), only one symbiosis-related transporter from the HAK family has been characterized so far in tomato plants, SlHAK10 (Liu et al. 2019) (Fig. 1a). On the fungal side, it has been hypothesized that transporters and channels could be involved in K acquisition by extra-radical hyphae and unload from the arbuscules, respectively, but only a couple of putative candidate proteins (RiHAK and RiSKC) have been identified so far, and none have been functionally characterized yet (Garcia and Zimmermann 2014) (Fig. 1a). The identification and characterization of several other candidate genes combined with the use of proxies to track K movement from fungal hyphae to the host plant has the potential to improve our understanding of K acquisition mechanisms and enhance strategies for optimizing nutrient use efficiency in agricultural systems, ultimately contributing to sustainable crop production and soil health management.

## 2.2 *Ectomycorrhizal Fungi*

Ectomycorrhizal (ECM) fungi form symbiotic associations with approximately 95% of the fine roots of evergreen and deciduous trees in temperate and boreal forests, are known to enhance tree nutrition and water uptake (Lee et al. 2010; Becquer et al. 2019), and improve host plant defenses against biotic and abiotic stresses (Sousa et al. 2011; Onwuchekwa et al. 2014). ECM fungi include a diverse range of fungal species within the phyla Ascomycota and Basidiomycota, belonging to over 70 distinct lineages from diverse ancestral clades (Ryberg and Matheny 2012; Tedersoo and Smith 2013; Martin et al. 2016). The earliest evolutions of ECM symbioses are estimated to have occurred around 150 MYA, coinciding with the appearance of the *Pinaceae*, which is a family of host plant species associating with members of many ancient ECM lineages (Hibbett and Matheny 2009). Different ECM fungal species can employ various strategies for nutrient scavenging (Rineau et al. 2012; Schmalenberger et al. 2015) and can have diverse influence on plant nutrient status and growth. Additionally, different ECM fungal species can respond differently to nutrient availability and other conditions in the soil environment (Rosling et al. 2004; Rosenstock et al. 2016). However, all ECM lineages have evolved convergently to share similar morphological characteristics and general symbiotic functions (Tedersoo and Smith 2013), improving host root access to soil nutrients and water in exchange for carbon supplied to the root-associated fungi by photosynthates translocated from aboveground plant tissue (Becquer et al. 2019). The exchange of resources in ECM symbioses occurs in the intercellular apoplastic space in the cortex of plant roots, often in the fine short-root tips, which are major uptake sites for root nutrient and water acquisition from the soil. Fungal hyphae penetrate the root epidermis and extend between epidermal cells to form a network known as the Hartig net that can occupy the apoplastic space in the cortex to varying depths depending on the host and fungal species (Sa et al. 2019). The exterior of the root is sheathed by a mantle of fungal hyphae, which can in some cases present a barrier preventing direct apoplastic root uptake from the surrounding environment (Bücking et al. 2002), bringing soil nutrient and water transport under the control of the fungal partner. The extraradical hyphae of ECM fungi can explore a much greater volume of soil than the plants alone (Genney et al. 2006), scavenging resources from beyond the rhizosphere and transporting them toward the Hartig net in exchange for carbon released by the host root. This symbiotic root interaction can have many benefits for the health and growth of tree, especially an enhanced nutritional status due to the improvement of mineral nutrient transport to the roots via fungal hyphae (Bidartondo et al. 2001; Casieri et al. 2013; Courty et al. 2016).

Few studies have investigated the role of ECM fungi in tree K nutrition, despite the vital importance of K for plant health and the crucial role of ECM fungi in tree mineral nutrient uptake. Yuan et al. (2004) demonstrated the ability of ECM fungi to absorb K from vermiculite and phlogopite. Eucalyptus trees in this study that were colonized by the same fungi displayed higher levels of K than non-colonized plants with access to the same K sources. However, the authors note that they cannot conclude that the increase in K uptake by the plant is a direct consequence of the fungal absorption of the mineral. A potentially confounding factor is that an increase in "root" surface area from fungal hyphae could increase K uptake, independent of a fungus' K uptake. Similarly, Wallander and Wickman (1999) used biotite and microcline as sole K sources and found that all *Pinus sylvestris* plants, colonized and non-colonized, were able to acquire K from the biotite. The primary sources sustaining K nutrition in forest soils are mineral particles, wherein K is either trapped in interlayer spaces in secondary minerals such as vermiculite or embedded in the crystal matrix of primary minerals, such as K feldspars, biotite, or muscovite (Markewitz and Richter 2000). As stated above, for K to be available for plant or fungal uptake, it must be released from these sources into the soil solution. Aside from abiotic weathering processes that release soluble K ions into the soil over time, some ECM fungi have been shown to release K from biotite and muscovite (Van Schöll et al. 2006; Schmalenberger et al. 2015), which may then become available for fungal and plant use. The small diameter of extraradical hyphae in relation to plant roots allows fungal exploration and K foraging in micropores and soil aggregates that are inaccessible to plant roots, thus potentially unlocking additional K resources for plant uptake via symbiotic transport. More recently, it has also been demonstrated that the hyphae of the ECM fungus *Paxillus ammoniavirescens* accumulate various K chemistries, including complexation with organic ions potentially involved in solubilization of mineral nutrients, suggesting the importance of fungal K nutrition for the weathering of other minerals in the environment (Richardson et al. 2024). However, the mechanisms involved in this production of organic K in ECM fungi have not been identified so far.

The direct transport of K via ECM hyphae from a distant source to colonized loblolly pine was demonstrated by Frank and Garcia (2021), who applied Rb as a proxy for K to a compartment adjacent to inoculated roots, which could only be accessed by fungal hyphae, as described above. Among the ECM fungi tested in this study, *P. ammoniavirescens* and *Hebeloma cylindrosporum* increased the Rb and K concentrations in the shoots of inoculated loblolly pine (*Pinus taeda*) seedlings when supplied with these elements in the adjacent fungal compartment, demonstrating the ability of these fungi to allocate K to the plant root from distant soil resources (Frank and Garcia 2021). Concerning the molecular mechanisms governing K transport in ECM symbiosis, almost nothing is known on the plant side (Guerrero-Galán et al. 2018c). Only one putative Shaker-like channel (PsSKOR) was reported upregulated in *Pinus sylvestris* roots colonized by *Cenococcum geophilum*, suggesting a possible involvement of this transport protein in ECM symbiosis (Peter et al. 2016). On the fungal side, several fungal K transporters (HcTrk1/2, HcHAK) and channels (HcTOK1/2.1/2.2 and HcSKC) were identified and/or characterized in the model fungus *H. cylindrosporum* and shown to be involved in K acquisition from the soil or transfer to the host plant (Corratgé et al. 2007; Garcia et al. 2014, 2020; Guerrero-Galán et al. 2018a, b; Houdinet et al. 2023) (Fig. 1b). In addition to the enhanced K nutrition provided to loblolly pine by these ECM interactions, the shoot Na concentrations were reduced by inoculation with *P. ammoniavirescens*, indicating that this fungus can also play a role in regulating the K:Na homeostasis in loblolly pine (Frank and Garcia 2021; Rose et al. 2024). ECM symbioses are vital for forest health, productivity, and resilience, particularly under conditions of environmental stress such as drought or soil degradation. Understanding ECM-mediated K dynamics can inform sustainable forestry practices, maintain native fungal communities, and improve plantation yields, contributing to ecological balance and long-term soil fertility (Hackman et al. 2022).

## 2.3 Endophytic and Rhizospheric Fungi

Many other fungi live close to plant roots, in the rhizosphere, and several of them also internally colonize roots. These endophytic fungi inhabit plant tissue without causing any symptoms of disease to the host cells. Contrary to AM and ECM fungi, these fungi do not belong to specific clades but can be found within various groups and do not form specific structures in and on roots as described above. They do not have any obligate biotrophic life stages and can live at least part of their life cycle away from the plant (Rodriguez et al. 2009). Endophytic fungi have been found to associate with almost all plant species, and some are capable of internal migration within plant tissue to inhabit foliage, stems, and bark, as well as roots (Behie et al. 2013). The ability of endophytes to transfer mineral nutrients to plants is a relatively recent discovery, and the mechanisms of this transfer are still unknown in most cases. *Heteroconium chaetospira* is an example of a fungal endophyte that can transfer N from decomposing organic matter to its host plant (Usuki and Narisawa 2007). *Colletotrichum tofieldiae* (Hiruma et al. 2016) has demonstrated the ability to transfer P to the non-mycorrhizal host, *Arabidopsis thaliana*, promoting its growth under low-P conditions. Regarding K nutrition, *Serendipita indica* has been described to improve the K content in shoots of maize and tomato plants growing under salt stress (Liu et al. 2018; Ghorbani et al. 2019). However, its role in plant K nutrition remained unclear until a recent study from Conchillo et al. (2021), which revealed that while *S. indica* does not improve *A. thaliana* growth or K accumulation during K shortage, the fungus benefits from increased colonization under these conditions. Indeed, functional characterization of the fungal K transporters SiHAK1 suggests that the fungus prioritizes its own K acquisition during symbiosis and therefore limits its transfer to the colonized plant (Fig. 1c).

Additionally, some root rhizospheric fungi that do not invade plant roots have been reported as K solubilizers, transforming insoluble K minerals into forms accessible to plants, such as *Fusarium haematococcum* (Koshila Ravi et al. 2022) or members of the *Aspergillus* genus. For example, the fungus *Aspergillus niger* was able to solubilize K from alkaline ultramafic rock powder in small-scale fermentations, with efficiency decreasing at larger volumes due to oxygen transfer limitations (Lopes-Assad et al. 2010). Besides *A. niger*, two other *Aspergillus* species (*Aspergillus terreus* and *Aspergillus violaceofuscus*) also produce organic acids to solubilize significant K from mica and K aluminum silicate (Muthuraja and Muthukumar 2021). Inoculation with these fungi improved maize growth and K uptake, particularly with *A. violaceofuscus* showing the most promise as a bio-inoculant for enhancing crop growth across soil types. More recently, three K-solubilizing *Mortierella* strains (2K3, 2K4, and 2K12) were isolated from poplar rhizosphere soil, with strain 2K4 exhibiting the highest K-dissolving ability (Zhu et al. 2024). Further analysis revealed that 2K4 secreted large amounts of organic acids, mainly succinic acid, and increased polysaccharide content, which correlated with K release. Pot experiments demonstrated that 2K4 promoted maize seedling growth, suggesting its potential for use in microbial fertilizers and ecological restoration. Indeed, although a limited number of studies have investigated the impact of K-solubilizing fungi on plant K nutrition, these fungi may have the potential to be used in sustainable agriculture, reducing our dependency on chemical fertilizers, and improving soil fertility. The insights gained from their mechanisms might guide the development of bioinoculants to enhance soil K cycling, K-use efficiency, and plant productivity.

## 3 Plant Potassium Uptake Facilitated by Bacteria

### *3.1 Associative and Endophytic Bacteria*

Besides fungi, many bacterial species act on soil minerals to release bound K, making it available to plant roots. Phylogenetically diverse bacterial communities, either associative or endophytic, release K from soil rocks. Generally, species or isolates of *Acidothiobacillus*, *Agrobacterium*, *Bacillus*, *Burkholderia*, *Enterobacter*, *Flavobacterium*, *Klebsiella*, *Microbacterium*, *Myroides*, *Paenibacillus*, *Pantoea*, *Pseudomonas*, and *Rhizobium* have been described as K-solubilizing microbes (Setiawati and Mutmainnah 2016; Sarikhani et al. 2018). To dissolve the minerals and release K, these bacteria produce and excrete organic acids, such as citric acid, oxalic acid, tartaric acid, or protons, into the environment. Acidolysis and chelation are common mechanistic processes by which these acids dissolve or leach K from rock minerals into the soil (Meena et al. 2014; Sarikhani et al. 2018; Haro and Benito

2019). The recent genome sequencing of the K-solubilizing bacterium *Bacillus aryabhattai* SK1–7 revealed that K solubilization is enhance through organic acid secretion and extracellular polymeric substance (EPS) production (Chen et al. 2022). Genes such as *ackA* (acetic acid synthesis), *gltA* (citric acid synthase), *mdh* (malate dehydrogenase), and *ppc* (PEPC) were found upregulated under insoluble K conditions, boosting the TCA cycle to generate acids like malic, citric, and gluconic acid, which lower soil pH and dissolve K minerals. Additionally, *epsB*, a key regulator of EPS synthesis, is significantly upregulated, facilitating bacterial adhesion to mineral particles and enhancing K release through acid hydrolysis, chelation, and mineral dissolution (Chen et al. 2022).

Several studies have demonstrated that the application of K-solubilizing bacteria alone, or in combination with other bacteria, to media supplemented with mineral-bound K, significantly increased plant K acquisition and overall plant growth. Plants inoculated with several species in association with feldspar had 50% higher K content than non-inoculated plants, which clearly demonstrates the effectiveness or efficiency of this bacteria for enhancing plant K uptake (Han et al. 2006; Basak and Biswas 2009). Different bacteria on different plant species have pronounced effects on plant K content and plant growth (Sarikhani et al. 2018), for example, *Bacillus edaphicus* on cotton and rapeseed (Sheng 2005) and *Bacillus mucilaginous* on pepper, cucumber, Sundar grass, and eggplant (Sarikhani et al. 2018). Similarly, tomato plants inoculated with bacterial isolates of *Pseudomonas* genus showed an increase of shoot K concentrations and plant biomass when compared to control plants. This bacterial isolate solubilizes K from biotite and muscovite and subsequently makes it available to plants. The elevated level of tissue K was directly related to K-solubilizing bacteria because the growth substrate was lacking in plant available K (Sarikhani et al. 2018).

More recently, a consortium of K-solubilizing bacteria (*Enterobacter hormaechei*, *Citrobacter braakii*, *Pseudomonas putida*, *Erwinia iniecta*, and *Pantoea agglomerans*) has been tested for their ability to enhance K uptake and improve K/Na ratios in rice grown under saline-sodic conditions (Nawaz et al. 2023). This bacterial consortium significantly improved plant growth, K assimilation, root and shoot K/Na ratios, and grain yield, demonstrating the importance of studying more complex bacterial communities than individual species. Indeed, to establish a stable soil ecosystem, diverse soil-dwelling organisms interact with each other and with their environments. Indeed, the observed effects likely result from synergistic interactions among bacterial species, where some may solubilize K through organic acid or enzyme production, while others enhance nutrient transport or mitigate stress. The consortium's functional diversity enables efficient resource partitioning, reducing competition and maximizing K uptake. Additionally, certain bacteria might contribute to soil structure improvement by producing exopolysaccharides or biofilms (Costa et al. 2018) and enhancing soil aggregation and nutrient availability. Furthermore, some bacteria may produce phytohormones like auxins or gibberellins (Orozco-Mosqueda et al. 2023), enhancing root growth and nutrient absorption, further improving plant health under saline-sodic conditions. Consequently, inoculation of even more complex consortia composed of soil microbes from different functional niches is not only highly beneficial to the plants but also to these microbes for their growth and abundance (Shayanthan et al. 2022; Martins et al. 2023; Jing et al. 2024). It has been shown that K-solubilizing, P-solubilizing, and N-fixing bacteria work together in combination with mycorrhizal fungi to deliver nutrients to the most valuable host for them (Wu et al. 2005; Basak and Biswas 2009). Despite the coordination among beneficial microbes, we have underestimated the potential of mineral solubilizing soil microbes. The global agricultural community has widely adopted the use of N-fixing bacteria to reduce N fertilizer inputs but has only begun to explore the broader potential of these microbes for optimizing nutrient management.

## 3.2 *Impact of Improved Potassium Nutrition on Nitrogen-Fixing Symbioses*

N is one of the most limiting nutrients for plant growth and development (Glass et al. 2002). Many plants, including numerous agricultural crop species, have a high demand for N. To fulfill this requirement, legumes establish symbiotic associations with N-fixing bacteria, commonly referred to as rhizobia bacteria (Mendoza-Suárez et al. 2020). The bacteria reside in host root nodules, which are specialized structures providing a protected environment for bacteria to convert atmospheric N into ammonium using a key enzyme, the nitrogenase (Udvardi and Poole 2013). It has been demonstrated that N-fixing bacteria need considerable amounts of K, P, and carbon for their functionality (Mengel et al. 1974; Lynd et al. 1981; Divito and Sadras 2014). For instance, Premaratne and Oertli (1994) and Sangakkara et al. (1996) showed that increasing external K concentrations improved the nodule size and numbers of soybean plants, although they did not observe any effect on the nitrogenase activity. However, more recently, Høgh-Jensen (2003) provided evidence that deprivation of K in the plant tissue was responsible for the decrease in nitrogenase activity of white clover plants.

As mentioned above, inoculating plants with multiple microbes has the potential to reveal synergistic effects on plant nutrition. Basak and Biswas (2009) investigated this phenomenon using muscovite mica as a K source to grow Sudan grass, a legume plant. In their study, plants were inoculated with either K-solubilizing bacteria (*Bacillus mucilaginosus*), N-fixing bacteria (*Azotobacter chroococcum*), both species, or left uninoculated. The results showed that dual inoculation significantly increased K and N concentrations in both plants and the growth substrate compared to single inoculations or uninoculated controls. These findings highlight the role of K-solubilizing bacteria in enhancing K availability, which subsequently improved plant N uptake and overall growth performance. This beneficial effect of multiple microbial inoculations on plant fitness is attributed to the compatibility among microbes, where some facilitate the release of K and P from minerals, while others enhance N fixation (Wu et al. 2005). Other studies indicated that improved K levels enhance photosynthesis, enabling plants to allocate more carbon to their roots and the symbiotic or associative microbes (Geiger and Conti 1983; Kanai et al. 2011). N-fixing rhizobia and other root-associated microbes benefit from increased photosynthate availability, which boosts their abundance and functional activity (Kafle et al. 2019a). Beyond these indirect effects, K plays a direct role in N fixation processes, including its involvement in nitrogenase enzyme functionality and the transport of fixed N to the plant's roots (Duke et al. 1980; Lynd et al. 1981; Divito and Sadras 2014). Collectively, improved K availability in soil and plant tissues positively influences N acquisition in legume plants (Duke and Collins 1985; Divito and Sadras 2014). Therefore, to maximize the potential of N fixation, it is essential to promote the presence of P- and K-solubilizing bacteria in conjunction with enhancing fungal diversity. A holistic understanding of the interconnected nature of biological components is crucial, as each unit plays a vital role in maintaining the balance and functionality of the system.

## 4 Conclusion

Despite its importance, K availability in soils is often limited due to its tendency to bind with mineral complexes, rendering it inaccessible to plants. While traditional solutions like chemical fertilizers have been widely used to address K deficiencies, these methods are not sustainable and can lead to environmental degradation. Here, we highlighted the critical role of soil microbial communities, including bacteria, endophytic fungi, and mycorrhizal fungi, in enhancing K bioavailability. These microbes contribute to the solubilization of mineral-bound K, enabling plants to access this crucial nutrient while also fostering symbiotic relationships that improve overall soil fertility and plant health. Although the potential of these microbial interactions has been recognized for over a century, their application in natural and agricultural ecosystems remains underexplored. Future research must focus on understanding the synergistic relationships among soil microbes, plants, and K availability to harness these natural processes effectively. Integrating microbial solutions with sustainable agricultural practices offers a promising pathway to improve nutrient management, reduce dependency on chemical fertilizers, and enhance crop productivity in an environmentally responsible manner.

## References

Adeyemi NO, Atayese MO, Sakariyawo OS et al (2021) Arbuscular mycorrhizal fungi species differentially regulate plant growth, phosphorus uptake and stress tolerance of soybean in lead contaminated soil. J Plant Nutr 44:1633–1648. https://doi.org/10.1080/01904167.2021.1871748

Alfaro MA, Jarvis SC, Gregory PJ (2004) Factors affecting potassium leaching in different soils. Soil Use Manag 20:182–189. https://doi.org/10.1111/j.1475-2743.2004.tb00355.x

Armengaud P, Breitling R, Amtmann A (2010) Coronatine-insensitive 1 (COI1) mediates transcriptional responses of *Arabidopsis thaliana* to external potassium supply. Mol Plant 3:390–405. https://doi.org/10.1093/mp/ssq012

Basak BB, Biswas DR (2009) Influence of potassium solubilizing microorganism (*Bacillus mucilaginosus*) and waste mica on potassium uptake dynamics by sudan grass (*Sorghum vulgare* Pers.) grown under two Alfisols. Plant Soil 317:235–255. https://doi.org/10.1007/s11104-008-9805-z

Becquer A, Guerrero-Galán C, Eibensteiner JL et al (2019) The ectomycorrhizal contribution to tree nutrition. Adv Bot Res 89:77–126. https://doi.org/10.1016/BS.ABR.2018.11.003

Begum N, Qin C, Ahanger MA et al (2019) Role of arbuscular mycorrhizal fungi in plant growth regulation: implications in abiotic stress tolerance. Front Plant Sci 10:1068. https://doi.org/10.3389/fpls.2019.01068

Behie SW, Padilla-Guerrero IE, Bidochka MJ (2013) Nutrient transfer to plants by phylogenetically diverse fungi suggests convergent evolutionary strategies in rhizospheric symbionts. Commun Integr Biol 6:e22321. https://doi.org/10.4161/cib.22321

Benito B, Haro R, Amtmann A et al (2014) The twins K+ and Na+ in plants. J Plant Physiol 171:723–731. https://doi.org/10.1016/j.jplph.2013.10.014

Bidartondo MI, Ek H, Wallander H, Söderström B (2001) Do nutrient additions alter carbon sink strength of ectomycorrhizal fungi? New Phytol 151:543–550. https://doi.org/10.1046/j.1469-8137.2001.00180.x

Bücking H, Kuhn AJ, Schröder WH, Heyser W (2002) The fungal sheath of ectomycorrhizal pine roots: an apoplastic barrier for the entry of calcium, magnesium, and potassium into the root cortex? J Exp Bot 53:1659–1669

Campanelli A, Ruta C, De Mastro G, Morone-Fortunato I (2013) The role of arbuscular mycorrhizal fungi in alleviating salt stress in *Medicago sativa* L. var. icon. Symbiosis 59:65–76. https://doi.org/10.1007/s13199-012-0191-1

Casieri L, Lahmidi NA, Doidy J et al (2013) Biotrophic transportome in mutualistic plant – fungal interactions. Mycorrhiza 23:597–625. https://doi.org/10.1007/s00572-013-0496-9

Chen M, Arato M, Borghi L et al (2018) Beneficial services of arbuscular mycorrhizal fungi – from ecology to application. Front Plant Sci 9:1270. https://doi.org/10.3389/fpls.2018.01270

Chen Y, Yang H, Shen Z, Ye J (2022) Whole-genome sequencing and potassium-solubilizing mechanism of *Bacillus aryabhattai* SK1-7. Front Microbiol 12:722379. https://doi.org/10.3389/fmicb.2021.722379

Conchillo LB, Haro R, Benito B (2021) K+ nutrition exchange in the *Serendipita-Arabidopsis* symbiosis: study of the fungal K+ transporters involved. Front Ecol Evol 9:789371. https://doi.org/10.3389/fevo.2021.789371

Corratgé C, Zimmermann S, Lambilliotte R et al (2007) Molecular and functional characterization of a Na(+)-K(+) transporter from the Trk family in the ectomycorrhizal fungus *Hebeloma cylindrosporum*. J Biol Chem 282:26057–26066. https://doi.org/10.1074/jbc.M611613200

Costa OYA, Raaijmakers JM, Kuramae EE (2018) Microbial extracellular polymeric substances: ecological function and impact on soil aggregation. Front Microbiol 9:1636. https://doi.org/10.3389/fmicb.2018.01636

Courty P-E, Doidy J, Garcia K et al (2016) The transportome of mycorrhizal systems. In: Molecular mycorrhizal symbiosis. Wiley, Hoboken, pp 239–256

Divito GA, Sadras VO (2014) How do phosphorus, potassium and sulphur affect plant growth and biological nitrogen fixation in crop and pasture legumes? A meta-analysis. Field Crop Res 156:161–171. https://doi.org/10.1016/j.fcr.2013.11.004

Duke SH, Collins M (1985) Role of potassium in legume dinitrogen fixation. In: Potassium in agriculture. ASA-CSSA-SSSA, Madison, pp 443–465

Duke SH, Collins M, Soberalske RM (1980) Effects of potassium fertilization on nitrogen fixation and nodule enzymes of nitrogen metabolism in alfalfa. Crop Sci 20:213–219. https://doi.org/10.2135/cropsci1980.0011183X002000020016x

Estrada B, Aroca R, Maathuis FJM et al (2013) Arbuscular mycorrhizal fungi native from a Mediterranean saline area enhance maize tolerance to salinity through improved ion homeostasis. Plant Cell Environ 36:1771–1782. https://doi.org/10.1111/pce.12082
Fellbaum CR, Gachomo EW, Beesetty Y et al (2012) Carbon availability triggers fungal nitrogen uptake and transport in arbuscular mycorrhizal symbiosis. Proc Natl Acad Sci USA 109:2666–2671. https://doi.org/10.1073/pnas.1118650109
Frank HER, Garcia K (2021) Benefits provided by four ectomycorrhizal fungi to *Pinus taeda* under different external potassium availabilities. Mycorrhiza 31:755–766. https://doi.org/10.1007/s00572-021-01048-z
Frey SD (2019) Mycorrhizal fungi as mediators of soil organic matter dynamics. Annu Rev Ecol Evol Syst 50:237–259. https://doi.org/10.1146/annurev-ecolsys-110617-062331
Garcia K, Ané J-M (2017) Polymorphic responses of *Medicago truncatula* accessions to potassium deprivation. Plant Signal Behav 12:e1307494. https://doi.org/10.1080/15592324.2017.1307494
Garcia K, Zimmermann SD (2014) The role of mycorrhizal associations in plant potassium nutrition. Front Plant Sci 5:1–9. https://doi.org/10.3389/fpls.2014.00337
Garcia K, Delteil A, Conéjéro G et al (2014) Potassium nutrition of ectomycorrhizal *Pinus pinaster*: overexpression of the *Hebeloma cylindrosporum* HcTrk1 transporter affects the translocation of both K(+) and phosphorus in the host plant. New Phytol 201:951–960. https://doi.org/10.1111/nph.12603
Garcia K, Doidy J, Zimmermann SD et al (2016) Take a trip through the plant and fungal transportome of mycorrhiza. Trends Plant Sci 21:937–950. https://doi.org/10.1016/j.tplants.2016.07.010
Garcia K, Guerrero-Galán C, Frank HER et al (2020) Fungal Shaker-like channels beyond cellular K+ homeostasis: a role in ectomycorrhizal symbiosis between *Hebeloma cylindrosporum* and *Pinus pinaster*. PLoS One 15:e0242739
Geiger DR, Conti TR (1983) Relation of increased potassium nutrition to photosynthesis and translocation of carbon. Plant Physiol 71:141–144. https://doi.org/10.1104/pp.71.1.141
Genney DR, Anderson IC, Alexander IJ (2006) Fine-scale distribution of pine ectomycorrhizas and their extramatrical mycelium. New Phytol 170:381–390. https://doi.org/10.1111/j.1469-8137.2006.01669.x
George E, Marschner H, Jakobsen I (1995) Role of arbuscular mycorrhizal fungi in uptake of phosphorus and nitrogen from soil. Crit Rev Biotechnol 15:257–270. https://doi.org/10.3109/07388559509147412
Ghorbani A, Omran VOG, Razavi SM et al (2019) *Piriformospora indica* confers salinity tolerance on tomato (*Lycopersicon esculentum* Mill.) through amelioration of nutrient accumulation, K+/Na+ homeostasis and water status. Plant Cell Rep 38:1151–1163. https://doi.org/10.1007/s00299-019-02434-w
Glass ADM, Britto DT, Kaiser BN et al (2002) The regulation of nitrate and ammonium transport systems in plants. J Exp Bot 53:855–864. https://doi.org/10.1093/jexbot/53.370.855
Goulding K, Murrell TS, Mikkelsen RL et al (2021) Outputs: potassium losses from agricultural systems. In: Murrell TS, Mikkelsen RL, Sulewski G et al (eds) Improving potassium recommendations for agricultural crops. Springer International Publishing, Cham, pp 75–97
Guerrero-Galán C, Delteil A, Garcia K et al (2018a) Plant potassium nutrition in ectomycorrhizal symbiosis: properties and roles of the three fungal TOK potassium channels in *Hebeloma cylindrosporum*. Environ Microbiol 20:1873–1887. https://doi.org/10.1111/1462-2920.14122
Guerrero-Galán C, Garcia K, Houdinet G, Zimmermann SD (2018b) HcTOK1 participates in the maintenance of K+ homeostasis in the ectomycorrhizal fungus *Hebeloma cylindrosporum*, which is essential for the symbiotic K+ nutrition of *Pinus pinaster*. Plant Signal Behav 13:e1480845. https://doi.org/10.1080/15592324.2018.1480845
Guerrero-Galán C, Houdinet G, Calvo-Polanco M et al (2018c) The role of plant transporters in mycorrhizal symbioses. Adv Bot Res 87:303–342. https://doi.org/10.1016/BS.ABR.2018.09.012

Hackman JJ, Rose BD, Frank HER et al (2022) NPK fertilizer use in loblolly pine plantations: who are we really feeding? For Ecol Manag 520:120393. https://doi.org/10.1016/j.foreco.2022.120393

Han HS, Supanjani, Lee KD (2006) Effect of co-inoculation with phosphate and potassium solubilizing bacteria on mineral uptame and growth of pepper and cucumber. Plant Soil Environ 52:130–137

Haro R, Benito B (2019) The role of soil fungi in K(+) plant nutrition. Int J Mol Sci 20:3169. https://doi.org/10.3390/ijms20133169

Hibbett DS, Matheny PB (2009) The relative ages of ectomycorrhizal mushrooms and their plant hosts estimated using Bayesian relaxed molecular clock analyses. BMC Biol 7:13. https://doi.org/10.1186/1741-7007-7-13

Hiruma K, Gerlach N, Sacristán S et al (2016) Root endophyte *Colletotrichum tofieldiae* confers plant fitness benefits that are phosphate status dependent. Cell 165:464–474. https://doi.org/10.1016/j.cell.2016.02.028

Høgh-Jensen H (2003) The effect of potassium deficiency on growth and N2-fixation in Trifolium repens. Physiol Plant 119:440–449. https://doi.org/10.1034/j.1399-3054.2003.00189.x

Houdinet G, Guerrero-Galán C, Rose BD et al (2023) Secrets of the fungus-specific potassium channel TOK family. Trends Microbiol 31:511–520. https://doi.org/10.1016/j.tim.2022.11.007

Jing J, Garbeva P, Raaijmakers JM, Medema MH (2024) Strategies for tailoring functional microbial synthetic communities. ISME J 18:wrae049. https://doi.org/10.1093/ismejo/wrae049

Kafle A, Cooney DR, Shah G, Garcia K (2022) Mycorrhiza-mediated potassium transport in Medicago truncatula can be evaluated by using rubidium as a proxy. Plant Sci 322:111364 https://doi.org/10.1016/j.plantsci.2022.111364

Kafle A, Garcia K (2022) Cesium could be used as a proxy for potassium in mycorrhizal *Medicago truncatula*. Plant Signal Behav 17:2134676. https://doi.org/10.1080/15592324.2022.2134676

Kafle A, Garcia K, Peta V et al (2019a) Beneficial plant microbe interactions and their effect on nutrient uptake, yield, and stress resistance of soybeans. In: Soybean – biomass, yield and productivity. IntechOpen, Rijeka

Kafle A, Garcia K, Wang X et al (2019b) Nutrient demand and fungal access to resources control the carbon allocation to the symbiotic partners in tripartite interactions of *Medicago truncatula*. Plant Cell Environ 42:270–284. https://doi.org/10.1111/pce.13359

Kaiser C, Kilburn MR, Clode PL et al (2015) Exploring the transfer of recent plant photosynthates to soil microbes: mycorrhizal pathway vs direct root exudation. New Phytol 205:1537–1551. https://doi.org/10.1111/nph.13138

Kanai S, Moghaieb RE, El-Shemy HA et al (2011) Potassium deficiency affects water status and photosynthetic rate of the vegetative sink in green house tomato prior to its effects on source activity. Plant Sci 180:368–374. https://doi.org/10.1016/j.plantsci.2010.10.011

Keller-Pearson M, Bortolazzo A, Willems L et al (2023) A dual transcriptomic approach reveals contrasting patterns of differential gene expression during drought in arbuscular mycorrhizal fungus and carrot. Mol Plant-Microbe Interact 36:821–832. https://doi.org/10.1094/MPMI-04-23-0038-R

Koshila Ravi R, Prema Sundara Valli P, Muthukumar T (2022) Physiological characterization of root endophytic *Fusarium haematococcum* for hydrolytic enzyme production, nutrient solubilization and salinity tolerance. Biocatal Agric Biotechnol 43:102392. https://doi.org/10.1016/j.bcab.2022.102392

Lanfranco L, Bonfante P, Genre A (2016) The mutualistic interaction between plants and arbuscular mycorrhizal fungi. Microbiol Spectr 4:10. https://doi.org/10.1128/microbiolspec.funk-0012-2016

Lee SH, Calvo-Polanco M, Chung GC, Zwiazek JJ (2010) Role of aquaporins in root water transport of ectomycorrhizal jack pine (*Pinus banksiana*) seedlings exposed to NaCl and fluoride. Plant Cell Environ 33:769–780. https://doi.org/10.1111/j.1365-3040.2009.02103.x

Liu A, Hamel C, Hamilton RI et al (2000) Acquisition of Cu, Zn, Mn and Fe by mycorrhizal maize (*Zea mays* L.) grown in soil at different P and micronutrient levels. Mycorrhiza 9:331–336. https://doi.org/10.1007/s005720050277

Liu F, Xu Y, Han G et al (2018) Identification and functional characterization of a maize phosphate transporter induced by mycorrhiza formation. Plant Cell Physiol 59:1683–1694

Liu J, Liu J, Liu J et al (2019) The potassium transporter SlHAK10 is involved in mycorrhizal potassium uptake. Plant Physiol 180:465–479. https://doi.org/10.1104/pp.18.01533

Lopes-Assad ML, Avansini SH, Rosa MM et al (2010) The solubilization of potassium-bearing rock powder by *Aspergillus niger* in small-scale batch fermentations. Can J Microbiol 56:598–605. https://doi.org/10.1139/W10-044

Lynd JQ, Odell GV Jr, McNew RW (1981) Soil potassium effects on nitrogenase activity with associated nodule components of hairy vetch at anthesis. J Plant Nutr 4:303–318. https://doi.org/10.1080/01904168109362920

Markewitz D, Richter DD (2000) Long-term soil potassium availability from a Kanhapludult to an aggrading loblolly pine ecosystem. For Ecol Manag 130:109–129. https://doi.org/10.1016/S0378-1127(99)00175-9

Martin F, Kohler A, Murat C et al (2016) Unearthing the roots of ectomycorrhizal symbioses. Nat Rev Microbiol 14:760–773

Martins SJ, Pasche J, Silva HAO et al (2023) The use of synthetic microbial communities to improve plant health. Phytopathology 113:1369–1379. https://doi.org/10.1094/PHYTO-01-23-0016-IA

Meena VS, Maurya BR, Verma JP (2014) Does a rhizospheric microorganism enhance $K^+$ availability in agricultural soils? Microbiol Res 169:337–347. https://doi.org/10.1016/j.micres.2013.09.003

Mendoza-Suárez MA, Geddes BA, Sánchez-Cañizares C et al (2020) Optimizing *Rhizobium*-legume symbioses by simultaneous measurement of rhizobial competitiveness and N2 fixation in nodules. Proc Natl Acad Sci USA 117:9822–9831. https://doi.org/10.1073/pnas.1921225117

Mengel K, Haghparast MR, Koch K (1974) The effect of potassium on the fixation of molecular nitrogen by root nodules of *Vicia faba*. Plant Physiol 54:535–538. https://doi.org/10.1104/pp.54.4.535

Mikkelsen RL, Roberts TL (2021) Inputs: potassium sources for agricultural systems. In: Murrell TS, Mikkelsen RL, Sulewski G et al (eds) Improving potassium recommendations for agricultural crops. Springer International Publishing, Cham, pp 47–73

Muentz A (1890) Sur la decomposition des roches et la formation de la terre arable. C R Acad Sci 110:1370–1372

Muthuraja R, Muthukumar T (2021) Isolation and characterization of potassium solubilizing *Aspergillus* species isolated from saxum habitats and their effect on maize growth in different soil types. Geomicrobiol J 38:672–685. https://doi.org/10.1080/01490451.2021.1928800

Nawaz A, Qamar ZU, Marghoob MU et al (2023) Contribution of potassium solubilizing bacteria in improved potassium assimilation and cytosolic K+/Na+ ratio in rice (*Oryza sativa* L.) under saline-sodic conditions. Front Microbiol 14:1196024. https://doi.org/10.3389/fmicb.2023.1196024

Ngwene B, Gabriel E, George E (2013) Influence of different mineral nitrogen sources (NO(3)(−)-N vs. NH(4)(+)-N) on arbuscular mycorrhiza development and N transfer in a *Glomus intraradices*–cowpea symbiosis. Mycorrhiza 23:107–117. https://doi.org/10.1007/s00572-012-0453-z

Onwuchekwa NE, Zwiazek JJ, Quoreshi A, Khasa DP (2014) Growth of mycorrhizal jack pine (*Pinus banksiana*) and white spruce (*Picea glauca*) seedlings planted in oil sands reclaimed areas. Mycorrhiza 24:431–441. https://doi.org/10.1007/s00572-014-0555-x

Orozco-Mosqueda MDC, Santoyo G, Glick BR (2023) Recent advances in the bacterial phytohormone modulation of plant growth. Plants (Basel) 12:606. https://doi.org/10.3390/plants12030606

Pallon J, Wallander H, Hammer E et al (2007) Symbiotic fungi that are essential for plant nutrient uptake investigated with NMP. Nucl Instrum Methods Phys Res B 260:149–152. https://doi.org/10.1016/j.nimb.2007.02.018

Pérez-Tienda J, Corrêa A, Azcón-Aguilar C, Ferrol N (2014) Transcriptional regulation of host $NH_4^+$ transporters and GS/GOGAT pathway in arbuscular mycorrhizal rice roots. Plant Physiol Biochem 75:1–8. https://doi.org/10.1016/j.plaphy.2013.11.029

Peter M, Kohler A, Ohm RA, Kuo A, Krützmann J, Morin E, Arend M, Barry KW, Binder M, Choi C, Clum A, Copeland A, Grisel N, Haridas S, Kipfer T, LaButti K, Lindquist E, Lipzen A, Maire R, Meier B, Mihaltcheva S, Molinier V, Murat C, Pöggeler S, Quandt CA, Sperisen C, Tritt A, Tisserant E, Crous PW, Henrissat B, Nehls U, Egli S, Spatafora JW, Grigoriev IV, Martin FM (2016) Ectomycorrhizal ecology is imprinted in the genome of the dominant symbiotic fungus Cenococcum geophilum. Nat Commun 7:12662. https://doi.org/10.1038/ncomms12662

Plassard C, Becquer A, Garcia K (2019) Phosphorus transport in mycorrhiza: how far are we? Trends Plant Sci 24:794–801. https://doi.org/10.1016/j.tplants.2019.06.004

Pratt PF (1965) Potassium. In: Methods of soil analysis. American Society of Agronomy, Madison, pp 1022–1030

Premaratne KP, Oertli JJ (1994) The influence of potassium supply on nodulation, nitrogenase activity and nitrogen accumulation of soybean (*Glycine max* L. Merrill) grown in nutrient solution. Fertil Res 38:95–99. https://doi.org/10.1007/BF00748769

Rich MK, Vigneron N, Libourel C et al (2021) Lipid exchanges drove the evolution of mutualism during plant terrestrialization. Science 372:864–868. https://doi.org/10.1126/science.abg0929

Richardson JA, Rose BD, Garcia K (2024) X-ray fluorescence and XANES spectroscopy revealed diverse potassium chemistries and colocalization with phosphorus in the ectomycorrhizal fungus *Paxillus ammoniavirescens*. Fungal Biol 128:2054–2061. https://doi.org/10.1016/j.funbio.2024.08.004

Rineau F, Roth D, Shah F et al (2012) The ectomycorrhizal fungus *Paxillus involutus* converts organic matter in plant litter using a trimmed brown-rot mechanism involving Fenton chemistry. Environ Microbiol 14:1477–1487. https://doi.org/10.1111/j.1462-2920.2012.02736.x

Rodriguez RJ, White JF Jr, Arnold AE, Redman RS (2009) Fungal endophytes: diversity and functional roles. New Phytol 182:314–330. https://doi.org/10.1111/j.1469-8137.2009.02773.x

Rose BD, Dellinger MA, Larmour CP et al (2024) The ectomycorrhizal fungus *Paxillus ammoniavirescens* influences the effects of salinity on loblolly pine in response to potassium availability. Environ Microbiol 26:e16597. https://doi.org/10.1111/1462-2920.16597

Rosenstock NP, Berner C, Smits MM et al (2016) The role of phosphorus, magnesium and potassium availability in soil fungal exploration of mineral nutrient sources in Norway spruce forests. New Phytol 211:542–553. https://doi.org/10.1111/nph.13928

Rosling A, Lindahl BD, Finlay RD (2004) Carbon allocation to ectomycorrhizal roots and mycelium colonising different mineral substrates. New Phytol 162:795–802. https://doi.org/10.1111/j.1469-8137.2004.01080.x

Ruytinx J, Kafle A, Usman M et al (2020) Micronutrient transport in mycorrhizal symbiosis; zinc steals the show. Fungal Biol Rev 34:1–9. https://doi.org/10.1016/j.fbr.2019.09.001

Ryberg M, Matheny PB (2012) Asynchronous origins of ectomycorrhizal clades of *Agaricales*. Proc R Soc B 279:2003–2011. https://doi.org/10.1098/rspb.2011.2428

Sa G, Yao J, Deng C et al (2019) Amelioration of nitrate uptake under salt stress by ectomycorrhiza with and without a Hartig net. New Phytol 222:1951–1964. https://doi.org/10.1111/nph.15740

Sangakkara R, Hartwig UA, Nösberger J (1996) Growth and symbiotic nitrogen fixation of Vicia faba and Phaseolus vulgaris as affected by fertiliser potassium and temperature. J Sci Food Agric 70:315–320. https://doi.org/10.1002/(SICI)1097-0010(199603)70:3<315::AID-JSFA504>3.0.CO;2-B

Santander C, Sanhueza M, Olave J et al (2019) Arbuscular mycorrhizal colonization promotes the tolerance to salt stress in lettuce plants through an efficient modification of ionic balance. J Soil Sci Plant Nutr 19:321–331. https://doi.org/10.1007/s42729-019-00032-z

Santander C, Aroca R, Cartes P, Vidal G, Cornejo P (2021) Aquaporins and cation transporters are differentially regulated by two arbuscular mycorrhizal fungi strains in lettuce cultivars growing under salinity conditions. Plant Physiol Biochem 158:396–409. https://doi.org/10.1016/j.plaphy.2020.11.025

Sarikhani MR, Oustan S, Ebrahimi M, Aliasgharzad N (2018) Isolation and identification of potassium-releasing bacteria in soil and assessment of their ability to release potassium for plants. Eur J Soil Sci 69:1078–1086. https://doi.org/10.1111/ejss.12708

Sarwat M, Hashem A, Ahanger MA et al (2016) Mitigation of NaCl stress by arbuscular mycorrhizal fungi through the modulation of osmolytes, antioxidants and secondary metabolites in mustard (*Brassica juncea* L.) plants. Front Plant Sci 7:2040

Scheloske S, Maetz M, Schneider T et al (2004) Element distribution in mycorrhizal and nonmycorrhizal roots of the halophyte Aster tripolium determined by proton induced X-ray emission. Protoplasma 223:183–189. https://doi.org/10.1007/s00709-003-0027-1

Schmalenberger A, Duran AL, Bray AW et al (2015) Oxalate secretion by ectomycorrhizal *Paxillus involutus* is mineral-specific and controls calcium weathering from minerals. Sci Rep 5:12187. https://doi.org/10.1038/srep12187

Setiawati TC, Mutmainnah L (2016) Solubilization of potassium containing mineral by microorganisms from sugarcane rhizosphere. Agric Agric Sci Procedia 9:108–117. https://doi.org/10.1016/j.aaspro.2016.02.134

Shayanthan A, Ordoñez PAC, Oresnik IJ (2022) The role of synthetic microbial communities (SynCom) in sustainable agriculture. Front Agron 4:896307. https://doi.org/10.3389/fagro.2022.896307

Sheng XF (2005) Growth promotion and increased potassium uptake of cotton and rape by a potassium releasing strain of *Bacillus edaphicus*. Soil Biol Biochem 37:1918–1922. https://doi.org/10.1016/j.soilbio.2005.02.026

Smith SE, Read D (2008) Mycorrhizal symbiosis, 3rd edn. Academic Press, London

Smith SE, Jakobsen I, Grønlund M, Smith FA (2011) Roles of arbuscular mycorrhizas in plant phosphorus nutrition: interactions between pathways of phosphorus uptake in arbuscular mycorrhizal roots have important implications for understanding and manipulating plant phosphorus acquisition. Plant Physiol 156:1050–1057. https://doi.org/10.1104/pp.111.174581

Sousa NR, Franco AR, Ramos MA et al (2011) Reforestation of burned stands: the effect of ectomycorrhizal fungi on *Pinus pinaster* establishment. Soil Biol Biochem 43:2115–2120. https://doi.org/10.1016/j.soilbio.2011.06.013

Spatafora JW, Chang Y, Benny GL et al (2016) A phylum-level phylogenetic classification of zygomycete fungi based on genome-scale data. Mycologia 108:1028–1046. https://doi.org/10.3852/16-042

Tang B, Man J, Lehmann A, Rillig MC (2023) Arbuscular mycorrhizal fungi benefit plants in response to major global change factors. Ecol Lett 26:2087–2097. https://doi.org/10.1111/ele.14320

Tedersoo L, Smith ME (2013) Lineages of ectomycorrhizal fungi revisited: foraging strategies and novel lineages revealed by sequences from belowground. Fungal Biol Rev 27:83–99. https://doi.org/10.1016/j.fbr.2013.09.001

Tester M, Davenport R (2003) Na+ tolerance and Na+ transport in higher plants. Ann Bot 91:503–527. https://doi.org/10.1093/aob/mcg058

Thioub M, Ewusi-Mensah N, Sarkodie-Addo J, Adjei-Gyapong T (2019) Arbuscular mycorrhizal fungi inoculation enhances phosphorus use efficiency and soybean productivity on a Haplic Acrisol. Soil Tillage Res 192:174–186. https://doi.org/10.1016/j.still.2019.05.001

Tisdale SL, Nelson WL, Beaton JD (1985) Soil fertility and fertilizers, 4th edn. Macmillan Publishing Company, New York, pp 188–239

Troufflard S, Mullen W, Larson TR et al (2010) Potassium deficiency induces the biosynthesis of oxylipins and glucosinolates in *Arabidopsis thaliana*. BMC Plant Biol 10:172. https://doi.org/10.1186/1471-2229-10-172

Udvardi M, Poole PS (2013) Transport and metabolism in legume-rhizobia symbioses. Annu Rev Plant Biol 64:781–805. https://doi.org/10.1146/annurev-arplant-050312-120235

Usuki F, Narisawa K (2007) A mutualistic symbiosis between a dark septate endophytic fungus, *Heteroconium chaetospira*, and a nonmycorrhizal plant, Chinese cabbage. Mycologia 99:175–184. https://doi.org/10.1080/15572536.2007.11832577

Van Schöll L, Hoffland E, Van Breemen N (2006) Organic anion exudation by ectomycorrhizal fungi and *Pinus sylvestris* in response to nutrient deficiencies. New Phytol 170:153–163. https://doi.org/10.1111/j.1469-8137.2006.01649.x

Wallander H, Wickman T (1999) Biotite and microcline as potassium sources in ectomycorrhizal and non-mycorrhizal *Pinus sylvestris* seedlings. Mycorrhiza 9:25–32. https://doi.org/10.1007/s005720050259

Wang Y, Wu W-H (2017) Regulation of potassium transport and signaling in plants. Curr Opin Plant Biol 39:123–128. https://doi.org/10.1016/j.pbi.2017.06.006

Watts-Williams SJ, Cavagnaro TR (2018) Arbuscular mycorrhizal fungi increase grain zinc concentration and modify the expression of root ZIP transporter genes in a modern barley (*Hordeum vulgare*) cultivar. Plant Sci 274:163–170. https://doi.org/10.1016/J.PLANTSCI.2018.05.015

Watts-Williams SJ, Patti AF, Cavagnaro TR (2013) Arbuscular mycorrhizas are beneficial under both deficient and toxic soil zinc conditions. Plant Soil 371:299–312. https://doi.org/10.1007/s11104-013-1670-8

Watts-Williams SJ, Tyerman SD, Cavagnaro TR (2017) The dual benefit of arbuscular mycorrhizal fungi under soil zinc deficiency and toxicity: linking plant physiology and gene expression. Plant Soil 420:375–388. https://doi.org/10.1007/s11104-017-3409-4

White PJ, Karley AJ (2010) Potassium. In: Hell R, Mendel R-R (eds) Cell biology of metals and nutrients. Springer, Berlin, Heidelberg, pp 199–224

Wu SC, Cao ZH, Li ZG et al (2005) Effects of biofertilizer containing N-fixer, P and K solubilizers and AM fungi on maize growth: a greenhouse trial. Geoderma 125:155–166. https://doi.org/10.1016/j.geoderma.2004.07.003

Ye Z, Zeng J, Li X et al (2017) Physiological characterizations of three barley genotypes in response to low potassium stress. Acta Physiol Plant 39:232. https://doi.org/10.1007/s11738-017-2516-4

Yuan L, Huang J, Li X, Christie P (2004) Biological mobilization of potassium from clay minerals by ectomycorrhizal fungi and eucalypt seedling roots. Plant Soil 262:351–361

Zhu R, Jin L, Sang Y et al (2024) Characterization of potassium-solubilizing fungi, Mortierella spp., isolated from a poplar plantation rhizosphere soil. Arch Microbiol 206:157. https://doi.org/10.1007/s00203-024-03912-w

Zörb C, Senbayram M, Peiter E (2014) Potassium in agriculture--status and perspectives. J Plant Physiol 171:656–669. https://doi.org/10.1016/j.jplph.2013.08.008

# Revisiting Paradigms Related to Root Hydraulic Limitation Under Drought

**Bernhard Schuldt, Mutez Ahmed, Brendan Choat, Sylvain Delzon, Steven Jansen, Martyna M. Kotowska, Christoph Leuschner, Laurent J. Lamarque, Torben Lübbe, Peter Petrik, José M. Torres-Ruiz, and Scott McAdam**

## Contents



Communicated by Ulrich Lüttge

B. Schuldt (✉) and P. Petrik
Institute of Forest Botany and Forest Zoology, Technical University of Dresden (TUD), Tharandt, Germany
e-mail: bernhard.schuldt@plant-ecology.de; peter.petrik@tu-dresden.de

M. Ahmed
Root-Soil Interaction, TUM School of Life Sciences, Technical University of Munich, Freising, Germany
e-mail: mutez.ahmed@tum.de

B. Choat
Hawkesbury Institute for the Environment, Western Sydney University, Richmond, NSW, Australia
e-mail: B.Choat@westernsydney.edu.au

S. Delzon
University of Bordeaux, INRAE, BIOGECO, Pessac, France

S. Jansen
Institute of Botany, Ulm University, Ulm, Germany
e-mail: steven.jansen@uni-ulm.de

M. M. Kotowska
School of Natural Sciences, Macquarie University, Sydney, NSW, Australia
e-mail: martyna.kotowska@mq.edu.au

C. Leuschner and T. Lübbe
Plant Ecology, Albrecht von Haller Institute for Plant Sciences, University of Goettingen, Goettingen, Germany
e-mail: cleusch@gwdg.de; torben.luebbe@biologie.uni-goettingen.de



Progress in Botany (2026) 86: 151–180, https://doi.org/10.1007/124_2025_93,
Published online: 05 September 2025

**Abstract** In vascular plants, the long-distance water transport between organs occurs under tension. Because it is difficult to study a system under negative pressure without measurement artefacts, the field of plant hydraulics has been afflicted with various controversies. One such debate relates to the relative vulnerability of roots versus aboveground organs within woody angiosperms; specifically, whether coarse roots are far more vulnerable to drought-induced dysfunction than stems or branches. It has been proposed that the wider and longer xylem vessels observed in roots compared with aboveground organs are associated with higher vulnerability to xylem embolism, suggesting a trade-off between xylem safety and hydraulic efficiency at the whole-plant level. Since roots were thought to embolize first and to recover steadily from embolism, they were assumed to act as 'hydraulic fuses' that limit tension and embolism in proximal parts of the xylem network. Moreover, the 'hydraulic fuse' concept was consistent with the idea that roots were the sole source of abscisic acid and that chemical root-shoot signals were required to induce stomatal closure. Here, we re-evaluate several paradigms that still persist in the literature and show that they are outdated and require reinterpretation based on recent evidence. The surveyed paradigms include (i) a high root vulnerability to embolism formation, (ii) a within-tree trade-off between xylem safety and hydraulic efficiency, (iii) the assumption that roots act as hydraulic fuses, (iv) the occurrence of embolism repair in roots, and (v) the root-to-shoot abscisic acid signalling. Based on available evidence, we show that none of these classic hypotheses hold, except for an updated hydraulic fuse hypothesis, which could be attributed to a loss of contact at the soil-root interface when the cortex shrinks during moderate soil drought.

L. J. Lamarque
Natural Resources Canada, Canadian Forest Service, Laurentian Forestry Centre, Québec, QC, Canada

Département des Sciences de l'environnement, Université du Québec à Trois-Rivières, Trois-Rivières, QC, Canada

J. M. Torres-Ruiz
Instituto de Recursos Naturales y Agrobiologıa (IRNAS), Consejo Superior de Investigaciones Cientıficas (CSIC), Seville, Spain

S. McAdam
Department of Botany and Plant Pathology, Purdue University, West Lafayette, IN, USA
e-mail: smcadam@purdue.edu

**Keywords** Embolism refilling, Embolism resistance, Hydraulic architecture, Measurement artefact, Plant hydraulics, Root, r-Shaped vulnerability curve, Wood anatomy

## 1 Long-Lasting Paradigms in the Field of Plant Hydraulics

Plant water relations play a major role both in developing mechanistic models of plant growth and yield and in predicting future species distribution and plant responses to climate change (Jansen and Schenk 2015; Sack et al. 2016; Sanchez-Martinez et al. 2023; Torres-Ruiz et al. 2024). Hence, it is surprising that our understanding of plant hydraulics is still incomplete even 130 years after the cohesion-tension theory was first proposed (Askenasy 1895; Dixon and Joly 1895). According to the cohesion-tension theory, transpiring surfaces of leaves generate a negative pressure that drives the ascent of sap from the roots to the leaves within the hydraulic systems of plants (Schenk et al. 2015). Although this theory has frequently been questioned (e.g., Canny 2001; Zimmermann et al. 2002), it has resisted several challenges and is now accepted as the driving mechanism behind plant water transport (Meinzer et al. 2001; Angeles 2004). The passive nature of the physical process that drives the ascent of sap demands that hundreds of water molecules are lost by transpiration through the leaf stomata in order to take up a single molecule of $CO_2$ for photosynthesis (Venturas et al. 2017). Accordingly, vascular water transport through dead xylem cells has been identified as the 'backbone of terrestrial plant productivity' (Brodribb 2009), and the field of plant hydraulics has emerged as a major component of plant water relations research (Venturas et al. 2017).

Because a long-distance water transport system under tension (negative pressure) is liable to failure by air invasion, it is inherently challenging to study without the introduction of artefacts. This has led to repeated controversies within the field of plant hydraulics, with many debates relating to the methodology used to measure the vulnerability of plants to drought. One of these debates has centred on roots being by far the most vulnerable organ along the hydraulic pathway (e.g., Sperry and Saliendra 1994; Alder et al. 1996). These early studies from the 1990s made use of positive pressure in custom-made pressure sleeves for constructing xylem vulnerability curves in angiosperms. Since the development of newer technologies for assessing embolism formation in xylem, there has emerged a general agreement that the majority of these early results have been misinterpreted (Choat et al. 2010; Ennajeh et al. 2011; Martin-StPaul et al. 2014; Cochard et al. 2010, 2015). Hence, a growing body of studies applying non-invasive techniques indicate that roots are not generally much more vulnerable to embolism formation than stems or branches (Skelton et al. 2017; Losso et al. 2019; Peters et al. 2020). The historical assumption of high root vulnerability to drought-induced hydraulic failure contributed to several further paradigms including the idea that there is a trade-off between xylem safety and hydraulic efficiency within trees (Sperry and Saliendra 1994; Kavanagh et al.

1999), the hydraulic fuse theory (Jackson et al. 2000), embolism repair in roots (McCully et al. 1998; Buchard et al. 1999; Domec et al. 2006) or the root-to-shoot abscisic acid (ABA) signalling to close stomata during drought (Zhang and Davies 1987). Here, we review the paradigm of high root vulnerability to hydraulic dysfunction in vessel-bearing woody plants in the light of recent evidence and discuss the interrelated set of now outdated paradigms that still persist in the literature.

## 2 Hydraulic Adjustments Along the Flow Path for Water

As early as in the seventeenth century, Nehemiah Grew (1641–1712) noted that xylem conduits were wider in roots than in stems (Baas 1982), while Sanio (1872) observed that conduits in branches were smaller than in trunks (Aloni 1987). Since then, numerous studies found xylem conduits to taper along the hydraulic pathway from roots to stems and further to the leaves, regardless of organ diameter (e.g., Fegel 1941; Carlquist 1975; Denne 1972; Zimmermann and Potter 1982; Gasson 1985; Aloni 1987; Tyree and Zimmermann 2002; McElrone et al. 2004; Choat et al. 2005; Mencuccini et al. 2007; Lintunen and Kalliokoski 2010; Anfodillo et al. 2013; Lübbe et al. 2022; but see Zotz et al. 1997; Schuldt et al. 2013; Kotowska et al. 2015). For example, in adult trees of four forest-grown diffuse-porous temperate tree species, vessel diameters were on average three-times larger in similar-sized coarse roots than in branches (Fig. 1). Consistent with a vessel diameter difference, the root system of various woody angiosperms was found to be more vulnerable to embolism formation than the stems (saplings) or branches (adults) of these species, especially when the air injection (Cochard et al. 1992) or centrifuge techniques (Alder et al. 1997; Cochard 2002) were used (Table 1), with stems in saplings being morphologically comparable to branches of mature trees. The anticipated axial increase in xylem safety from deep roots upwards (e.g., McElrone et al. 2004) was interpreted as an advantage when growing tall. Indeed, the ability to maintain an intact water column is constrained by gravitational force and friction as tree height increases (Woodruff and Meinzer 2011). With increasing tree height and flow path length, the hydraulic constraints imposed by gravity reduce the xylem water potential by 0.01 MPa per metre (Scholander et al. 1965), which causes a linear decline in water potential in the absence of transpiration (Koch et al. 2004; Ishii et al. 2008; Ambrose et al. 2016). Based on this increase in gravitational force and friction with height, Burgess et al. (2006) suggested that a vertical hydrostatic gradient would cause a progressive increase in xylem safety and a concomitant decrease in xylem efficiency. Because the risk of hydraulic failure increases with reduced water potential, it seemed reasonable to assume that xylem conduits taper in order to reduce the risk of embolism formation with increasing height. At first glance, this is supported by the high number of studies claiming that roots are the most vulnerable organ along the flow path (Table 1). However, the reliability of the majority of these results needs to be questioned because measurements can be

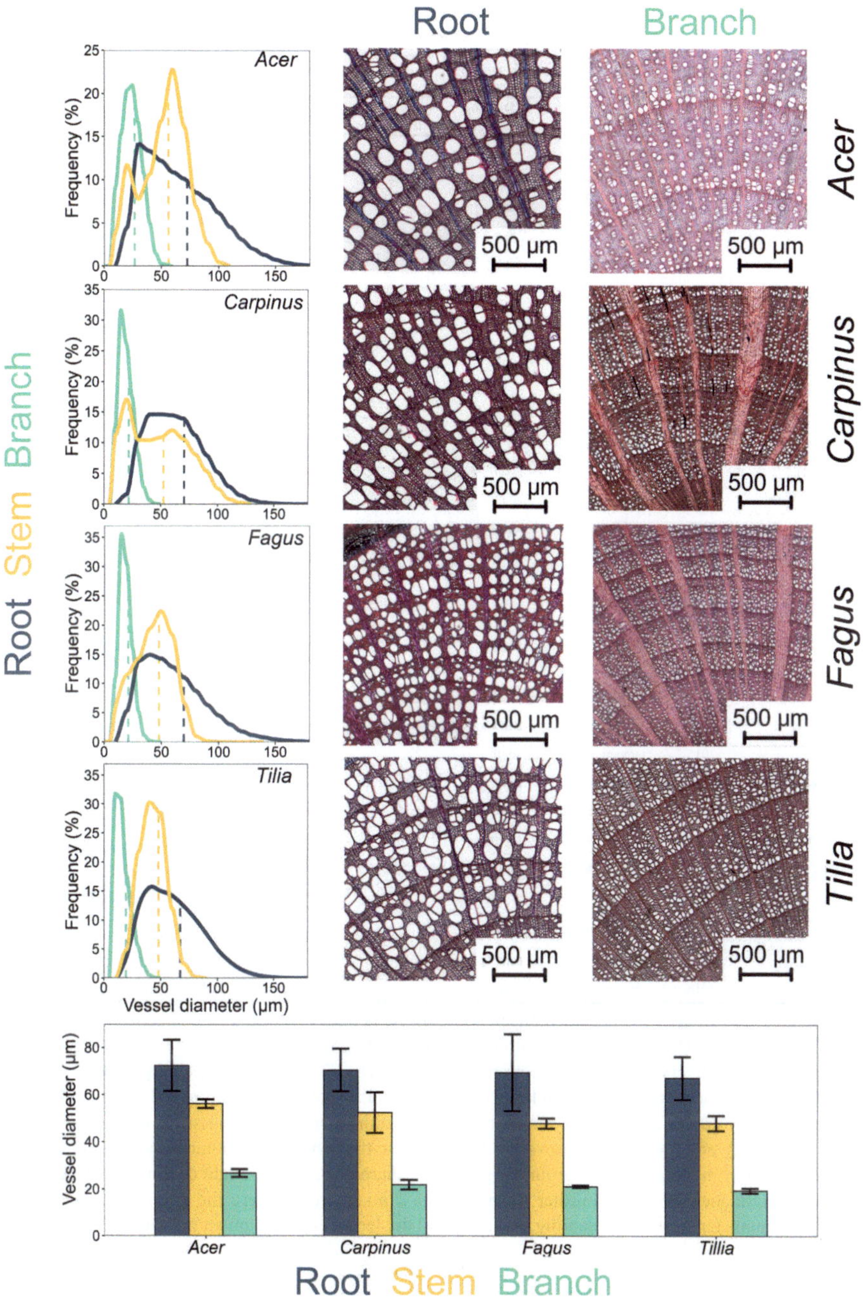


**Fig. 1** Wood anatomical sections of coarse root, stem, and branch wood for mature trees (tree height range: 24.3–33.7 m) of four temperate broad-leaved tree species (right), vessel diameter distribution (left), and average vessel diameter variation across root, stem, and branch wood (bottom). The species studied were *Acer pseudoplatanus*, *Carpinus betulus*, *Fagus sylvatica*, and *Tilia cordata*. (For methodological details, please refer to Lübbe et al. (2022))

**Table 1** Literature review of empirical data on root xylem embolism resistance of woody angiosperms by the group of sampled plants, applied method and, if available, the shape of fitted vulnerability curve ($VC_{Root}$, $VC_{Stem}$, $VC_{Branch}$)

| Authors | Year | Climate zone | Develop. stage | Segment length (cm) | Method | $VC_{Root}$ | $VC_{Stem}$ | $VC_{Branch}$ | $\Delta_{P50}$ |
|---|---|---|---|---|---|---|---|---|---|
| Cochard et al. | 2002 | temperate | sapling | – | AI | S | S | – | ↗ |
| Domec et al. | 2006 | seasonally dry | young tree | 15-40 | AI | R/L | – | – | ↔ |
| Hultine et al. | 2006 | semi-arid | adult | – | AI | R | R | | ↘ |
| Johnson et al. | 2018 | semi-arid | adult | 90, 60 | AI | R, S | – | R, S | ↘ |
| Johnson et al. | 2016 | temperate | adult | 36, 18 | AI | L, R | – | L, S | ↓ |
| Maherali et al. | 2006 | temperate | young tree | 14, 24-59 | AI | R/(S) | (R)/S | – | ↘ |
| Matzner et al. | 2001 | mediterranean | adult | 20 | AI | R | R/S | – | ↘ |
| McElrone et al. | 2004 | temperate | adult | 14 | AI | R | R/S | – | ↘ |
| Mencuccini & Comstock | 1997 | arid | sapling | 20 | AI | R | S | S | ↓ |
| Martinez-Vilata et al. | 2002 | mediterranean | adult | 20-45 | AI, BD | R/(S) | S | – | ↓ |
| Tsuda & Tyree | 1997 | temperate | sapling | 2 | BD | S | R/S | – | ↗ |
| Alder et al. | 1996 | temperate | adult | 21-40 | BD, AI | R | – | S | ↓ |
| Hacke & Sauter | 1996 | temperate | adult | 4-6, 12-15 | BD, AI, AE | R | S | S | ↓ |
| Hukin et al. | 2005 | arid | sapling | 2 | CSEM, AI | (S) | (S) | – | ↗ |
| Domec et al. | 2010 | temperate | adult | 14 | SC | R | – | R/S | ↓ |
| Fickle et al. | 2023 | semi-arid | adult | 14 | SC | R | – | R/L | ↓ |
| Hacke et al. | 2000 | semi-arid | adult | 14 | SC | R | – | (S) | ↓ |
| Pratt et al. | 2007 | semi-arid | adult | 14, 27 | SC | R/(L)/(S) | (L)/S | – | ↓ |
| Pratt et al. | 2015 | semi-arid | adult | 14 | SC | R | – | – | – |
| Sperry & Hacke | 2002 | arid | adult | 14 | SC | R/(L) | – | – | – |
| Davis et al. | 2002 | mediterranean | adult | 26 | FC | (S) | – | S | ↔ |
| Jansen et al. | 2015 | temperate | adult | 27 | FC | R | – | S | ↓ |
| Lübbe et al. | 2022 | temperate | adult | 100, 27 | FC | S | – | S | ↔ |
| Losso et al. | 2019 | temperate | seedling | – | µCT | S | S | – | ↔ |
| Peters et al. | 2020 | subtropical | sapling | – | µCT | S | S | – | ↔ |
| Skelton et al. | 2017 | subtropical | sapling | – | µCT, OV | S | S | – | ↔ |
| Harrison Day et al. | 2024 | temperate | sapling | – | OV | S | S | – | ↔ |
| Rodriguez-Dominguez et al. | 2018 | mediterranean | seedling | – | OV | S | S | – | ↗ |
| Wu et al. | 2020 | temperate | sapling | – | PN | (S) | (S) | – | ↔ |
| Creek et al. | 2018 | temperate | sapling | – | RK | R/S | R/S | – | (↔) |

The studies are sorted by methods, which are abbreviated as follows: AE ultrasound acoustic emissions, AI air injection, BD bench dehydration, SC standard-centrifuge technique, FC flow-centrifuge technique, CSEM cryo-scanning electron microscopy, µCT micro-CT technique, OV optical vulnerability method, RK rehydration kinetics technique, PN pneumatic technique. The general shape of vulnerability curves across species and treatments within each study is classified as sigmoidal (S), exponential (R), and linear (L). Arrows for $\Delta P_{50}$ indicate the general direction of change in $P_{50}$ along the flow path from roots to above-ground organs (i.e., sharp decline (↓), slight decline (↘), no change (↔) or slight increase (↗)). Further given are the climatic zone where the study took place, the developmental stage (seedling, sapling, or adult) and the length of the segment measured if given.

misinterpreted due to various artefacts, such as the occurrence of long vessels in roots (Fig. 2).

## 3 Measuring Root Embolism Resistance in Angiosperms Correctly

Over the last 30 years, numerous studies have attempted to measure embolism resistance in the root xylem of woody angiosperms (Table 1). However, the reliability of most of these findings is under question, as vulnerability curves (VCs) established for long-vesselled plant segments may be subject of both sampling (e.g., Wheeler et al. 2013; Torres-Ruiz et al. 2015) and measurement artefacts (e.g., Jansen et al. 2015). This so-called 'vessel-length artefact', now better known as the 'open-vessel artefact', i.e., the occurrence of open-cut conduits in the xylem segments, may favour exponential (r-shaped) VCs that overestimate the vulnerability to embolism up to orders of magnitude (Choat et al. 2010, 2016; Cochard et al. 2010, 2013, 2015; Ennajeh et al. 2011; Martin-StPaul et al. 2014; Torres-Ruiz et al. 2014, 2017). This is particularly the case with air-injection or centrifugation techniques, when cut-open vessels traverse the centre of the plant segment or even its whole length (Torres-Ruiz et al. 2014). In contrast to these error-prone methods for long-vesselled woody segments such as roots (e.g., Hajek et al. 2014) or tropical tree species and lianas (e.g., De Guzman et al. 2017), more recent studies acknowledging potential sampling or measurement artefacts did not confirm earlier results that roots are by far the most vulnerable organ in trees (Table 1). In the following, we will focus on the centrifugation technique as it is still claimed that root embolism resistance can be accurately measured in angiosperms on 14 cm long root segments using a short-diameter rotor (Ø 15 cm) with this device (Pratt et al. 2015; Fickle et al. 2023). The same limitation also applies to the air injection technique when using segments of insufficient length. To test the effect of root segment length on the shape of the vulnerability curve using the flow-centrifuge technique, we inserted roots from four diffuse-porous temperate tree species in either a large-diameter rotor (Ø 100 cm) or a medium-diameter rotor (Ø 30 cm). According to our vessel length measurements (Fig. 2), short root segments of 28 cm length showed a considerable fraction of open-cut vessels. Therefore, in agreement with Pratt et al. (2015) and Fickle et al. (2023), all VCs measured on short root segments showed a steep decrease in conductivity at moderate xylem pressures corresponding to r-shaped VCs, whereas all VCs measured on long root segments without cut-open vessels showed sigmoidal (s)-shaped VCs (Fig. 3). These results provide substantial support for a clear effect of the presence of open vessels on root embolism resistance when measuring segments of insufficient length using the centrifuge method, as the water potential at 50% loss of hydraulic conductance ($P_{50}$-value) would be highly overestimated. In fact, the $P_{50}$-values measured in 100 cm long coarse root segments with s-shaped VCs were

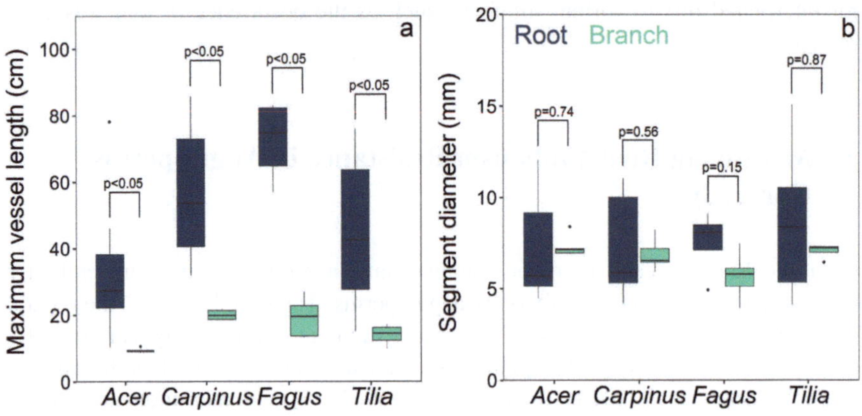


**Fig. 2** (**a**) Maximum xylem vessel length estimated with the air-injection method at 0.1 MPa air pressure inserted against the flow direction in (**b**) equally sized roots and branches in four temperate broad-leaved tree species (*Acer pseudoplatanus*; *Carpinus betulus*; *Fagus sylvatica*; *Tilia cordata*). (For methodological details, please refer to Lübbe et al. (2022))

on average −2.0 to −2.9 MPa lower (more negative) than corresponding values derived from 28 cm long segments with r-shaped VCs, which would lead to massive overestimation of root sensitivity to embolism, in *Acer pseudoplatanus*, for example, by up to 1,450% (−0.2 and −3.1 MPa in short and long root segments, respectively; Fig. 4). The divergence in the shape of VCs derived from medium-diameter (30 cm; r-shaped) or large-diameter (100 cm; s-shaped) rotors can be explained by predicted differences in cut-open vessels, as maximum root vessel length exceeds 28 cm in all four species (Fig. 2). Although mechanisms are not fully understood, recreating the natural conditions under which water is transported under tension in living plants is not possible in centrifuge experiments for cut-open vessels. This is because, during the first spins of the centrifuge, water escapes through the open ends of the vessels, even when water reservoirs are installed at the sample ends. Consequently, the higher the number of open vessels in a segment, the greater the overestimation of vulnerability to cavitation because bordered pits are excluded from the measurements. It seems likely that many studies in the past that have assessed root xylem vulnerability to embolism using air injection or centrifugation techniques have not used root segments long enough to exclude cut-open vessels, and this factor has been largely ignored in the interpretation of the results. Indeed, most of the data for angiosperm root segments exhibited r-shaped VCs, in marked contrast to the above-ground organs studied (Table 1). For conifer roots with relatively short tracheids, on the other hand, r-shaped VCs have been related to enhanced flexibility of the pit margo, causing the 'torus aspiration artefact' by sealing of the pit aperture when the pressure gradient between the sample ends is high (Beikircher et al. 2010; Bouche et al. 2015). In angiosperms, several recent studies using five independent methods

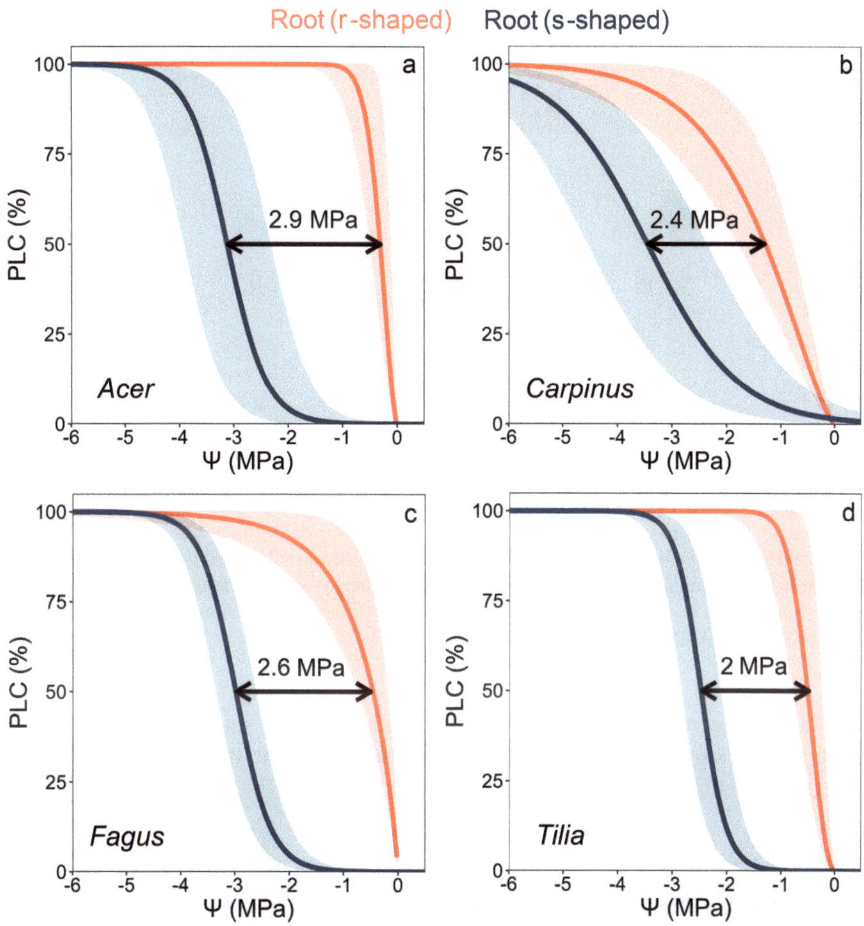


**Fig. 3** Averaged vulnerability curves for coarse roots measured on short segments (28 cm, red curves, r-shaped) or long segments (100 cm, blue curves, s-shaped) with the flow-centrifuge method for four temperate broad-leaved tree species ((**a**) *Acer pseudoplatanus*; (**b**) *Carpinus betulus*; (**c**) *Fagus sylvatica*; (**d**) *Tilia cordata*) with xylem pressure (Ψ) plotted against percentage loss of hydraulic conductivity (PLC). The shaded areas indicate the 95% confidence intervals for each curve. (For methodological details, please refer to Lübbe et al. (2022))

for estimating xylem vulnerability to embolism formation and acknowledging the 'open-vessel artefact' have challenged the classic paradigm of high root vulnerability to drought-induced hydraulic failure (Table 1). These studies used the micro-CT technique and the optical method, both direct and non-invasive methods applied on intact plants (Skelton et al. 2017; Rodriguez-Dominguez et al. 2018; Losso et al. 2019; Peters et al. 2020; Harrison Day et al. 2024a), the large-diameter flow-centrifuge technique (Lübbe et al. 2022), the rehydration kinetics technique (Creek et al. 2018), and the pneumatic method (Wu et al. 2020). Although roots appeared to

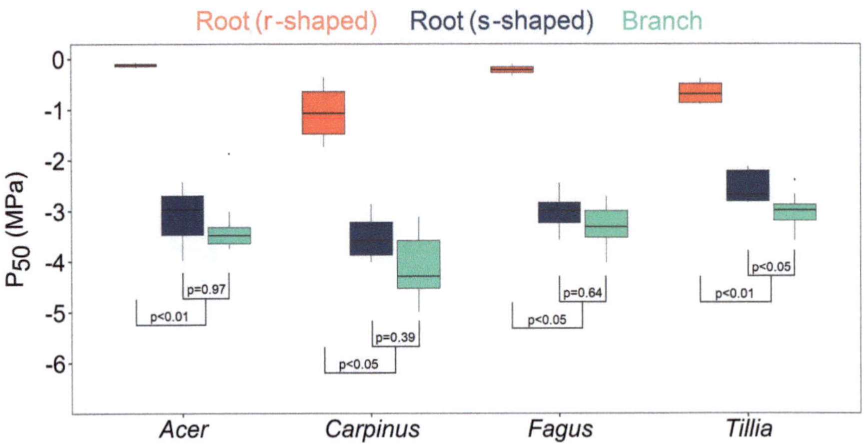


**Fig. 4** $P_{50}$-values for short coarse root segments (length 28 cm; orange) estimated from r-shaped vulnerability curves (VCs), long coarse root segments (length 100 cm; blue) estimated from s-shaped VCs, and sun-canopy branches (green) from mature field-grown trees of four temperate broad-leaved tree species (*Acer pseudoplatanus*; *Carpinus betulus*; *Fagus sylvatica*; *Tilia cordata*). For short root segments, 4–8 vulnerability curves (r-shaped) were measured per species (*Acer*: $n = 8$; *Carpinus*: $n = 4$; *Fagus*: $n = 4$; *Tilia*: $n = 8$). (For further methodological details, please refer to Lübbe et al. (2022))

be slightly more vulnerable to embolism formation in some studies, differences were generally not significant in most of them (please refer to the studies listed in Table 1 and compare Fig. 4).

## 4 Within-Tree Trade-Off Between Xylem Safety and Efficiency

Due to the biased assumption of high root vulnerability to embolism, i.e., higher $P_{50}$-values, another long-lasting paradigm in the field of plant hydraulics postulates a strong trade-off between xylem safety and hydraulic efficiency at the tree level (e.g., Sperry and Saliendra 1994; Kavanagh et al. 1999). Across species, a relationship between hydraulically-weighted conduit diameter ($D_h$) as a measure of hydraulic efficiency and the $P_{50}$-value can be suggested in angiosperms (Lens et al. 2022; Isasa et al. 2023). However, Tyree and Sperry (1989) already have argued that vessel diameter is not directly related to the mechanism of embolism formation but could be related to pit properties. While vessel diameters can differ several-fold between roots and branches (Fig. 1), pit membrane thickness as a determinant of embolism resistance in angiosperms (e.g., Li et al. 2016; Kaack et al. 2021; Isasa et al. 2023) shows considerable convergence between roots and stems in saplings or branches in

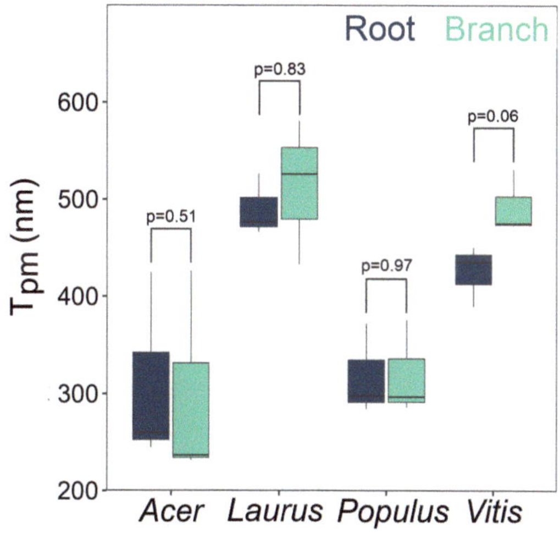


**Fig. 5** Interconduit pit membrane thickness ($T_{pm}$, nm) in small-diameter roots and branches in a mature tree (*Acer pseudoplatanus*) or roots and stems of saplings (*Laurus nobilis, Populus tremula, Vitis vinifera cv. Cabernet sauvignon*). (For methodological details, please refer to Kotowska et al. (2019))

adult trees (Fig. 5). This might explain why the safety-efficiency trade-off is virtually absent at the intra-specific level in angiosperms, either across individuals within given organs (e.g., Hajek et al. 2016; Isasa et al. 2023) or across organs within individuals (e.g., Lübbe et al. 2022). Therefore, inconsistent claims in the literature regarding the efficiency-safety trade-off may be due to the range of trait values covered and the level of data aggregation (Isasa et al. 2023; see also Lens et al. 2023). It is therefore not surprising that for the adult trees of the four temperate broadleaved species mentioned above, no trade-off between xylem efficiency and safety is observed after excluding the 'open-vessel artefact' in the roots (Fig. 6). Although recent studies acknowledging the 'open-vessel artefact' do not support the paradigm that roots are orders of magnitude more vulnerable than stems, for some species roots may appear to be slightly more vulnerable to embolism formation although differences commonly are non-significant (e.g., Losso et al. 2019; Lübbe et al. 2022; and partly Peters et al. 2020; Harrison Day et al. 2024a, b). Since the thickness of the pit membrane is more or less comparable between organs (Fig. 5), other factors must be involved. Lübbe et al. (2022) attributed this finding to differences in the vessel lumen-to-sapwood area ratio, i.e., the total area occupied by vessels per cross-sectional xylem area; the higher the vessel lumen fraction, the lower the resistance to drought-induced xylem embolism. Indeed, a growing number of studies report this relationship between the vessel lumen-to-sapwood area ratio and embolism resistance, both within (Weithmann et al. 2022; Waite et al. 2024) and across species (Lübbe et al. 2022; Avila et al. 2023). With an increase in the vessel lumen-to-sapwood area ratio, it seems reasonable to assume that the total wall area per cross-section likewise increases. The total vessel wall area, on the other hand, is associated with the total number of intervessel pits per cross-section (cf., Kaack et al. 2021). This could provide an explanation for the observed relationship between

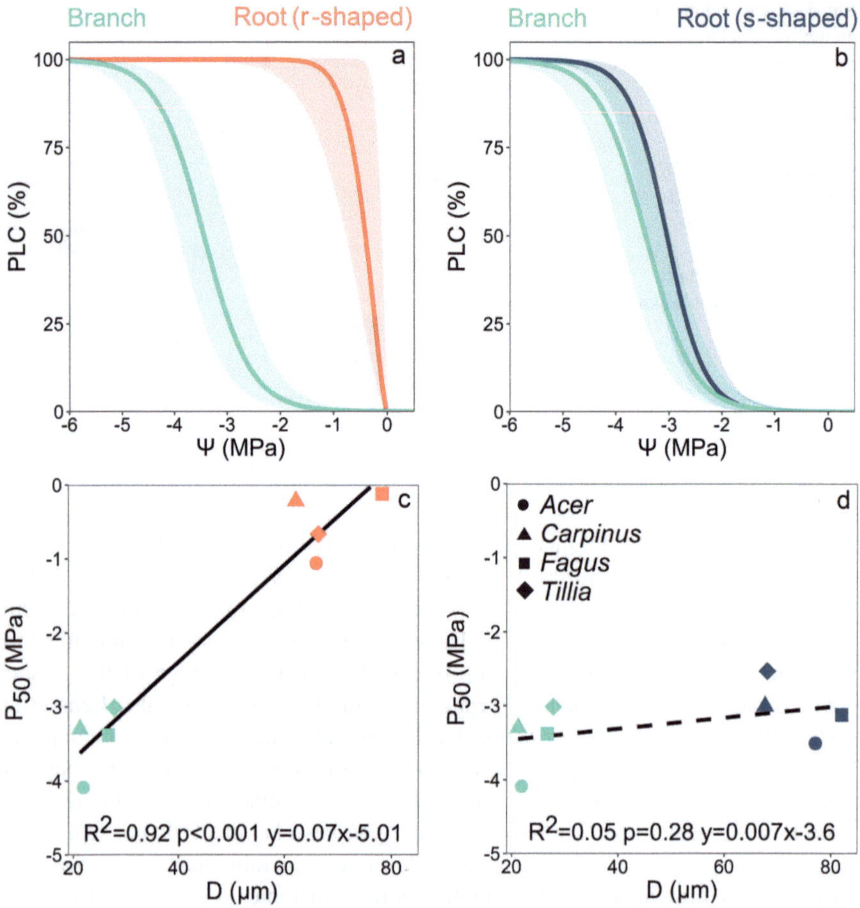


**Fig. 6** Branch embolism resistance (green) and root xylem embolism resistance estimated from (**a**) r-shaped vulnerability curves (VC) from 28 cm long root segments (orange) or (**b**) s-shaped vulnerability curves from 100 cm long root segments (blue), and the estimated water potential at 50% loss of hydraulic conductance ($P_{50}$) in relationship to vessel diameter for s-shaped (**c**) and s-shaped (**d**) root VCs for four broad-leaved temperate tree species (*Acer pseudoplatanus*, *Carpinus betulus*, *Fagus sylvatica*, and *Tilia cordata*). All vulnerability curves have been measured with the flow-centrifuge technique. (For methodological details, please refer to Lübbe et al. (2022))

vessel lumen-to-sapwood area ratio and embolism resistance, which is consistent with the pit area hypothesis (Sperry et al. 2006). Testing these structure-functional ideas would require quantitative information about vessel dimensions and interconnectivity, and how the 3D-vessel network is affected by the occurrence of tracheids as unicellular conduits (Johnson et al. 2023).

# 5 Hydraulic Fuse Hypothesis

According to his 'segmentation hypothesis', Zimmermann (1983) assumed that woody plants might restrict hydraulic failure to redundant organs that are readily replaced, analogous to an electric circuit (Sperry et al. 2002). Building on this concept, Jackson et al. (2000) speculated that small roots, which were thought to show high vulnerability to embolism, might act as 'hydraulic fuses' during soil drying, localizing failure to relatively 'expendable' units compared to tap roots, stems, and branches. Unfortunately, root diameters are rarely given in older plant hydraulic studies, but we assume that the 'small roots' mentioned (Jackson et al. 2000; citing Hacke et al. 1996) refer to woody coarse roots (Table 2). However, the 'hydraulic fuse hypothesis' is used here as a synonym of the 'hydraulic vulnerability segmentation hypothesis' (HVSH), which is a modified version of Zimmermann's 'segmentation hypothesis' by Tyree and Ewers (1991). The HVSH suggests that more distal portions of trees such as root systems are more vulnerable to drought-induced embolism formation than aboveground stems. For example, Hacke and Sperry (2001) argued that small woody roots in shallow and dry soil layers are less embolism resistant than roots from deeper soil layers, enabling water uptake from soil layers where water status is more favourable once small roots have become heavily embolized. The authors concluded that 'Zimmermann's segmentation concept, which originally applied for aboveground xylem, may be even more applicable and relevant to the root system'. This high embolism vulnerability of roots was interpreted as an advantage in terms of carbon investment, because small-diameter roots can be replaced cost-efficiently, avoiding embolism formation in stems and branches (Sperry et al. 2002). Yet, the construction costs of lignified small-diameter roots in terms of carbon and nutrients may not be much smaller than for twigs and leaves (Gonzalez-Benecke et al. 2010), and the loss of roots is directly related to the loss of absorption capacity for nutrients and water. Furthermore, it would take days to weeks for new roots to develop, which may be too long to ensure a supply of water and nutrients. Instead, the inhibition of xylem water flow following moderate soil drought seems rather short, according to a field study monitoring sap flow in small-diameter roots; continuous measurements with miniature sap flow gauges showed that root water flow ceased during prolonged periods of soil drought but resumed

**Table 2** Root classification using the diameter ranges after Sutton and Tinus (1983)

<table>
<tr><th>Root diameter</th><th>Classification</th><th></th></tr>
<tr><td>Ø >50 mm</td><td>Strong roots</td><td rowspan="5">Ø >2 mm (coarse roots)</td></tr>
<tr><td>Ø 20–50 mm</td><td>Very large roots</td></tr>
<tr><td>Ø 10–20 mm</td><td>Large roots</td></tr>
<tr><td>Ø 5–10 mm</td><td>Medium roots</td></tr>
<tr><td>Ø 2–5 mm</td><td>Small roots</td></tr>
<tr><td>Ø 1–2 mm</td><td>Fine roots</td><td rowspan="2">Ø ≤2 mm (fine roots)</td></tr>
<tr><td>Ø ≤1 mm</td><td>Very fine roots</td></tr>
</table>

again within 3–4 h after soil rewetting in temperate oak and beech trees (Coners and Leuschner, unpublished). Because roots are not more vulnerable to embolism formation compared to aboveground organs as discussed above, other mechanisms must be responsible for this observation, such as the disconnection of the capillary continuum from the soil to the root as speculated by Körner (2019). Meanwhile, a growing body of studies has shown that root cortex shrinkage during soil dehydration is responsible for a loss of connectivity at the soil-root-interface (e.g., Cuneo et al. 2016, 2021; Duddek et al. 2022; Barrientos-Sanhueza et al. 2023; Affortit et al. 2024; Yang et al. 2023). However, when the root 'hydraulic fuse hypothesis' was developed more than 20 years ago (Jackson et al. 2000), it was also assumed that embolism occurs daily in the root system (e.g., McCully et al. 1998; Buchard et al. 1999). If this assumption were true, there would be no need for root regrowth if the embolized root vessels could easily be refilled in support of the root 'hydraulic fuse hypothesis'. However, evidence for embolism refilling in roots is virtually absent, except for one micro-CT study on grapevine (Cuneo et al. 2018), a woody liana known to exert considerable root pressure (Tibbetts and Ewers 2000).

## 6 Root Cortex Shrinkage Responsible for Contact Loss at the Soil-Root Interface

During progressive soil drying, the outside xylem hydraulic conductivity within the root cortical cells declines much earlier before any significant embolism is observed in coarse roots (Cuneo et al. 2016, 2021; Duddek et al. 2022; Barrientos-Sanhueza et al. 2023; Affortit et al. 2024; Yang et al. 2023). In addition to this process, in which apoplastic and symplastic barriers are likely to play a main role, a physical disconnection at the soil-root interface and a formation of gaps may occur due to root cortex shrinkage (Koebernick et al. 2018; Duddek et al. 2022; Fig. 7). For instance, Koebernick et al. (2018) used X-ray CT to study the dynamics of root-soil contact during soil drying in broad bean (*Vicia faba* L.), an annual herbaceous plant in the legume family. The authors found that root-soil contact significantly decreased as the soil drying period progressed, beginning at soil matric potentials below −20 kPa, and there was no significant difference in root shrinkage between taproots and lateral roots. The authors also observed that upon re-irrigation of the soil, the roots expanded, partially or completely closing the gaps (Koebernick et al. 2018). It is important to note that the soil matric potentials reported by Koebernick et al. (2018) are significantly higher than the root water potentials at which Cole and Alston (1974) observed root shrinkage (between −0.5 and −1.0 MPa). Similarly, they are also much higher than the soil water potentials at which Huck et al. (1970) detected diameter changes in cotton (between −0.1 and −0.7 MPa). Even lower soil water potentials were recorded by North and Nobel (1997) for desert succulents. In a recent study, Duddek et al. (2022) leveraged high-resolution (actual pixel size: $0.65 \times 0.65\ \mu m^2$) synchrotron-based X-ray micro-CT to observe roots, root hairs,

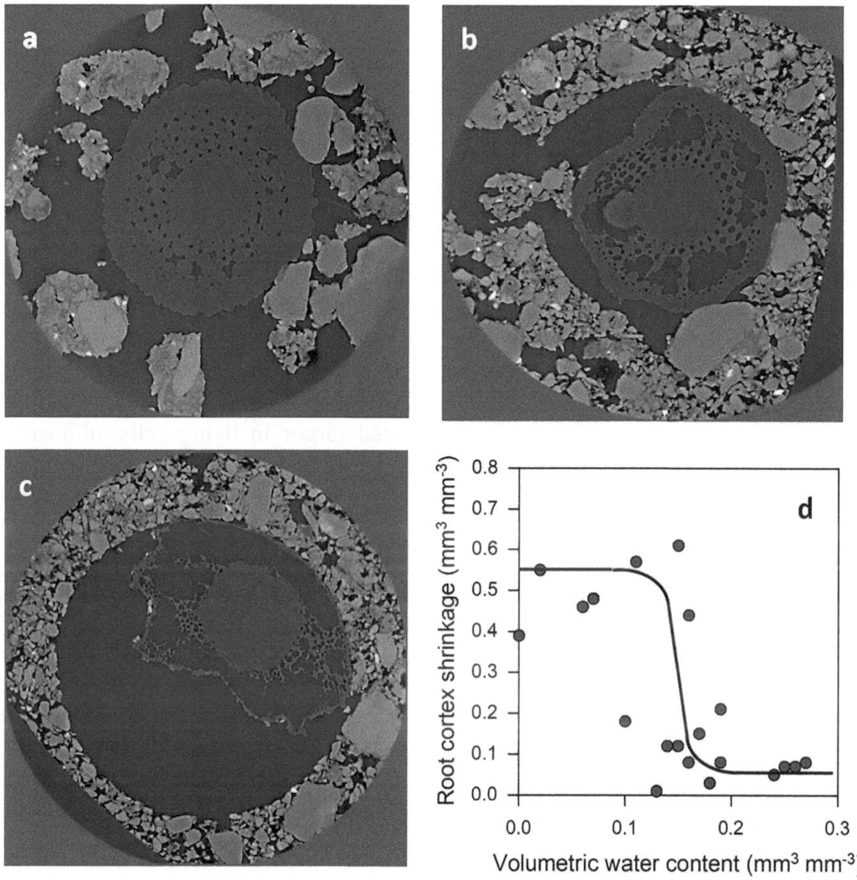


**Fig. 7** Synchrotron micro-CT images showing maize roots and root hair shrinkage, and the formation of air-filled gaps during soil drying (air is black, soil is light grey, and root tissue is dark grey; actual pixel size: 0.65 × 0.65 μm$^2$). (**a**) A maize root in a wet loamy soil showing a turgid root and shrunken root hairs; (**b**) and (**c**) a maize root in a dry soil showing an air-filled gap between the roots and the soil matrix; (**d**) relation between the volumetric soil water content and cortex shrinkage during soil drying. (For methodological details, please refer to Duddek et al. (2022))

and root-soil contact of maize (*Zea mays* L.), an herbaceous annual monocot. The authors found that, during soil drying, root hairs shrinkage in maize began at relatively high soil matric potentials (ranging from −10 to −310 kPa). Root hair shrinkage marks the initial step in a series of root responses to soil drying, followed by the formation of cortical lacunae and finally root shrinkage (note that xylem embolism was rarely observed in this study). In contrast to hair shrinkage, cortex drying and root shrinkage began at a more negative soil matric potential (below −1 MPa).

The latter leads to the creation of air-filled gaps at the root-soil interface. Together, these processes contribute to a gradual disconnection of roots from the soil, resulting in a significant reduction in root-soil contact during soil drying. Interestingly, Rodriguez-Dominguez and Brodribb (2020) demonstrated that, during moderate water stress in woody olive plants (*Olea europaea* L.), losses in root hydraulic conductance, primarily caused by cortex shrinkage and loss of contact at the root-soil interface, are significant enough to trigger stomatal closure before cavitation takes place.

In fact, both the loss in outside xylem hydraulic conductivity and the physical disconnection at the soil-root interface processes increase root hydraulic resistance and may cause a long-distance pressure-induced hydraulic signal from roots to leaves, causing a decline in leaf water potential, which induces ABA synthesis in foliar living cells when approaching the water potential at turgor loss point (Christmann et al. 2007). Depending also on the internal water storage capacity, reduced conductivity in the root cortex and disconnection from the soil will result in a decline in leaf water potential and thus cell turgor in living cells, ultimately inducing ABA synthesis in leaves resulting in stomatal closure. After rewetting, root functionality is restorable within hours (Ahmed unpublished; Koebernick et al. 2018), as long as roots were not embolized and only the cortex needs to be rehydrated in order to swell for restoring soil-root connectivity. Thus, fine roots showing a high root cortex fraction (cf., Li et al. 2022) could still be viewed as 'hydraulic fuses' because the shrinkage of the root cortex is responsible for the loss of contact at the soil-root-interface, thereby protecting the root xylem from tissue dehydration when the soil water potential would become more negative than the root water potential.

## 7 Refilling of Embolized Root Vessels

Once embolized, xylem can recover from conductivity loss over different time scales. In the short term, it has been suggested that embolism refilling may occur (Klein et al. 2018), while in the medium to long term, water transport capacity could be restored through the formation of new xylem tissue (Hammond and Adams 2019; Arend et al. 2022). In contrast to former studies on crop species biased by measurement artefacts (McCully et al. 1998; Buchard et al. 1999; Pate and Canny 1999), Harrison Day et al. (2024b) found no evidence that wheat roots refill embolized conduits after rehydration when using the optical vulnerability technique (Brodribb et al. 2017). In the late 1990s, the cryo-scanning electron microscopy (CSEM) technique was used to visualize embolized vessels, but this method is prone to a 'freezing artefact' when applied to tissue with water under tension. The CSEM technique induces artificial embolism formation in segments immersed in liquid nitrogen when the xylem sap is under negative pressure (Cochard et al. 2000; Umebayashi et al. 2016). Due to suboptimal cooling rates and a high amount of soluble gas in xylem sap, the instantaneous freezing of xylem sap generates artificial emboli (Schenk et al. 2016; Wightman 2022). It might therefore not be surprising that since then only a limited number of studies have demonstrated embolism refilling using in vivo imaging techniques. Refilling has foremost been observed in various grapevine varieties, once in roots (Cuneo et al. 2018) and several times in stems (e.g., Brodersen et al. 2010, 2018; Knipfer et al. 2015a; Secchi et al. 2021; Vuerich et al. 2023). When observed, embolism refilling in grapevines was always associated with positive root pressure and has not been observed under negative pressure (Charrier et al. 2016). In addition, various studies have failed to detect embolism refilling across a range of angiosperm (Choat et al. 2015, 2018; Knipfer et al. 2015b, 2018; Arend et al. 2022; Gauthey et al. 2022; Schreel et al. 2022). Embolism refilling could theoretically occur through positive xylem pressure before leaf-out, which is traditionally described as 'root pressure' (Schenk et al. 2021; Charrier et al. 2016). Most likely, the development of positive xylem pressure requires a combination of osmotic exudation and hydraulic pressure (Schenk et al. 2021), but when present it is insufficient to refill embolized vessels after extreme drought in upper canopy branches of mature trees (Arend et al. 2022). Overall, it seems unlikely that embolism refilling under substantial tension after drought is a common phenomenon in woody plants (Cochard et al. 2015; McDowell et al. 2019). Many former studies claiming that emboli are formed and refilled diurnally in angiosperm xylem were either prone to the 'freezing artefact' (Cochard et al. 2000) or the 'cutting artefact', i.e., artificial increases in the degree of embolism when samples were cut under tension (Wheeler et al. 2013; Torres-Ruiz et al. 2015). Altogether, this suggests that (i) embolism refilling under tension is not playing a major role in tree hydraulic recovery (Choat et al. 2016), (ii) embolism formation is rare under typical xylem conditions, and (iii) tree species rely primarily on new xylem formation rather than embolism repair mechanisms (Hammond and Adams 2019; Rehschuh et al. 2020; Arend et al. 2022; Gauthey et al. 2022). However, further belowground studies are needed to generally rule out the possibility of embolism refilling in roots whose xylem may periodically reach near-zero water potentials. Nevertheless, the incorrect assumption of high root vulnerability and frequent root embolism formation during moderate soil drought contributed to another still widely accepted paradigm, namely, that the phytohormone abscisic acid is readily produced in roots and afterwards translocated to the shoots in order to induce stomatal closure (e.g., Kaur et al. 2021; Li et al. 2020; Li and Liu 2021).

## 8 Root-to-Shoot Abscisic Acid Signalling

The suggestion that there might be early onset root embolism during soil drought, caused by a mild decrease in root water potential, has been linked to the induction of stomatal closure via increased levels of the phytohormone abscisic acid (ABA) in roots during drought (Lovisolo et al. 2008; Secchi et al. 2012). In fact, root-origin ABA was once believed to be the primary trigger for stomatal closure and the regulation of transpiration during drought (Zhang and Davies 1990; Davies and Zhang 1991), a paradigm that was highly complementary to the root 'hydraulic fuse hypothesis' (Jackson et al. 2000). ABA triggers stomatal closure by activating anion channels in the guard cell membrane, causing a loss of guard cell turgor pressure and the closing of the stomatal pore (Geiger et al. 2009). An increase in ABA levels during drought, sufficient to close stomata, was first suggested to have a root origin in the seminal work of Zhang et al. (1987). Zhang et al. (1987) investigated a small annual herbaceous plant (*Commelina communis* L., Fig. 8b) and found that ABA

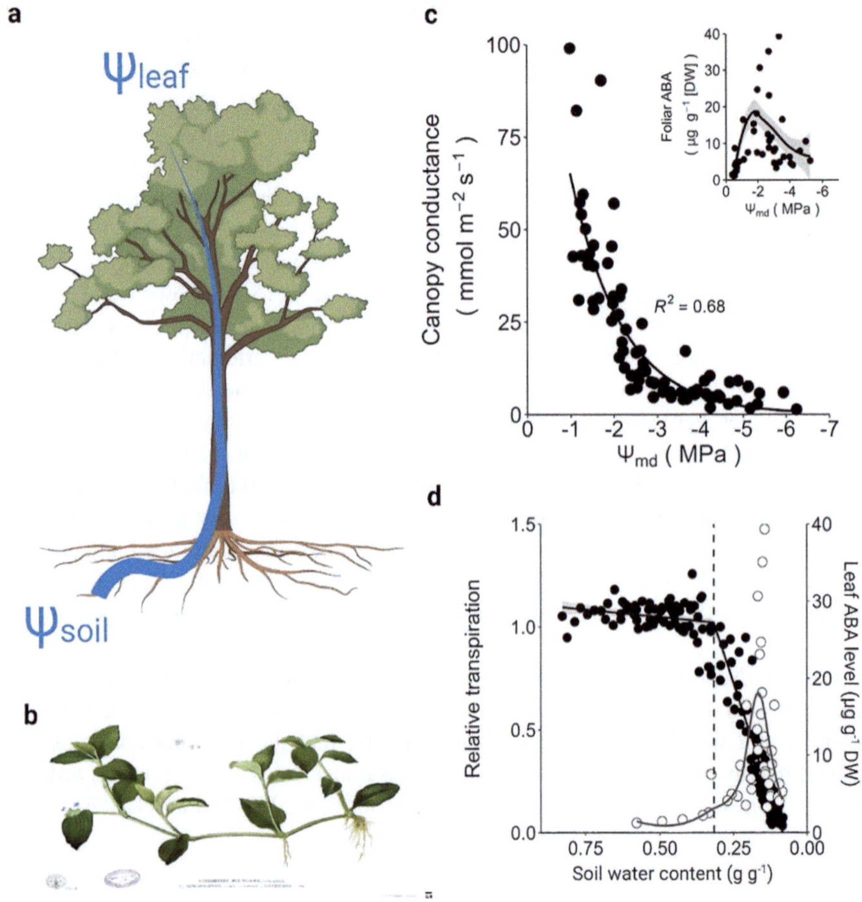


**Fig. 8** In the absence of embolism, there is a continuous hydraulic connection between the water in the soil and the leaf via water potential (Ψ) in both (**a**) trees and (**b**) the diminutive herb *Commelina communis*, the annual species studied by Zhang et al. (1987). (Data source: http://commons.wikimedia.org/wiki, File: commelina_communis_blanco1.13.jpg. This hydraulic connection is depicted in the tree as a blue line.) (**c**) In the tree species *Umbellularia californica* when midday leaf water potential ($\Psi_{md}$) declines during drought, canopy conductance declines and foliar ABA levels increase (depicted in the insert). (**d**) The increase in leaf ABA levels (open symbols) driven by declining Ψ in this species causes transpiration (closed symbols) to decline as soil water content declines. Please note that in this species, ABA levels peak after stomatal closure and decline when leaves reach very low Ψ at which point stomata transition to being passively closed by low leaf water status, for details see McAdam et al. (2024). (Data from panels (**c**) and (**d**) are taken from Manandhar et al. (2024a). Figure created in BioRender (https://BioRender.com/m29p596))

levels increased during dehydration in root tips and that ABA could be readily absorbed by roots and transported in the xylem to the leaves. This discovery was complemented by split-root and partial root zone drying experiments, which have found that ABA levels in leaf xylem sap increase and stomata close when water is withheld from half or parts of the root system (Lovisolo et al. 2002; Sobeih et al. 2004; Li et al. 2011). These discoveries, and the strong relationship between stomatal conductance and leaf xylem sap ABA concentration (Khalil and Grace 1993; Tardieu et al. 1993), led to models for whole plant transpiration based on stomatal regulation by root-sourced ABA (Tardieu and Davies 1993). The concept of a transmitted, long-distance molecule that moves from roots to stomata to trigger stomatal closure in response to declines in soil water status remains persistent in the literature (e.g., Schachtman and Goodger 2008; Kaur et al. 2021; Li et al. 2020; Li and Liu 2021; Liu et al. 2022; Castro-Valdecantos et al. 2023; Batool et al. 2024).

The earliest experiments suggesting root-sourced ABA might not provide the primary driver of stomatal closure during soil drought came from reciprocal grafting studies between ABA biosynthetic mutants and wild-type plants. Holbrook et al. (2002), followed by Christmann et al. (2007), found that the stomata of tomato and *Arabidopsis*, respectively, can close in plants with a wild-type scion grafted to a mutant rootstock, unable to synthesize ABA. These studies also found that ABA deficiency in roots had no effect on the stomata of a scion that can synthesize ABA. This work suggested that if ABA closed stomata during drought, then roots were not the primary source. Christmann et al. (2007) proposed that the signal for triggering stomatal closure when soil moisture declines is hydraulic and relatively instantaneous.

A hydraulic signal triggering leaf ABA biosynthesis close to stomata, when plant water status declines, is supported by evidence that ABA biosynthesis in leaves is triggered by a decrease in cell turgor (Pierce and Raschke 1980, 1981; McAdam and Brodribb 2016; Bacete et al. 2022). The expression of the gene encoding the enzyme responsible for catalysing the rate-limiting step in the ABA biosynthesis pathway, 9-cis-epoxycarotenoid cleavage dehydrogenase (NCED), increases in leaves of plants exposed to low soil moisture or high VPD (Qin and Zeevaart 1999; McAdam et al. 2016; Vo et al. 2025). This upregulation of NCED in leaves can occur within 5 min of a decline in leaf turgor and when stomata close on exposure to high VPD (Sussmilch et al. 2017). A number of studies have found that when leaf water status declines, at high VPD or when soil moisture declines, ABA levels increase, and stomatal closure is initiated (Sobeih et al. 2004; Bauerle et al. 2004; Tombesi et al. 2015; Binstock et al. 2024; Castro-Valdecantos et al. 2023; Manandhar et al. 2024a, b; Fig. 8c). Based on this evidence, roots provide a near-instantaneous hydraulic signal to the leaves via water potential; this occurs if leaf water potential declines close to turgor loss point and ABA biosynthesis is rapidly triggered in mesophyll cells leading to the closure of stomata (McAdam and Brodribb 2018; Cardoso et al. 2020). This model by which ABA closes stomata in seed plants when soil moisture declines is able to overcome one of the more problematic limitations to applying a root-sourced molecular regulation of stomatal aperture in response to declining soil moisture in trees: the time it takes for root-to-stomata transport of molecules. Unlike in diminutive agricultural herbs where drought-induced root-to-shoot signals can influence stomata (Dodd 2005), in trees the transpiration stream moves at a rate at least an order of magnitude slower in the transpiration stream than a signal needs to be transmitted physically from the roots to trigger stomatal closure when the soil dries. For example, in mature broad-leaved trees of 20 m or more in tree height, sap velocity typically peaks at around 15–30 cm $h^{-1}$ at midday (cf., Gimenez et al. 2019) and does not exceed 400 cm $day^{-1}$ (cf., Guillén et al. 2022). This slow transport from root to stomata is further reduced as the contact at the soil-root-interface is lost during soil drying or stomata closure. In contrast, a hydraulic signal driving a leaf origin of stomatal-acting ABA is not dependent on plant height.

There are some key questions that remain unanswered about the role of ABA in roots. Studies suggest that shoot-derived ABA can influence root phenotypes, including promoting root elongation while inhibiting adventitious and secondary root formation and growth (Sharp and LeNoble 2002; McAdam et al. 2016; Huntenburg et al. 2022). Roots are able to synthesize endogenous ABA, particularly when carotenoid precursor levels are high, although once depleted, ABA biosynthesis in roots in response to dehydration does not occur (Manzi et al. 2015). Whether this root-sourced ABA has a localized function influencing root growth, root water uptake or suberization (cf., Lu et al. 2022) remains an open area of investigation.

## 9 Conclusion and Outlook

Various methods applied over recent years, including non-invasive methods, have rejected the paradigm of high root vulnerability to embolism formation. By interpreting vulnerability curves accurately, the trade-off between xylem safety and hydraulic efficiency at the whole-plant level disappears. As root embolism repair appears to be unlikely, suberized roots cannot be considered as 'hydraulic fuses' that are readily embolized during moderate soil drought. Similarly, a chemical signal from the roots to the shoots when they reach critical hydraulic thresholds would take far too long for instantaneous stomatal closure in mature trees. Instead, a hydraulic signal is rapidly transmitted, triggering abscisic acid synthesis in the leaves once they approach the water potential at the turgor loss point. However, we argue that fine roots without secondary growth can still be considered as 'hydraulic fuses', because they lose contact with the soil once the cortex starts to shrink. As long as this shrinkage can be reversed by subsequent tissue swelling upon rewetting, root conductivity should be rapidly restorable after a rain event without additional carbon costs. Although embolism formation in suberized woody roots must be a rare phenomenon, we still lack field data on the extreme climatic and edaphic conditions under which it might occur and to what extent the structure of outside root xylem tissue influences this. Furthermore, one might speculate that the different above-ground stomatal responses may be related to differences in belowground responses across species. As more water-conservative species show earlier stomatal closure than less water-conservative ones, the fine roots of the former may lose contact with the soil earlier as soil drying progresses. Indeed, for two annual crop species, Yang et al. (2023) showed that rice, considered as drought-sensitive, has a higher proportion of root cortical tissue than wheat, which is considered to be more drought-resistant. Further studies should therefore test whether common fine root traits, such as mean root diameter or root tissue density, might be related to the root cortex fraction and thus to fine root shrinkage behaviour.

**Acknowledgements** We thank Sebastian Fuchs for help with root sampling in the Hainich National Park, Régis Burlett for help with Cavi1000 measurements, and Alexandra Weigelt and an anonymous reviewer for helpful and constructive comments that helped to improve the manuscript.

## References

Affortit P, Ahmed MA, Grondin A, Delzon S, Carminati A, Laplaze L (2024) Keep in touch: the soil–root hydraulic continuum and its role in drought resistance in crops (L Dupuy, Ed.). J Exp Bot 75:584–593

Alder NN, Sperry JS, Pockman WT (1996) Root and stem xylem embolism, stomatal conductance, and leaf turgor in Acer grandidentatum populations along a soil moisture gradient. Oecologia 105:293–301

Alder NN, Pockman WT, Sperry JS, Nuismer S (1997) Use of centrifugal force in the study of xylem cavitation. J Exp Bot 48:665–674

Aloni R (1987) Differentiation of vascular tissues. Annu Rev Plant Physiol Plant Mol Biol 38:179–204

Ambrose AR, Baxter WL, Wong CS, Burgess SSO, Williams CB, Næsborg RR, Koch GW, Dawson TE (2016) Hydraulic constraints modify optimal photosynthetic profiles in giant sequoia trees. Oecologia 182:713–730

Anfodillo T, Petit G, Crivellaro A (2013) Axial conduit widening in woody species: a still neglected anatomical pattern. IAWA J 34:352–364

Angeles G (2004) The cohesion-tension theory. New Phytol 163:451–452

Arend M, Link RM, Zahnd C, Hoch G, Schuldt B, Kahmen A (2022) Lack of hydraulic recovery as a cause of post-drought foliage reduction and canopy decline in European beech. New Phytol 234:1195–1205

Askenasy E (1895) Über das Saftsteigen. Verh Naturw Med Ver Heidelberg NF 5:325–345

Avila RT, Kane CN, Batz TA, Trabi C, Damatta FM, Jansen S, McAdam SAM (2023) The relative area of vessels in xylem correlates with stem embolism resistance within and between genera (J Martinez-Vilalta, Ed.). Tree Physiol 43:75–87

Baas P (1982) Systematic, phylogenetic and ecological wood anatomy. History and perspectives. In: Baas P (ed) New perspectives in wood anatomy. Nijhoff/Junk, The Hague, pp 23–58

Bacete L, Schulz J, Engelsdorf T, Bartosova Z, Vaahtera L, Yan G, Gerhold JM, Tichá T, Øvstebø C, Gigli-Bisceglia N et al (2022) THESEUS1 modulates cell wall stiffness and abscisic acid production in Arabidopsis thaliana. Proc Natl Acad Sci 119:1–38

Barrientos-Sanhueza C, Hormazabal-Pavat V, Cuneo IF (2023) Extreme drought enhances Opuntia ficus-indica fine root cells elasticity preventing permanent damage. Theor Exp Plant Physiol 35: 233–246

Batool A, Li S-S, Yue D-X, Ullah F, Zhao L, Cheng Z-G, Wang C, Duan H-X, Lv G-C, Haq Z et al (2024) Root-to-shoot signaling positively mediates source-sink relation in late growth stages in diploid and tetraploid wheat. BMC Plant Biol 24:492

Bauerle WL, Whitlow TH, Setter TL, Vermeylen FM (2004) Abscisic acid synthesis in Acer rubrum L. leaves—a vapor-pressure-deficit-mediated response. J Am Soc Hortic Sci 129:182–187

Beikircher B, Ameglio T, Cochard H, Mayr S (2010) Limitation of the Cavitron technique by conifer pit aspiration. J Exp Bot 61:3385–3393

Binstock BR, Manandhar A, McAdam SAM (2024) Characterizing the breakpoint of stomatal response to vapor pressure deficit in an angiosperm. Plant Physiol 194:732–740

Bouche PS, Jansen S, Cochard H, Burlett R, Capdeville G, Delzon S (2015) Embolism resistance of conifer roots can be accurately measured with the flow-centrifuge method. J Plant Hydraul 2: e002

Brodersen CR, McElrone AJ, Choat B, Matthews MA, Shackel KA (2010) The dynamics of embolism repair in xylem: in vivo visualizations using high-resolution computed tomography. Plant Physiol 154:1088–1095

Brodersen CR, Knipfer T, McElrone AJ (2018) In vivo visualization of the final stages of xylem vessel refilling in grapevine (*Vitis vinifera*) stems. New Phytol 217:117–126

Brodribb TJ (2009) Xylem hydraulic physiology: the functional backbone of terrestrial plant productivity. Plant Sci 177:245–251

Brodribb TJ, Carriqui M, Delzon S, Lucani C (2017) Optical measurement of stem xylem vulnerability. Plant Physiol 174:2054–2061

Buchard C, McCully M, Canny M (1999) Daily embolism and refilling of root xylem vessels in three dicotyledonous crop plants. Agronomy 2:97–106

Burgess SSO, Pittermann J, Dawson TE (2006) Hydraulic efficiency and safety of branch xylem increases with height in Sequoia sempervirens (D. Don) crowns. Plant Cell Environ 29:229–239

Canny MJ (2001) Contributions to the debate on water transport. Am J Bot 88:43–46

Cardoso AA, Gori A, Da-Silva CJ, Brunetti C (2020) Abscisic acid biosynthesis and signaling in plants: key targets to improve water use efficiency and drought tolerance. Appl Sci 10:6322

Carlquist S (1975) Ecological strategies of xylem evolution. University of California Press, Berkely

Castro-Valdecantos P, Puértolas J, Dodd IC (2023) Similar soil drying-induced stomatal closure in soybean genotypes varying in abscisic acid accumulation and stomatal sensitivity to abscisic acid. Funct Plant Biol 51:FP23012

Charrier G, Torres-Ruiz JM, Badel E, Burlett R, Choat B, Cochard H, Delmas CEL, Domec J-C, Jansen S, King A et al (2016) Evidence for hydraulic vulnerability segmentation and lack of xylem refilling under tension. Plant Physiol 172:1657–1668

Choat B, Ball MC, Luly JG, Holtum JAM (2005) Hydraulic architecture of deciduous and evergreen dry rainforest tree species from north-eastern Australia. Trees-Struct Funct 19:305–311

Choat B, Drayton WM, Brodersen CR, Matthews MA, Shackel KA, Wada H, McElrone AJ (2010) Measurement of vulnerability to water stress-induced cavitation in grapevine: a comparison of four techniques applied to a long-vesseled species. Plant Cell Environ 33:1502–1512

Choat B, Brodersen CR, McElrone AJ (2015) Synchrotron X-ray microtomography of xylem embolism in Sequoia sempervirens saplings during cycles of drought and recovery. New Phytol 205:1095–1105

Choat B, Badel E, Burlett R, Delzon S, Cochard H, Jansen S (2016) Noninvasive measurement of vulnerability to drought-induced embolism by X-ray microtomography. Plant Physiol 170:273–282

Choat B, Brodribb TJ, Brodersen CR, Duursma RA, López R, Medlyn BE (2018) Triggers of tree mortality under drought. Nature 558:531–539

Christmann A, Weiler EW, Steudle E, Grill E (2007) A hydraulic signal in root-to-shoot signalling of water shortage. Plant J 52:167–174

Cochard H (2002) A technique for measuring xylem hydraulic conductance under high negative pressures. Plant Cell Environ 25:815–819
Cochard H, Cruiziat P, Tyree MT (1992) Use of positive pressures to establish vulnerability curves – further support for the air-seeding hypothesis and implications for pressure-volume analysis. Plant Physiol 100:205–209
Cochard H, Bodet C, Ameglio T, Cruiziat P (2000) Cryo-scanning electron microscopy observations of vessel content during transpiration in walnut petioles. Facts or artifacts? Plant Physiol 124:1191–1202
Cochard H, Coll L, Le Roux X, Améglio T (2002) Unraveling the effects of plant hydraulics on stomatal closure during water stress in walnut. Plant Physiol 128(1):282–290. https://doi.org/10.1104/pp.010400
Cochard H, Herbette S, Barigah T, Badel E, Ennajeh M, Vilagrosa A, Mencuccini M (2010) Xylem hydraulic failures and recovery in Quercus pubescens and Quercus ilex. New Phytol 186 (2):285–297. https://doi.org/10.1111/j.1365-3040.2010.02163.x
Cochard H, Badel E, Herbette S, Delzon S, Choat B, Jansen S (2013) Methods for measuring plant vulnerability to cavitation: a critical review. J Exp Bot 64:4779–4791
Cochard H, Delzon S, Badel E (2015) X-ray microtomography (micro-CT): a reference technology for high-resolution quantification of xylem embolism in trees. Plant Cell Environ 38:201–206
Cole PJ, Alston AM (1974) Effect of transient dehydration on absorption of chloride by wheat roots. Plant Soil 40:243–247
Creek D, Blackman CJ, Brodribb TJ, Choat B, Tissue DT (2018) Coordination between leaf, stem, and root hydraulics and gas exchange in three arid-zone angiosperms during severe drought and recovery. Plant Cell Environ 41:2869–2881
Cuneo IF, Knipfer T, Brodersen CR, McElrone AJ (2016) Mechanical failure of fine root cortical cells initiates plant hydraulic decline during drought. Plant Physiol 172:1669–1678
Cuneo IF, Knipfer T, Mandal P, Brodersen CR, McElrone AJ (2018) Water uptake can occur through woody portions of roots and facilitates localized embolism repair in grapevine. New Phytol 218:506–516
Cuneo IF, Barrios-Masias F, Knipfer T, Uretsky J, Reyes C, Lenain P, Brodersen CR, Walker MA, McElrone AJ (2021) Differences in grapevine rootstock sensitivity and recovery from drought are linked to fine root cortical lacunae and root tip function. New Phytol 229:272–283
Davies WJ, Zhang J (1991) Root signals and the regulation of growth and development of plants in drying soil. Annu Rev Plant Physiol Plant Mol Biol 42:55–76
Davis CC, Endress PK, Baum DA (2002) Floral morphogenesis in Buxus and *Sarcococca* (Buxaceae): implications for understanding the evolution of perianth number in flowers. Am J Bot 89(5):820–830. https://doi.org/10.3732/ajb.89.5.820
De Guzman ME, Santiago LS, Schnitzer SA, Álvarez-Cansino L (2017) Trade-offs between water transport capacity and drought resistance in neotropical canopy liana and tree species (N Phillips, Ed.). Tree Physiol 37:1404–1414
Denne M (1972) A comparison of root-and shoot-wood development in conifer seedlings. Ann Bot 36:579–587
Dixon HH, Joly J (1895) On the ascent of sap. Philos Trans R Soc B 186:563–576
Dodd IC (2005) Root-to-shoot signalling: assessing the roles of 'up' in the up and down world of long-distance signalling in planta. Plant Soil 274:251–270
Domec JC, Scholz FG, Bucci SJ, Meinzer FC, Goldstein G, Villalobos-Vega R (2006) Diurnal and seasonal variation in root xylem embolism in neotropical savanna woody species: impact on stomatal control of plant water status. Plant Cell Environ 29:26–35
Domec JC, Schafer K, Oren R, Kim HS, McCarthy HR (2010) Variable conductivity and embolism in roots and branches of four contrasting tree species and their impacts on whole-plant hydraulic performance under future atmospheric CO2 concentration. Tree Physiol 30:1001–1015
Duddek P, Carminati A, Koebernick N, Ohmann L, Lovric G, Delzon S, Rodriguez-Dominguez CM, King A, Ahmed MA (2022) The impact of drought-induced root and root hair shrinkage on root–soil contact. Plant Physiol 189:1232–1236

Ennajeh M, Simões F, Khemira H, Cochard H (2011) How reliable is the double-ended pressure sleeve technique for assessing xylem vulnerability to cavitation in woody angiosperms? Physiol Plant 142:205–210

Fegel AC (1941) Comparative anatomy and varying physical properties of trunk, branch and root wood in certain northeastern trees. Bull. N.Y. State College

Fickle JC, Pratt RB, Jacobsen AL (2023) Xylem structure and hydraulic function in roots and stems of chaparral shrub species from high and low elevation in the Sierra Nevada, California. Physiol Plant 175:e13970

Gasson P (1985) Automatic-measurement of vessel lumen area and diameter with particular reference to pedunculate oak and common beech. IAWA Bull 6:219–237

Gauthey A, Backes D, Balland J, Alam I, Maher DT, Cernusak LA, Duke NC, Medlyn BE, Tissue DT, Choat B (2022) The role of hydraulic failure in a massive mangrove die-off event. Front Plant Sci 13:1–11

Geiger D, Scherzer S, Mumm P, Stange A, Marten I, Bauer H, Ache P, Matschi S, Liese A, Al-Rasheid KAS et al (2009) Activity of guard cell anion channel SLAC1 is controlled by drought-stress signaling kinase-phosphatase pair. Proc Natl Acad Sci 106:21425–21430

Gimenez BO, Jardine KJ, Higuchi N, Negrón-Juárez RI, Sampaio-Filho IJ, Cobello LO, Fontes CG, Dawson TE, Varadharajan C, Christianson DS et al (2019) Species-specific shifts in diurnal sap velocity dynamics and hysteretic behavior of ecophysiological variables during the 2015–2016 El Niño event in the Amazon forest. Front Plant Sci 10:1–16

Gonzalez-Benecke CA, Martin TA, Peter GF (2010) Hydraulic architecture and tracheid allometry in mature Pinus palustris and Pinus elliottii trees. Tree Physiol 30:361–375

Guillén LA, Brzostek E, McNeil B, Raczka N, Casey B, Zegre N (2022) Sap flow velocities of Acer saccharum and Quercus velutina during drought: insights and implications from a throughfall exclusion experiment in West Virginia, USA. Sci Total Environ 850:158029

Hacke U, Sauter JJ (1996) Drought-induced xylem dysfunction in petioles, branches, and roots of Populus balsamifera L and Alnus glutinosa (L) Gaertn. Plant Physiol 111:413–417

Hacke UG, Sperry JS (2001) Functional and ecological xylem anatomy. Perspect Plant Ecol Evol Syst 4:97–115

Hacke U, Sperry J, Pittermann J (2000) Drought experience and cavitation resistance in six shrubs from the Great Basin, Utah. Basic Appl Ecol 41:31–41

Hajek P, Leuschner C, Hertel D, Delzon S, Schuldt B (2014) Trade-offs between xylem hydraulic properties, wood anatomy and yield in Populus. Tree Physiol 34:744–756

Hajek P, Kurjak D, von Wühlisch G, Delzon S, Schuldt B (2016) Intraspecific variation in wood anatomical, hydraulic, and foliar traits in ten European beech provenances differing in growth yield. Front Plant Sci 7:791

Hammond WM, Adams HD (2019) Dying on time: traits influencing the dynamics of tree mortality risk from drought (S Sevanto, Ed.). Tree Physiol 39:906–909

Harrison Day BL, Johnson KM, Tonet V, Bourbia I, Blackman CJ, Brodribb TJ (2024a) A one-way ticket: wheat roots do not functionally refill xylem emboli following rehydration. Plant Physiol 196:2362–2373

Harrison Day BL, Brodersen CR, Brodribb TJ (2024b) Weak link or strong foundation? Vulnerability of fine root networks and stems to xylem embolism. New Phytol 244:1288–1302

Holbrook NM, Shashidhar VR, James RA, Munns R (2002) Stomatal control in tomato with ABA-deficient roots: response of grafted plants to soil drying. J Exp Bot 53:1503–1514

Huck MG, Klepper B, Taylor HM (1970) Diurnal variations in root diameter. Plant Physiol 45:529–530

Hukin D, Cochard H, Dreyer E, Thiec DL, Bogeat-Triboulot MB (2005) Cavitation vulnerability in roots and shoots: does Populus euphratica Oliv., a poplar from arid areas of Central Asia, differ from other poplar species? J Exp Bot 56:2003–2010

Hultine KR, Koepke DF, Pockman WT, Fravolini A, Sperry JS, Williams DG (2006) Influence of soil texture on hydraulic properties and water relations of a dominant warm-desert phreatophyte. Tree Physiol 26:313–323

Huntenburg K, Puértolas J, de Ollas C, Dodd IC (2022) Bi-directional, long-distance hormonal signalling between roots and shoots of soil water availability. Physiol Plant 174:1–9

Isasa E, Link RM, Jansen S, Tezeh FR, Kaack L, Sarmento Cabral J, Schuldt B (2023) Addressing controversies in the xylem embolism resistance–vessel diameter relationship. New Phytol 238: 283–296

Ishii HT, Jennings GM, Sillett SC, Koch GW (2008) Hydrostatic constraints on morphological exploitation of light in tall Sequoia sempervirens trees. Oecologia 156:751–763

Jackson RB, Sperry JS, Dawson TE (2000) Root water uptake and transport: using physiological processes in global predictions. Trends Plant Sci 5:482–488

Jansen S, Schenk HJ (2015) On the ascent of sap in the presence of bubbles. Am J Bot 102:1561–1563

Jansen S, Schuldt B, Choat B (2015) Current controversies and challenges in applying plant hydraulic techniques. New Phytol 205:961–964

Johnson DM, Wortemann R, McCulloh KA, Jordan-Meille L, Ward E, Warren JM, Palmroth S, Domec J-C (2016) A test of the hydraulic vulnerability segmentation hypothesis in angiosperm and conifer tree species (N Phillips, Ed.). Tree Physiol 36:983–993

Johnson DM, Domec J, Carter Berry Z, Schwantes AM, McCulloh KA, Woodruff DR, Wayne Polley H, Wortemann R, Swenson JJ, Scott Mackay D et al (2018) Co-occurring woody species have diverse hydraulic strategies and mortality rates during an extreme drought. Plant Cell Environ 41:576–588

Johnson KM, Everbach SR, Holbrook NM, Olson ME (2023) Evaluating Carlquist's Law from a physiological perspective. IAWA J 44(3–4):429–438. https://doi.org/10.1163/22941932-bja10134

Kaack L, Weber M, Isasa E, Karimi Z, Li S, Pereira L, Trabi CL, Zhang Y, Schenk HJ, Schuldt B et al (2021) Pore constrictions in intervessel pit membranes provide a mechanistic explanation for xylem embolism resistance in angiosperms. New Phytol 230:1829–1843

Kaur H, Kohli SK, Khanna K, Bhardwaj R (2021) Scrutinizing the impact of water deficit in plants: transcriptional regulation, signaling, photosynthetic efficacy, and management. Physiol Plant 172:935–962

Kavanagh KL, Bond BJ, Aitken SN, Gartner BL, Knowe S (1999) Shoot and root vulnerability to xylem cavitation in four populations of Douglas-fir seedlings. Tree Physiol 19:31–37

Khalil AAM, Grace J (1993) Does xylem sap ABA control the stomatal behaviour of water-stressed sycamore (Acer pseudoplatanus L.) seedlings? J Exp Bot 44:1127–1134

Klein T, Zeppel MJB, Anderegg WRL, Bloemen J, De Kauwe MG, Hudson P, Ruehr NK, Powell TL, von Arx G, Nardini A (2018) Xylem embolism refilling and resilience against drought-induced mortality in woody plants: processes and trade-offs. Ecol Res 33:839–855

Knipfer T, Fricke W, Shatil-Cohen A, Steele MR (2015a) Part of the xylem path in a grapevine is at least 10 times more resistant to embolism than a corresponding part in a congeneric, less drought-tolerant species. Plant Cell Environ 38(5):839–850. https://doi.org/10.1111/pce.12497

Knipfer T, Eustis A, Brodersen CR, Walker AM, McElrone AJ (2015b) Grapevine species from varied native habitats exhibit differences in embolism resistance. Tree Physiol 35(10):1083–1094. https://doi.org/10.1093/treephys/tpv040

Knipfer T, Barrios-Masias FH, Cuneo IF, Bouda M, Albuquerque CP, Brodersen CR, Kluepfel DA, McElrone AJ (2018) Variations in xylem embolism susceptibility under drought between intact saplings of three walnut species (J Martinez-Vilalta, Ed.). Tree Physiol 38:1180–1192

Koch GW, Sillett SC, Jennings GM, Davis SD (2004) The limits to tree height. Nature 428:851–854

Koebernick N, Schlüter S, Blaser SRGA, Vetterlein D (2018) Root-soil contact dynamics of Vicia faba in sand. Plant Soil 431:417–431

Körner C (2019) No need for pipes when the well is dry—a comment on hydraulic failure in trees. Tree Physiol 39:695–700

Kotowska MM, Hertel D, Abou Rajab Y, Barus H, Schuldt B (2015) Patterns in hydraulic architecture from roots to branches in six tropical tree species from cacao agroforestry and their relation to wood density and stem growth. Front Plant Sci 6:1–16

Kotowska AE, Treichel AJ, Kaufman PL (2019) The effects of age on the morphology of the optic nerve head in rhesus macaques. Graefes Arch Clin Exp Ophthalmol 257(5):931–943. https://doi.org/10.1007/s00468-019-01897-4

Lens F, Gleason SM, Bortolami G, Brodersen C, Delzon S, Jansen S (2022) Functional xylem characteristics associated with drought-induced embolism in angiosperms. New Phytol 236: 2019–2036

Lens F, Gleason SM, Bortolami G, Brodersen C, Delzon S, Jansen S (2023) Comparative anatomy vs mechanistic understanding: how to interpret the diameter-vulnerability link. IAWA J 44:368–380

Li S, Liu F (2021) Exogenous abscisic acid priming modulates water relation responses of two tomato genotypes with contrasting endogenous abscisic acid levels to progressive soil drying under elevated CO2. Front Plant Sci 12:1–16

Li B, Feng Z, Xie M, Sun M, Zhao Y, Liang L, Liu G, Zhang J, Jia W (2011) Modulation of the root-sourced ABA signal along its way to the shoot in Vitis ripariax Vitis labrusca under water deficit. J Exp Bot 62:1731–1741

Li S, Lens F, Espino S, Karimi Z, Klepsch M, Schenk HJ, Schmitt M, Schuldt B, Jansen S (2016) Intervessel pit membrane thickness as a key determinant of embolism resistance in angiosperm xylem. IAWA J 37:152–171

Li S, Li X, Wei Z, Liu F (2020) ABA-mediated modulation of elevated CO2 on stomatal response to drought. Curr Opin Plant Biol 56:174–180

Li Z, Wang S, Wang W, Gu J, Wang Y (2022) The hierarchy of protoxylem groupings in primary root and their plasticity to nitrogen addition in three tree species. Front Plant Sci 13:903318

Lintunen A, Kalliokoski T (2010) The effect of tree architecture on conduit diameter and frequency from small distal roots to branch tips in Betula pendula, Picea abies and Pinus sylvestris. Tree Physiol 30:1433–1447

Liu H, Song S, Zhang H, Li Y, Niu L, Zhang J, Wang W (2022) Signaling transduction of ABA, ROS, and Ca2+ in plant stomatal closure in response to drought. Int J Mol Sci 23:14824

Losso A, Bär A, Dämon B, Dullin C, Ganthaler A, Petruzzellis F, Savi T, Tromba G, Nardini A, Mayr S et al (2019) Insights from in vivo micro-CT analysis: testing the hydraulic vulnerability segmentation in Acer pseudoplatanus and Fagus sylvatica seedlings. New Phytol 221:1831–1842

Lovisolo C, Schubert A, Sorce C (2002) Are xylem radial development and hydraulic conductivity in downwardly-growing grapevine shoots influenced by perturbed auxin metabolism? New Phytol 156:65–74

Lovisolo C, Perrone I, Hartung W, Schubert A (2008) An abscisic acid-related reduced transpiration promotes gradual embolism repair when grapevines are rehydrated after drought. New Phytol 180:642–651

Lu H, Gao Q, Han J, Guo X, Wang Q, Altosaar I, Barberon M, Liu J, Gatehouse AMR, Shu Q-Y (2022) An ABA-serotonin module regulates root suberization and salinity tolerance. New Phytol 236:958–973

Lübbe T, Lamarque LJ, Delzon S, Torres Ruiz JM, Burlett R, Leuschner C, Schuldt B (2022) High variation in hydraulic efficiency but not xylem safety between roots and branches in four temperate broad-leaved tree species. Funct Ecol 36:699–712

Maherali H, Moura CF, Caldeira MC, Willson CJ, Jackson RB (2006) Functional coordination between leaf gas exchange and vulnerability to xylem cavitation in temperate forest trees. Plant Cell Environ 29:571–583

Manandhar A, Pichaco J, McAdam SAM (2024a) Abscisic acid increase correlates with the soil water threshold of transpiration decline during drought. Plant Cell Environ 47:5067–5075

Manandhar A, Rimer IM, Soares Pereira T, Pichaco J, Rockwell FE, McAdam SAM (2024b) Dynamic soil hydraulic resistance regulates stomata. New Phytol 244:147–158

Manzi M, Lado J, Rodrigo MJ, Zacarías L, Arbona V, Gómez-Cadenas A (2015) Root ABA accumulation in long-term water-stressed plants is sustained by hormone transport from aerial organs. Plant Cell Physiol 56:2457–2466

Martinez-Vilalta J, Prat E, Oliveras I, Pinol J (2002) Xylem hydraulic properties of roots and stems of nine Mediterranean woody species. Oecologia 133:19–29

Martin-StPaul NK, Longepierre D, Huc R, Delzon S, Burlett R, Joffre R, Rambal S, Cochard H (2014) How reliable are methods to assess xylem vulnerability to cavitation? The issue of 'open vessel' artifact in oaks. Tree Physiol 34:894–905

Matzner SL, Rice KJ, Richards JH (2001) Intra-specific variation in xylem cavitation in interior live oak (Quercus wislizenii A. DC.). J Exp Bot 52:783–789

McAdam SA, Brodribb TJ (2016) Linking turgor with ABA biosynthesis: implications for stomatal responses to vapor pressure deficit across land plants. Plant Physiol 171:2008–2016

McAdam SAM, Brodribb TJ (2018) Mesophyll cells are the main site of abscisic acid biosynthesis in water-stressed leaves. Plant Physiol 177:911–917

McAdam SAM, Sussmilch FC, Brodribb TJ (2016) Stomatal responses to vapour pressure deficit are regulated by high speed gene expression in angiosperms. Plant Cell Environ 39:485–491

McAdam SAM, Manandhar A, Kane CN, Mercado-Reyes JA (2024) Passive stomatal closure under extreme drought in an angiosperm species. J Exp Bot 75:6850–6855

McCully ME, Huang CX, Ling LEC (1998) Daily embolism and refilling of xylem vessels in the roots of field-grown maize. New Phytol 138:327–342

McDowell NG, Brodribb TJ, Nardini A (2019) Hydraulics in the 21st century. New Phytol 224: 537–542

McElrone AJ, Pockman WT, Martinez-Vilalta J, Jackson RB (2004) Variation in xylem structure and function in stems and roots of trees to 20 m depth. New Phytol 163:507–517

Meinzer FC, Clearwater MJ, Goldstein G (2001) Water transport in trees: current perspectives, new insights and some controversies. Environ Exp Bot 45:239–262

Mencuccini M, Comstock J (1997) Vulnerability to cavitation in populations of two desert species, Hymenoclea salsola and Ambrosia dumosa, from different climatic regions. J Exp Bot 48:1323–1334

Mencuccini M, Hölttä T, Petit G, Magnani F (2007) Sanio's laws revisited. Size-dependent changes in the xylem architecture of trees. Ecol Lett 10:1084–1093

North GB, Nobel PS (1997) Root-soil contact for the desert succulent *Agave deserti* in wet and drying soil. New Phytol 135:21–29

Pate JS, Canny MJ (1999) Quantification of vessel embolisms by direct observation: a comparison of two methods. New Phytol 141:33–43

Peters JMR, Gauthey A, Lopez R, Carins-Murphy MR, Brodribb TJ, Choat B (2020) Non-invasive imaging reveals convergence in root and stem vulnerability to cavitation across five tree species (H Griffiths, Ed.). J Exp Bot 71:6623–6637

Pierce M, Raschke K (1980) Correlation between loss of turgor and accumulation of abscisic acid in detached leaves. Planta 148:174–182

Pierce M, Raschke K (1981) Synthesis and metabolism of abscisic acid in detached leaves of Phaseolus vulgaris L. after loss and recovery of turgor. Planta 153:156–165

Pratt RB, Jacobsen AL, Ewers FW, Davis SD (2007) Relationships among xylem transport, biomechanics and storage in stems and roots of nine Rhamnaceae species of the California chaparral. New Phytol 174:787–798

Pratt RB, MacKinnon ED, Venturas MD, Crous CJ, Jacobsen AL (2015) Root resistance to cavitation is accurately measured using a centrifuge technique. Tree Physiol 35:185–196

Qin X, Zeevaart JAD (1999) The 9-cis-epoxycarotenoid cleavage reaction is the key regulatory step of abscisic acid biosynthesis in water-stressed bean. Proc Natl Acad Sci 96:15354–15361

Rehschuh R, Cecilia A, Zuber M, Faragó T, Baumbach T, Hartmann H, Jansen S, Mayr S, Ruehr N (2020) Drought-induced xylem embolism limits the recovery of leaf gas exchange in Scots pine. Plant Physiol 184:852–864

Rodriguez-Dominguez CM, Brodribb TJ (2020) Declining root water transport drives stomatal closure in olive under moderate water stress. New Phytol 225:126–134

Rodriguez-Dominguez CM, Carins Murphy MR, Lucani C, Brodribb TJ (2018) Mapping xylem failure in disparate organs of whole plants reveals extreme resistance in olive roots. New Phytol 218:1025–1035

Sack L, Ball MC, Brodersen C, Davis SD, Des Marais DL, Donovan LA, Givnish TJ, Hacke UG, Huxman T, Jansen S et al (2016) Plant hydraulics as a central hub integrating plant and ecosystem function: meeting report for 'Emerging Frontiers in Plant Hydraulics' (Washington, DC, May 2015). Plant Cell Environ 39:2085–2094

Sanchez-Martinez P, Mencuccini M, García-Valdés R, Hammond WM, Serra-Diaz JM, Guo W-Y, Segovia RA, Dexter KG, Svenning J-C, Allen C et al (2023) Increased hydraulic risk in assemblages of woody plant species predicts spatial patterns of drought-induced mortality. Nat Ecol Evol 7:1620–1632

Sanio K (1872) Über die Größe der Holzzellen bei der gemeinen der Kiefer (Pinus sylvestris). J Wissen Bot 8:401–420

Schachtman DP, Goodger JQD (2008) Chemical root to shoot signaling under drought. Trends Plant Sci 13:281–287

Schenk HJ, Steppe K, Jansen S (2015) Nanobubbles: a new paradigm for air-seeding in xylem. Trends Plant Sci 20:199–205

Schenk HJ, Espino S, Visser A, Esser BK (2016) Dissolved atmospheric gas in xylem sap measured with membrane inlet mass spectrometry. Plant Cell Environ 39:944–950

Schenk HJ, Jansen S, Hölttä T (2021) Positive pressure in xylem and its role in hydraulic function. New Phytol 230:27–45

Scholander PF, Bradstreet ED, Hemmingsen EA, Hammel HT (1965) Sap pressure in vascular plants: negative hydrostatic pressure can be measured in plants. Science 148:339–346

Schreel JDM, Brodersen C, De Schryver T, Dierick M, Rubinstein A, Dewettinck K, Boone MN, Van Hoorebeke L, Steppe K (2022) Foliar water uptake does not contribute to embolism repair in beech (*Fagus sylvatica* L.). Ann Bot 129:555–566

Schuldt B, Leuschner C, Brock N, Horna V (2013) Changes in wood density, wood anatomy and hydraulic properties of the xylem along the root-to-shoot flow path in tropical rainforest trees. Tree Physiol 33:161–174

Secchi F, Perrone I, Chitarra W, Zwieniecka A, Lovisolo C, Zwieniecki M (2012) The dynamics of embolism refilling in abscisic acid (ABA)-deficient tomato plants. Int J Mol Sci 14:359–377

Secchi F, Pagliarani C, Cavalletto S, Petruzzellis F, Tonel G, Savi T, Tromba G, Obertino MM, Lovisolo C, Nardini A et al (2021) Chemical inhibition of xylem cellular activity impedes the removal of drought-induced embolisms in poplar stems – new insights from micro-CT analysis. New Phytol 229:820–830

Sharp RE, LeNoble ME (2002) ABA, ethylene and the control of shoot and root growth under water stress. J Exp Bot 53:33–37

Skelton RP, Brodribb TJ, Choat B (2017) Casting light on xylem vulnerability in an herbaceous species reveals a lack of segmentation. New Phytol 214:561–569

Sobeih WY, Dodd IC, Bacon MA, Grierson D, Davies W (2004) Long-distance signals regulating stomatal conductance and leaf growth in tomato (Lycopersicon esculentum) plants subjected to partial root-zone drying. J Exp Bot 55:2353–2363

Sperry JS, Hacke UG (2002) Desert shrub water relations with respect to soil characteristics and plant functional type. Funct Ecol 16:367–378

Sperry JS, Saliendra NZ (1994) Intra-plant and inter-plant variation in xylem cavitation in Betula occidentalis. Plant Cell Environ 17:1233–1241

Sperry JS, Hacke UG, Oren R, Comstock JP (2002) Water deficits and hydraulic limits to leaf water supply. Plant Cell Environ 25:251–263

Sperry JS, Hacke UG, Pittermann J (2006) Size and function in conifer tracheids and angiosperm vessels. Am J Bot 93:1490–1500

Sussmilch FC, Brodribb TJ, McAdam SAM (2017) Up-regulation of NCED3 and ABA biosynthesis occur within minutes of a decrease in leaf turgor but AHK1 is not required. J Exp Bot 68: 2913–2918

Sutton RF, Tinus RW (1983) Root and root system terminology. For Sci 29:a0001–z0001

Tardieu F, Davies WJ (1993) Integration of hydraulic and chemical signalling in the control of stomatal conductance and water status of droughted plants. Plant Cell Environ 16:341–349

Tardieu F, Zhang J, Gowing DJG (1993) Stomatal control by both [ABA] in the xylem sap and leaf water status: a test of a model for draughted or ABA-fed field-grown maize. Plant Cell Environ 16:413–420

Tibbetts TJ, Ewers FW (2000) Root pressure and specific conductivity in temperate lianas: exotic *Celastrus orbiculatus* (Celastraceae) vs. native *Vitis riparia* (Vitaceae). Am J Bot 87:1272–1278

Tombesi S, Nardini A, Frioni T, Soccolini M, Zadra C, Farinelli D, Poni S, Palliotti A (2015) Stomatal closure is induced by hydraulic signals and maintained by ABA in drought-stressed grapevine. Sci Rep 5:12449

Torres-Ruiz JM, Cochard H, Mayr S, Beikircher B, Diaz-Espejo A, Rodriguez-Dominguez CM, Badel E, Fernández JE (2014) Vulnerability to cavitation in Olea europaea current-year shoots: further evidence of an open-vessel artifact associated with centrifuge and air-injection techniques. Physiol Plant 152:465–474

Torres-Ruiz JM, Jansen S, Choat B, McElrone AJ, Cochard H, Brodribb TJ, Badel E, Burlett R, Bouche PS, Brodersen CR et al (2015) Direct X-ray microtomography observation confirms the induction of embolism upon xylem cutting under tension. Plant Physiol 167:40–43

Torres-Ruiz JM, Cochard H, Choat B, Jansen S, López R, Tomášková I, Padilla-Díaz CM, Badel E, Burlett R, King A et al (2017) Xylem resistance to embolism: presenting a simple diagnostic test for the open vessel artefact. New Phytol 215:489–499

Torres-Ruiz JM, Cochard H, Delzon S, Boivin T, Burlett R, Cailleret M, Corso D, Delmas CEL, De Caceres M, Diaz-Espejo A et al (2024) Plant hydraulics at the heart of plant, crops and ecosystem functions in the face of climate change. New Phytol 241:984–999

Tsuda M, Tyree MT (1997) Whole-plant hydraulic resistance and vulnerability segmentation in Acer saccharinum. Tree Physiol 17:351–357

Tyree MT, Ewers FW (1991) The hydraulic architecture of trees and other woody plants. New Phytol 119:345–360

Tyree MT, Sperry JS (1989) Vulnerability of xylem to cavitation and embolism. Annu Rev Plant Physiol Plant Mol Biol 40:19–38

Tyree MT, Zimmermann MH (2002) Xylem structure and the ascent of sap. Springer, Berlin, Heidelberg

Umebayashi T, Ogasa MY, Miki NH, Utsumi Y, Haishi T, Fukuda K (2016) Freezing xylem conduits with liquid nitrogen creates artifactual embolisms in water-stressed broadleaf trees. Trees-Struct Funct 30:305–316

Venturas MD, Sperry JS, Hacke UG (2017) Plant xylem hydraulics: what we understand, current research, and future challenges. J Integr Plant Biol 59:356–389

Vo HM, Charleston MA, Brodribb TJ, Sussmilch FC (2025) Dehydration rapidly induces expression of NCED genes from a single subclade in diverse eudicots. Planta 261:46

Vuerich M, Petrussa E, Boscutti F, Braidot E, Filippi A, Petruzzellis F, Tomasella M, Tromba G, Pizzuto M, Nardini A et al (2023) Contrasting responses of two grapevine cultivars to drought: the role of non-structural carbohydrates in xylem hydraulic recovery. Plant Cell Physiol 64:920–932

Waite P-A, Kumar M, Link RM, Schuldt B (2024) Coordinated hydraulic traits influence the two phases of time to hydraulic failure in five temperate tree species differing in stomatal stringency (S Pfautsch, Ed.). Tree Physiol 44:tpae038

Weithmann G, Paligi SS, Schuldt B, Leuschner C (2022) Branch xylem vascular adjustments in European beech in response to decreasing water availability across a precipitation gradient. Tree Physiol 42:2224–2238

Wheeler JK, Huggett BA, Tofte AN, Rockwell FE, Holbrook NM (2013) Cutting xylem under tension or supersaturated with gas can generate PLC and the appearance of rapid recovery from embolism. Plant Cell Environ 36:1938–1949

Wightman R (2022) An overview of cryo-scanning electron microscopy techniques for plant imaging. Plants 11:1113

Woodruff DR, Meinzer FC (2011) Size-dependent changes in biophysical control of tree growth: the role of turgor. In: Meinzer FC, Lachenbruch B, Dawson TE (eds) Size- and age-related changes in tree structure and function. Springer, Dordrecht, pp 363–384

Wu M, Zhang Y, Oya T, Marcati CR, Pereira L, Jansen S (2020) Root xylem in three woody angiosperm species is not more vulnerable to embolism than stem xylem. Plant Soil 450:479–495

Yang Y, Ma X, Yan L, Li Y, Wei S, Teng Z, Zhang H, Tang W, Peng S, Li Y (2023) Soil–root interface hydraulic conductance determines responses of photosynthesis to drought in rice and wheat. Plant Physiol 194:376–390

Zhang J, Davies WJ (1987) Increased synthesis of ABA in partially dehydrated root tips and ABA transport from roots to leaves. J Exp Bot 38:2015–2023

Zhang J, Davies WJ (1990) Changes in the concentration of ABA in xylem sap as a function of changing soil water status can account for changes in leaf conductance and growth. Plant Cell Environ 13:277–285

Zhang J, Schurr U, Davies WJ (1987) Control of stomatal behaviour by abscisic acid which apparently originates in the roots. J Exp Bot 38:1174–1181

Zimmermann MH (1983) Xylem structure and the ascent of sap, 1st edn. Springer, Berlin

Zimmermann MH, Potter D (1982) Vessel-length distribution in branches, stem and roots of Acer rubrum L. IAWA Bull 3:103–109

Zimmermann U, Schneider H, Wegner LH, Wagner HJ, Szimtenings M, Haase A, Bentrup FW (2002) What are the driving forces for water lifting in the xylem conduit? Physiol Plant 114:327–335

Zotz G, Tyree MT, Patino S (1997) Hydraulic architecture and water relations of a flood-tolerant tropical tree, Annona glabra. Tree Physiol 17:359–365

# Compounding Effects of Heat, Drought, and Fire Increasingly Threaten Post-Calamity Recovery and Restoration of Forests in Biomes Worldwide

**Pierre L. Ibisch**

## Contents



**Abstract** Interest in restoration and reforestation is growing as forest destruction and degradation continue worldwide, and as increasingly severe climate-related disturbances such as fires, heat waves, and droughts affect their vitality or even cause local collapse. At the same time, there is increasing evidence that the resilience of forests is declining and, in the worst cases, that recovery from severe stand-replacing calamities is slowing down or failing altogether. The aim of this review is therefore to provide an overview of the extent to which the recent scientific literature assesses the problem of critically declining forest resilience after a calamity and the difficulties of recovery due to climate change. There is a wealth of compelling

Communicated by Christoph Leuschner

P. L. Ibisch (✉)
Centre for Econics and Ecosystem Management, Eberswalde University for Sustainable Development, Eberswalde, Germany
e-mail: pierre.ibisch@hnee.de

 
Progress in Botany (2026) 86: 181–209, https://doi.org/10.1007/124_2025_92,
Published online: 04 July 2025

literature on this topic, which is already a well-established field of research in some parts of the world. This is particularly true in western North America, and also in biomes in Australia where fire is a natural factor, but where the intensity and frequency of fire are increasing. There is evidence that fire in particular, but in systemic interaction with heat and drought, threatens the resilience of ecosystems that are in principle even fire-adapted. Particularly, the non-linearly increasing atmospheric water demand, the vapour pressure deficit, becomes a major risk to ecosystem health and recovery with rising temperatures. The problem of potential failure of recovery appears to be spreading to more and more biomes, already affecting boreal and occasionally temperate regions. It is the feedbacks with vegetation degradation and self-reinforcing and expanding droughts and heat waves that are of concern. What needs to be quickly understood is that the ability of damaged forest ecosystems to recover can be further reduced by post-calamity management (e.g. salvage logging). To better assess the challenge and effectiveness of restoration efforts, it is important to collect more data on failures of natural recovery and reforestation, and to learn from these results, especially failures and problems.

**Keywords** Climate change, Ecosystem restoration, Forest resilience, Wildfire

## 1 Introduction and Scope of Review

While the loss of forests due to land use continues in many biomes, and while forest utilisation and various disasters such as droughts or fires contribute to the weakening and degradation of forest ecosystems (e.g. Ceccherini et al. 2020; Vancutsem et al. 2021; McDowell et al. 2020), there is general agreement that forests need more decisive conservation and support. It is no longer just about maintaining largely intact forests, but also about restoring them. At the global level, the corresponding target was developed under the Global Biodiversity Framework of the Convention on Biological Diversity and ratified by a majority of countries. Consequently, endeavours to restore forests have been formally recognised as a national objective in numerous countries, with legal provisions being established to support these initiatives. China alone has implemented several large-scale afforestation programmes, such as the “Three North” Shelterbelt Development Programme, which have contributed substantially to the implementation of the country’s greening policies (e.g. Zhang et al. 2016). The EU Nature Restoration Law is an initiative designed to halt and reverse biodiversity loss across Europe. It aims to achieve this by establishing legally binding targets and deadlines for the restoration of ecosystems throughout the continent. The overarching objective of the EU Nature Restoration Law is to complement existing EU legislation and strategies. The initiative was established following a protracted negotiation process, and it entails the introduction of binding international legislation, accompanied by the adoption of explicit targets and timeframes for the comprehensive restoration of landscapes; the overarching objective of this initiative is the restoration of degraded ecosystems by the year 2050 (Hering et al. 2023; Cliquet et al. 2024; Aggestam 2024). The new initiatives are stimulating interest in restoration concepts, but they may also face enormous practical obstacles.

Amongst the many challenges arising, there is the issue what the baseline is and what nature is actually to be restored – natural ecosystems or cultural landscapes. The question is, of course, fundamentally legitimate. While conservation efforts often aim to restore a (presumed) pristine state, or at least to approach differently defined baselines, ethical issues and concerns about socio-economic feasibility aside, many ecological challenges arise. These include the alteration of the natural environment or site potential by human use, the limited spatial possibilities for restoring extensive landscape ecosystems, and many inhibiting influences that make a return to any reconstructed or projected past ecosystem state seem impossible. Whereas in the past it seemed possible to restore natural ecosystems or productive plantations by planting the tree species that (presumably) grew in the area before the disaster or forest loss, and thus to initiate recovery, human-induced climate change is now expected to become a definite game changer. In Germany, for example, extremely hot and dry summers in recent years have caused millions of hectares of forests (or plantations) to die. These large-scale calamities, caused by insufficient water supply, heat, pests and diseases, and even forest fires, imply extensive restoration needs.

The aforementioned situation indicates that forest conservation is confronted with unparalleled challenges. This assertion is equally applicable to restoration measures, encompassing both active reforestation and the promotion of natural succession. What will be and which forest we want to promote so that it has a future in the age of climate change is a discussion in its own right – it is also about discourse on proactive forest design to enable the forest to adapt to future climate conditions. One approach being discussed, among others, is assisted migration, that is, the transfer of organisms to other biomes to pre-empt or facilitate the expected biome shift (Chakraborty et al. 2024). There is also a debate, sometimes quite heated, on active or passive measures for adaption to and mitigation of climate change (e.g. Jandl et al. 2019; Erb et al. 2022). This often involves a mixture of scientific positions on how vegetation is likely to develop, with tangible forestry-economic interests, which are also pursued by practitioners and landowners. Interwoven with this discussion is the question of the various ecosystem services provided by forests and which silvicultural or nature conservation strategies should be pursued in order to ensure not only provisioning, but also regulating, cultural, or health-promoting ecosystem services.

All this outlines an important field for applied and, as far as possible, knowledge-based discussion. In the maelstrom of interests between commodity markets, forest owners, and forest practitioners, the potential of ecosystems is often in danger of being marginalised, and it is important not to lose sight of what is actually feasible in terms of natural and ecological conditions. Important questions are as follows: What does recent research say about the impact of climate change on forest restoration and recovery after severe (stand-replacing) disturbances? Are the problems intensifying and spreading globally, and how is this challenge being addressed? Therefore, the aim of this review is to provide an overview of the extent to which the scientific literature describes the problem of critically decreasing resilience of forests after calamities and the difficulties of recovery due to climate change. This is not so much about the resilience of individual tree species to climate change-related disturbances, for which there is a rapidly growing knowledge base. While these do play an important role when considering the resilience of ecosystems, the focus here is on forest development as such and, above all, on the crucial question of where and under what circumstances there is a risk of a tipping point towards a non-forest scenario because trees may no longer grow at all.

This paper is not intended to be a quantitative bibliographic analysis, but rather an exploration of recently documented issues and related research. As there have been some previous reviews on the topic of reduced forest resilience (Anderson-Teixeira et al. 2013; Ibáñez et al. 2019), the focus was on the most recent literature and on recovery after severe calamities. The dominance of published evidence from the western USA and Australia was duly reflected, but special efforts were made to find literature from all other continents as well. Mainly empirical studies were considered that reported actual changes, especially if they explicitly referred to climate change. Modelling-based projections or analyses that primarily dealt with the mechanisms of forest damage in the context of calamities mostly were not considered.

## 2 Forest Resilience and Climate Crisis: General Considerations

Resilience has been conceptualised in terms of (1) persistence (maintaining function despite disturbance), (2) recovery (returning to a pre-disturbance state), and (3) adaptability (adjusting to new conditions) (Strickland et al. 2024). Persistence tends to be associated with resisting without suffering significantly from external shocks, whereas resilience in the strict sense (from the Latin *resilire*) tends to refer to "bouncing back" to a previous state after being damaged or reduced in functionality. In the context of the climate crisis, and in line with Strickland et al. (2024), resilience is understood here as a concept that focuses on the dynamic maintenance of key characteristics and functions of forest ecosystems while enhancing their adaptive capacity. Nikinmaa et al. (2020) concluded in a review that a clear majority of studies applied the concept of engineering resilience, quantifying resilience as the recovery time after a disturbance. Falk et al. (2019) rightly point out that the different conceptualisations of resilience can coexist in the same system, but at different scales such as individual species in the wider ecosystem. Resilience also operates at different timescales, with humans often focusing on short-term processes that make it easier to judge whether a system is recovering critical features and functions (Falk et al. 2019). Rapid environmental change makes returning to a past state highly unlikely, so this paper focuses on a critical loss of short-term resilience, which means that an ecosystem does not recover following a severe disturbance and easily returns to an ecosystem state characterised by typical forest structures and properties.

In the last two decades, the effects of climate change on tree and forest ecosystem vitality was in the foreground – both in terms of currently observable impacts and potential future ones that can be derived from models. But for some time now, it has become increasingly clear that not only the health and existence of trees and forests, but also the recovery of damaged forests may suffer, at least in some regions of the world. This is a real double effect and potentially fatal combination of climate change, which first leads to increased damage and calamities – and then also impairs recovery. From a forest ecology and management perspective, the loss of resilience of forest ecosystems would, in the worst case, lead to a biome shift towards non-forest systems.

An early review of this topic was published over a decade ago (Anderson-Teixeira et al. 2013). On the one hand, the literature reviewed at the time suggested that increased atmospheric $CO_2$ levels, in combination with higher temperatures, may even have a beneficial effect on the dynamics of forest regeneration – provided water is available. On the other hand, it became increasingly clear that drought events negatively affect the growth and vitality of forests and also have an influence on the recruitment and survival rate of seedlings. Surpassing critical thresholds can lead to forests no longer being able to return to their previous state after a disturbance. Logically, the reactions of forests vary depending on their initial state, for example, in terms of the age and species composition of the trees. Allen et al. (2015) showed that some factors of global change could reduce vulnerability, while there is increasing evidence of a significant and probably underestimated risk associated with the effects of hotter droughts on global tree mortality and forest dieback. The comprehensive review by Ibáñez et al. (2019) investigated natural conditions that would hinder resilience sensu recovery. Among others, local moisture conditions were found to be relevant in drier regions, and regional moisture in boreal forests. Logically, the severity of the disturbance was associated with a lower recovery of the remaining adult trees.

It is widely accepted that climate change has a detrimental effect on forest ecosystems. The effects of climate change, which were already roughly predicted or feared decades ago, can now be observed. For a considerable period, climate change appeared to exert a favourable influence on global forest productivity, particularly in regions where water was not a limiting factor. However, this trend was not uniform, with some regions experiencing negative impacts due to resource limitations and increased disturbances (Boisvenue and Running 2006). Temperature increases and alterations in cloud cover have been observed to result in shifts in species distributions, predominantly towards higher elevations, which in turn have implications for biodiversity and ecological functions (Mata-Guel et al. 2023).

The intensification of abiotic disturbances, such as fire, drought, and wind, along with biotic disturbances from insects and pathogens, has been further exacerbated by climate change (Canelles et al. 2021). It is anticipated that these disturbances will be particularly pronounced in coniferous and boreal forests (Seidl et al. 2017). A discernible trend in the aggregate forest vulnerability is evident, which is principally driven by a warming-induced attenuation in plant defence mechanisms in response to insect outbreaks, particularly at higher latitudes (Forzieri et al. 2021). Temperate forests are also increasingly affected by extreme weather events such as droughts, storms, and floods, leading to phenomena like dieback and shifts in phenology (Wegler and Kuenzer 2024). Substantial droughts have been observed in the Amazon Basin over the past few decades, leading to reductions in photosynthesis and altered carbon allocation within trees. This has been indicated to result in stress and decreased vitality, consequently leading to increased tree mortality and reduced $CO_2$ uptake (Doughty et al. 2015). For a considerable period, an escalation in drought-fire interactions has been observed to result in elevated levels of tree mortality (Brando et al. 2014). In numerous forests, canopy dieback has become an increasingly prevalent phenomenon, driven by climate change (e.g. Andrus et al. 2024; Boehmer and Galvin 2024). Interactions between climate change impacts and local factors, such as forest management, have been demonstrated (Sangüesa-Barreda et al. 2015; Petritan et al. 2021; Blumröder et al. 2021).

There is an observed increase in tree mortality due to climate extremes such as heat and drought. These events are becoming more frequent and severe, affecting forests that were previously considered less vulnerable (Allen et al. 2010; Hartmann et al. 2022). Common species that currently dominate native vegetation across large ranges are threatened by rapid environmental changes implying substantial forest changes in the future (e.g. European beech, *Fagus sylvatica*; Leuschner 2020; Martinez del Castillo et al. 2022; Leuschner et al. 2023; Klesse et al. 2024). But also, tree species that until recently have been well adapted to heat and drought might be more vulnerable than expected by forest managers (e.g. Douglas fir, *Pseudotsuga menziesii*; Leuschner and Meinzer 2024).

The declining resilience of forests is also a growing concern in the context of scenarios that rely on forest sequestration of greenhouse gases. The fact that more and more forests are becoming, at least temporarily, a source of greenhouse gases, clouds the outlook for forest-based mitigation options, especially afforestation and reforestation (Windisch et al. 2025). So what we are discussing here is the fear of large-scale vegetation-atmosphere-climate feedback loops, which could make runaway climate scenarios more plausible.

The current levels of greenhouse gases in the atmosphere are without precedent in recent Earth history, with concentrations not observed for the past 800,000 years. However, analogous concentrations have been observed in much earlier geological periods, specifically during the Pliocene and Eocene epochs. During the Miocene, $CO_2$ levels were generally below 500 ppm, with transient reductions during periods of rapid cooling (Pearson and Palmer 2000; Guillermic et al. 2021). The current concentration of $CO_2$ in the atmosphere is above 420 ppm, representing a 52% increase compared to pre-industrial levels (around 278 ppm in 1750) (Friedlingstein et al. 2025). Should emissions continue to rise unchecked, $CO_2$ levels could reach 450 ppm by 2035, with even much higher levels possible by the end of the century under extreme scenarios. This trajectory could potentially result in climates reminiscent of the Eocene by 2,150, if high emission pathways are followed (Burke et al. 2018). The last 10 years include all 10 warmest years in the instrumental record, and

2024 was the hottest year on Earth since 1850, beating the previous record set in 2023 (Rohde 2025). Berkeley Earth estimates that the global mean temperature in 2023 will be 1.62 ± 0.06°C (2.91 ± 0.11°F) higher than the average temperature from 1850 to 1900, a time span often used as a pre-industrial baseline for global temperature targets (Rohde 2025).

The Berkeley Earth temperature report for 2024 also underlines two important facts: first, that since 1970, global warming has increased at a largely linear rate; and second, that the 2023/24 peak appears to deviate significantly from past trends (Rohde 2025). It appears that the implementation of new processes, which may involve feedbacks, could further push climate dynamics. For instance, research has shown that a reduction in albedo resulting from a decrease in low cloud coverage leads to additional warming that is not directly linked to greenhouse gases (Goessling et al. 2025).

The pace and magnitude of global warming – and the change in many dependent climatic parameters – alone justifies calling it a climate crisis. This does not imply that it is a temporary period in the short term, but rather that it underlines the real intensification of conditions for life on Earth, which is adapted to the Holocene climate. Strictly speaking, of course, it is a climate change-induced *biosphere crisis* that is already unfolding.

There is a growing realisation that extreme events, and even more so the chain reaction and interaction of various extremes in so-called compound events, are of central importance for the effects of climate change on humans and ecosystems. Conventional risk assessments typically examine only individual drivers and/or hazards, which can lead to an underestimation of risks, since extreme events are often linked and dependent in space and/or time (Zscheischler et al. 2018). The frequency and severity of compound events appear to be increasing (e.g. Ridder et al. 2022; Romanou et al. 2024; Tripathy and Mishra 2023; Wang et al. 2024; Zhou et al. 2024; Schillerberg and Tian 2024), making reliable projections of future impacts increasingly difficult.

## 3 Recent Evidence of a Decline in Post-Wildfire Resilience

Increasing evidence suggests that post-calamity resilience may be declining, at least in some ecoregions worldwide. The main relevant types of disturbances are drought – often in combination with heat – and wildfires. It may not be surprising that fire in particular can put forests in a state from which the system cannot easily recover. In the case of intense and widespread fires, not only is the above-ground biomass burned or killed, but the topsoil and humus layer are usually destroyed as well. In addition, there are severe microclimatic changes above and in the soil, as well as potential post-fire erosion. However, in a global study of post-fire vegetation productivity recovery (from 2004 to 2021), Xu et al. (2024) found evidence of relatively effective resilience in about 87% of burned areas, with vegetation returning to pre-fire productivity levels within 2 years. A slowing or reduction in resilience is expected and has been reported particularly in drier areas. Certain biomes are specifically affected.

In the following, the latest evidence of a critical loss of resilience is summarised and grouped geographically.

## 3.1 North America

Stevens-Rumann et al. (2018) show, based on a broad database of 52 wildfires from the US Rocky Mountains, how forest recovery in the twenty-first century has significantly worsened compared to previous decades. This is not least related to lower seedling densities. In particular, dry forests at the climatic limit of their range are candidates for being replaced by non-forest ecosystems after fires. For some time now, there have been indications of a biome change, especially in areas of the southwestern United States. One example is the Klamath Region of northern California and southwestern Oregon, where coniferous forests are being replaced by broadleaved trees and shrubs (Tepley et al. 2017). This study also yielded the interesting finding that the recruitment of conifers must occur in a relatively short period after the fire (about 4 years after the fire) in order to establish effective regeneration. With increasing aridity, the problem of insufficient seed availability arises.

Coop et al. (2020) reviewed evidence on the loss of forest resilience following high-severity fires in western North America. There are three main factors that may impede forest recovery. When large areas of forest are burned, there may be a loss of seed trees and a subsequent lack of seed. In addition, the establishment of new trees can be affected by a hotter and drier local or landscape climate as a result of the loss of functional vegetation, water holding capacity, and evaporation. In line with Landesmann et al. (2021) and Tepley et al. (2017), he also states that there are likely to be positive fire-vegetation feedbacks that increase the risk or severity of subsequent fires, that is, the problem of short-term reburning. These factors, alone or in combination, can lead to a biome shift away from the pre-fire forest ecosystem, in the worst case even to non-forest vegetation. In the case of reburns, increased carbon loss combined with sparse tree regeneration can delay the (simulated) recovery of carbon for centuries (*Pinus contorta* stands in Greater Yellowstone; Turner et al. 2019). Donato et al. (2016) provided evidence that after the big Yellowstone fires of 1988, after 24 years, there was robust self-replacement of Douglas-fir forests under mesic conditions, which recovered to a greater extent than dry stands not adjacent to the lower ecotone, while dry stands near the ecotone to non-forest did not regenerate at all. In fact, thousands of hectares of lower montane forest have been converted to grassland or sage steppe for at least several decades. Coop (2023) described post-fire ecosystems following both forest vegetation such as ponderosa pine, aspen, and mixed conifer forests, as well as shrub and herb-dominated plant communities. The latter would include native, widespread ruderal and exotic species. The risk of forest conversion to non-forest vegetation logically is also influenced by burn spatial patterns, insolation, and microhabitat conditions, for example, related to topography or exposition. In this context, forest refugia are of special importance (e.g. Coop et al. 2019).

Based on remote sensing data, Shrestha et al. (2024) find mixed patterns of post-fire recovery across forest types in the western USA. They examined all wildfires between 1986 and 2017 and found that, in some areas, the leaf area index of the corresponding sites had recovered to only 60–70% of pre-fire levels by 25 years after the fire. The lowest rates of recovery were observed at higher elevations, and a very important finding is that full recovery was rarely observed when annual precipitation was less than 500 mm and average summer temperatures were above 15–20°C. Zahura et al. (2024) similarly found that climate variables, particularly precipitation and temperature, were the most important drivers, with altitude being the most relevant topographical variable. Sorenson et al. (2025) concluded that the effectiveness of post-fire tree planting is determined by interacting microsite conditions and climate that control plantlet establishment, survival, and growth in the Sierra Nevada, but biotic interactions would strongly influence whether trees or shrubs would later emerge as dominant life forms and which tree species would become most abundant. Young et al. (2019) in northern California for 2004–2016 found that post-fire drought was a particularly important factor inhibiting tree recruitment, with one severe drought year following fire being more critical than several moderately dry years (compare Fig. 1).

**Fig. 1** Slope in potentially forested area after repeated burns in California (Sequoia National Forest). Harsh microclimatic conditions on hot days combine with soil erosion during rainfall. The recovery potential of the forest may be critically reduced or even lost for a long period of time. (Photo: P.L. Ibisch)

Climate change is a key driver not only of increasing wildfire risk itself – and of more frequent, more intensive, and more extended fires (e.g. Jones et al. 2022; Richardson et al. 2022; Brown et al. 2023; Jain et al. 2024; Kirchmeier-Young et al. 2024; Wang et al. 2025), but also of post-fire factors that hamper recovery. In vulnerable ecosystems such as those found in western North America, climate change increasingly results in extreme fire spread events with rapid growth rates and large spatial extent, but also higher burn severity and greater distances of severely burned areas from live tree seed sources (McFarland et al. 2025). This will result in limited tree regeneration and protracted forest conversion. The intensification of heat, drought, and fire will, as Rodman et al. (2020) put it, "snuff out" post-fire forest recovery. For the southern Rocky Mountains, they expect that the increasing climatic water deficit and an increased frequency of extreme droughts will cause a significant reduction in seedling densities after fires. Logically, this risk will be particularly elevated on the "trailing edge" of the forests (Parks et al. 2019).

The geographic focus of reports of weakened or failed post-fire recovery is obviously the southwestern United States. But there are also reports from other biomes and continents. For example, Baltzer et al. (2021) analysed forest resilience in the boreal biome in western North America, with reference to black spruce (*Picea mariana*) and jack pine (*Pinus banksiana*) forests, among others. In 18% of wildfire sites (fires from 1989 to 2014), spruce regeneration failed completely. The authors point out that this is a snapshot in time, and that resilience could quickly erode as moisture deficits and fire activity increase, pushing the system towards a tipping point unprecedented in millennia.

## 3.2 Australia

From Australia, the continent that has attracted much attention due to the recent prevalence of megafires, there are reports of significant biomass losses but also relatively rapid recovery rates (Ermitão et al. 2024). In southeastern eucalypt forests, approximately 97% of all burns observed between 2001 and 2014 recovered to pre-fire leaf area index within 6 years (Rifai et al. 2024). Also after large biomass losses in 2019, a wet 2020 supported large vegetation gains (Qin et al. 2022). However, there are some areas of concern, particularly in areas of severe drought (Gibson and Hislop (2022). In southeastern Australia, it was noted that the recovery time of live tree carbon to pre-fire levels increases with fire severity (Volkova et al. 2025). Another finding is that prolonged drought hinders recovery, leading to ongoing tree mortality, which means that the corresponding highly productive temperate forests are at risk of reduced carbon carrying capacity and resilience. In the same region, increased fire frequency has been observed to shift the species composition and abundance of plant functional groups not only of the current vegetation, but also of the soil seed bank towards a decrease in clonal resprouters and an increase in grasses and other herbaceous species (Duivenvoorden et al. 2024).

Fairman et al. (2019), also in southeastern Australia, found that in shrubby dry mixed eucalypt forest, short-interval wildfires increased tree "escape size" and reduced resprouting success, particularly of medium-size trees, which combined with reduced seedling recruitment leads to structural and demographic challenges for even the most fire-tolerant forests. Similarly, Anderson et al. (2025), based on studies of *Eucalyptus marginata* forests in southwestern Australia, in areas affected by drought, heatwave, and wildfire after 2010, conclude that the increased frequency of disturbances of various types also inevitably increases the likelihood of compounding effects leading to reduced resilience in forest ecosystems with resprouting trees adapted to wildfire. Walden et al. (2023), having studied eucalypt forests in the same ecoregion, concluded that "there is early evidence of a structural destabilisation of these forests under a more fire-prone, hotter and drier future climate".

Le Breton et al. (2022) bluntly state that the 2019–2020 wildfires not only broke global records in their extent and severity, but that ecosystem collapse may be inevitable if management cannot control the spread of future megafires that affect vegetation too frequently. The megafires affected over 5.5 million hectares in southeastern Australia, and a third of the vegetation would have burned too early. The "vegetation squeeze" caused by megafires means that time windows for vegetation recovery are shrinking below critical thresholds.

### 3.3 Mediterranean Europe

As seen in North America and Australia, the subtropical "Mediterranean" regions appear to be particularly prone to combined events of heat, drought, and fire (compare Ermitão et al. 2024), and the corresponding vulnerability of forests is also found in Mediterranean Europe. It is no coincidence that this is a region where forestry has introduced Australian eucalypts, creating high-risk anthropogenic fire ecosystems. In Portugal, where *Eucalyptus globulus* has become common in plantations, this fire-resistant species suffers from a high risk of top kill (Catry et al. 2013). In the Atlantic region of north-west Spain, Cruz et al. (2024) showed that native broadleaved *Quercus robur* forests were more resilient to fire than eucalypt and pine plantations. *Pinus pinaster* plantations were particularly vulnerable, with a post-fire increase in grasses, herbs, and spiny shrubs. The broadleaved forest recovered its vertical structure 1 year after the fire and maintained the distribution of plant functional types, both in the overstorey and in the understorey.

Referring to France and the 2022 fire season, Vallet et al. (2023) share concerns about low resistance and low post-fire growth rates in temperate and Atlantic pine forests. Rodrigues et al. (2023) analyse the record-breaking forest damage of the exceptional fire season of 2022 and state that this year "may be a spyglass into the 'new normal' for fire activity under global warming". In an analysis of the post-fire trajectories of Mediterranean pine forests, Viana-Soto et al. (2020) highlight that the recovery rates of pine forests were highly sensitive to the post-fire climate in the medium and long term, and to the severity of the fire in the short term. They also refer to evidence for the conversion from forest to shrubland on extremely dry and south-exposed slopes (*Pinus nigra* in northeast Spain; Martín-Alcón and Coll 2016) and in severely burned areas (*Pinus halepensis* in southeast Spain; González-De Vega et al. 2016).

Baudena et al. (2020) modelled that future aridity could push fire-prone ecosystems in the Mediterranean past a tipping point, with open, highly flammable shrublands stably replacing closed forests. While water stress alone would not hamper forest development after disturbance, the combination with poor post-fire recovery could be critical. Not only do the authors suggest that feedbacks involving the flammability of open shrublands and recurrent fires would impede forest recovery, they also warn that "drought-induced reductions in post-fire responses may drive Mediterranean vegetation transitions even more abruptly and irreversibly than previously predicted."

In eastern Spain, Salesa et al. (2022) conclude that the combination of drought and fire can severely compromise ecosystem resilience. Increasingly prolonged summer droughts would result in a persistent drought legacy that would affect post-fire recovery, which is usually mediated by seed species whose populations become established in the first year after fire. Paneghel et al. (2024) also see that "post-fire auto-successional dynamics" of obligate seeder species may be hampered by high temperatures, shortening the temporal window of opportunities for effective seedling establishment. In their case study in northeastern Spain, they predict a threat to post-fire resilience from a potential demographic collapse of Aleppo pine.

## 3.4 Mediterranean Vegetation in South America

The first reports on the interaction of drought effects with wildfires leading to reduced forest resilience in South America are also available from Mediterranean forests. Hernández-Duarte et al. (2024) investigated the effects of a mega-drought in central Chile since 2010, which even intensified into a hyper-drought in 2019, with record-low precipitation and high temperatures that hindered post-fire forest recovery. Analysing short-term (5-year) post-fire vegetation recovery in various types of sclerophyll forests, thorny forest, and shrubland, between 1987 and 2022, from 42 wildfires, they summarise that extreme and prolonged droughts showed a decrease in recovery in all forest types. Hyper-drought conditions with extremely low soil moisture availability would hinder both vegetative resprouting and seed germination.

## 3.5 Boreal Eurasia

With warming rates above the global average, the Eurasian boreal regions already include some of the hottest and driest regions of the entire biome (Burrell et al. 2021). The boreal forests of Eurasia and North America together have become one of the world's most important "pyromes"; fire emissions have almost tripled between 2001 and 2023, and this is related to increased vegetation productivity on the one hand, and an increase in fire-promoting weather conditions and reduced soil moisture on the other hand (Jones et al. 2024). In a review of post-fire regeneration at higher latitudes (above 55°N; according to the authors, particularly influenced by studies from the USA and Canada), it increased significantly with latitude. However, there is also clear and alarming evidence of a decrease in post-fire forest resilience in Siberian boreal forests, particularly in the southern belt of the biome (Ruggirello et al. 2023).

Burrell et al. (2022) published results showing that 1.2% of the Eurasian boreal zone would be at high or extreme risk of fire-induced forest loss, with a further 11% of areas at moderate risk. Failure of post-fire recruitment, particularly in the forests of southern Siberia, Russian Federation, could lead to a regional tipping point, resulting in the rapid replacement of large areas of forest by non-forest ecosystems (Burrell et al. 2021). The Zabaikal region east of Lake Baikal and the forests between Krasnoyarsk and Irkutsk, as well as the forests west of Yakutsk, are likely hotspots of post-fire recruitment failure and forest loss, and large areas of Siberia appear to be unstudied in this regard (e.g. Shvetsov et al. 2019; Burrell et al. 2022). Shvetsov et al. (2019) found that, depending on fire intensity, 43–56% of study sites in the southern Zabaikal taiga region showed poor or no regeneration. In the same region, Sun et al. (2021), investigating NDVI-based greenness of post-fire sites, found that a prolonged dry spring would have a strong effect consistently throughout the early developmental stages from the first 5 to 15 years post-fire.

In the European part of the southern taiga, in the middle Volga region, forests are characterised by species such as pine (*Pinus sylvestris*), birch (*Betula pendula*), spruce (*Picea abies*), lime (*Tilia cordata*), and aspen (*Populus tremula*). Here, the substantial variability in post-fire recovery from 2000 to 2023 is more strongly related to land surface temperature than to precipitation variability, the former accounting for approx. 65% of the variance (Kurbanov et al. 2024). Only 60% of the areas show a good recovery.

## 3.6 Temperate Central Europe

Possibly, the first report of variability in post-fire recovery and short-term failure in the temperate broadleaved forest biome comes from north-eastern Germany. Here, forest fires are actually associated with coniferous plantations of the native pine (*Pinus sylvestris*), which often grows on sandy soils and in a relatively arid climate.

**Fig. 2** Scientific photo monitoring in the second, third, and fourth post-fire years (Y2, Y3, Y4) of two burned areas in the same region southwest of Berlin, Germany. The salvaged sites in Treuenbrietzen (left; burned in 2018) show effective regeneration, especially with aspen (*Populus tremula*), despite intensive management. The sites in Jüterbog (right; burned in 2019) exhibit only poor tree recruitment even 4 years after the fire (Jouy et al. 2025; Photos: T. Geisel, PYROPHOB project)

A recent increase in fire risk appears to be linked to increasingly hot and dry weather, especially after 2018.

At a 2018 fire site in a region southwest of Berlin, there was a rapid recovery driven mainly by aspen (*Populus tremula*) (Blumröder et al. 2022; Schüle et al. 2023; Heinken et al. 2024; Jouy et al. 2025). At a 2019 fire site, only about 10 km away but surrounded by a more open and warmer landscape, short-term recovery was poor. This may be related to the legacy of the previous drought, as well as pronounced dry spells and higher temperatures, which hindered seeding and seedling establishment (Jouy et al. 2025; Fig. 2). The case illustrates that forest management has the potential to accelerate forest degradation and loss: Large-scale cultivation of more flammable conifer species in a region with broadleaved forest potential, harvesting practices that increase maximum temperatures and drought on hot days (Blumröder et al. 2022) as well as salvage logging post-fire, leading to even higher temperature extremes and vapour pressure deficit extremes on cleared areas (Rönnefarth et al. in prep.), can lead to more intense calamities and poor post-fire recovery. Furthermore, under certain conditions, such as sandy soils with poor water holding capacity and a significant management legacy, temperate forests may be closer to tipping points than one might imagine in the event of extreme compound events (Fig. 2).

# 4 Recovery After Windthrows and Hot Drought Calamities Without Fire

While there is clearly a lot of evidence for the loss of forest recovery after fires with the risk of biome change, often strongly influenced by climatic factors, there are not yet many reports on weather events alone that lead to large-scale mortality. However, it is becoming increasingly clear that heat and drought extremes, as particularly direct consequences of climate change, are having an ever greater impact on the health of forest ecosystems. Globally, there is evidence of changes in forests due to warming and drought, such as reduced biomass, smaller tree diameters, and changing composition (e.g. Zhou et al. 2013). For the northern hemisphere, Zhang et al. (2025) report a general decline in forest resistance to flash droughts. Reduced growth and persistent legacy effects following calamities have been documented (Anderegg et al. 2015; Johnstone et al. 2016; Pardos et al. 2021). Huang and Zhai (2025) note that little research has focused on comparing the recovery time between general drought and combined dry-hot events, and conducted a first global assessment of recovery from different types of drought between 1982 and 2016. They found that compound dry-hot events did indeed mostly cause longer recovery times, and that post-drought moisture and long-term precipitation were the key factors determining recovery time.

There is evidence of a significant decline in seedlings and saplings following hot droughts in Texas, USA (Lawal et al. 2025). Crockett and Hurteau (2024) show how reduced seedling survival rates under heat and drought will pose future demographic challenges for conifers in the southwestern US. Referring to parts of southern Europe and North Africa, Lloret et al. (2022) point out that short-term forest resilience is generally defined by pre-drought conditions and previous management legacies, as well as the impact of drought on the dominant species both before and after the drought. Gazol et al. (2018) document how post-drought development in forests of northeastern Spain leads to tree species replacement.

In central Europe, a series of heat and drought compound events from 2018 onwards (Xoplaki et al. 2025) caused significant damage to forests and plantations, with diverse recovery responses, especially in broadleaved forests (e.g. Schuldt et al. 2020; Beloiu et al. 2020, 2022; Seidl et al. 2024; Knutzen et al. 2025). Cerioni et al. (2024) found that temperate forests in Europe, after large and severe disturbance events, have a good potential for recovery (structurally better than compositionally). However, the models indicated that higher altitudes and salvage logging hindered regeneration after windthrow, especially for late successional species, while pioneer species were negatively impacted by increasing summer drought. In Germany alone, about two million hectares of forest have been damaged since 2018, and large parts of them have been salvaged and replanted. So far, no studies seem to have been carried out on the decline or lack of recovery of (reforested) clearcuts.

Perhaps it is not too surprising that reforestation failures are scarcely documented worldwide, although they do not seem to be rare. One exception is the study by Del Campo et al. (2021) of a project in Mediterranean drylands. Weather conditions, including extreme drought after planting, are the main cause of failure in Mediterranean drylands (e.g. Del Campo et al. 2020). They report a short-term mortality rate of 53%, rising to 83% after 10 years (dieback of certain species such as *Pinus pinaster* and *Quercus faginea* > 94%, after 10 years). Initially, soil moisture and weather conditions also led to problematic technical decisions.

## 5 Interacting Heat, Drought, and Ecosystem Degradation: Consequences for Vegetation Recovery

As noted above, the unfolding climate crisis is not just about shifting averages, but more and more about increasing volatility, extremes, and compounding events. In addition to these factors and stressors affecting the health and recovery prospects of forest ecosystems, there are risks associated with feedback loops and non-linear responses, especially when global climate change interacts with changes in land use, meso-, and microclimate. Looking at climate change and (the management of) ecosystems, often the focus is on water availability and soil drought. Nevertheless, there are multiple stressors also involving extreme heat and dry air. To date, the vulnerability of plants and ecosystems to extreme heat and atmospheric drought appears to be underestimated (Breshears et al. 2021; Grossiord et al. 2020; Novick et al. 2024). This is even more true for the systemic interaction of these climate-induced stressors as a major risk to forest conservation and management.

However, there is increasing research on systemic processes that exacerbate drought and vegetation damage through feedbacks involving extreme heat, high vapour pressure deficit, and plant death, explaining the non-linear unfolding of ecosystem tipping. In this context, it is important to recognise how soil and vegetation desiccation leads to increased sensible heat, which translates into more intense drought (e.g. Miralles et al. 2014, 2019; Schumacher et al. 2022). The systemic feedbacks can be active at local and even large-scale regional scales. For example, there is evidence that the "record-breaking spring heatwave" of 2023 in the western Mediterranean was amplified by long-term drought (Lemus-Canovas et al. 2024). Dirmeyer et al. (2021) concluded that the 2018 heatwave and drought were exacerbated by unprecedented low soil moisture across much of Europe. Another example is the record-breaking 2021 heatwave in western North America, which can be partly explained by a soil moisture-atmosphere feedback (Zhang et al. 2023). This feedback also contributes to the projected increase in the likelihood of future heat extremes similar to 2021.

While soil heat may exceed air temperature extremes in its detrimental effects (García-García et al. 2023), the non-linearly increasing atmospheric water demand, the vapour pressure deficit (VPD), becomes a major risk to ecosystem health and recovery with rising temperatures (Treydte et al. 2024). Extreme VPD values are becoming more frequent, and by the end of the century, this could be a key factor in the existence of forests. In the USA, the number of days with VPD above 9 kPa could increase by 10 days along the California coast, 30–40 days in the Northwest and Midwest, and 100 days in the Central Valley of California compared to the historical period (Gamelin et al. 2022).

In this context, it is worth highlighting that VPD also is a key predictor of wildfires, driving dead fine fuel moisture content, as one of the most important determinants of fire behaviour, more than temperature or relative humidity (Rodrigues et al. 2024). Unfortunately, high VPD "removes the barriers to the development of large forest fires" even in mesic forests, which dry slowly and therefore have good resistance to fire (Cawson et al. 2024). Thus, both fire weather and flammability are expected to expand into the temperate biome (see, e.g. Maringer et al. 2020; de Dios et al. 2021).

The barely existing and at least delayed recovery of a burnt area in pine plantations in northeast Germany after a fire in 2019 and, above all, the widespread absence of aspen, which had provided abundant regeneration in the region the year before after another fire, can presumably be linked to a previous period of pronounced drought and higher temperatures (Jouy et al. 2025). Consequently, VPD was critically high at the fire site. A feedback loop involving dry sandy soils, anomalously low precipitation, heat, fire, and a resulting high VPD, as well as the absence of tree germination, could even inhibit forest recovery in temperate latitudes (Rönnefarth et al. in prep.).

## 6 Management Practices of Forest Ecosystem Restoration in the Climate Crisis

The spread of the phenomenon that forests do not (easily) recover after severe disturbances should not only motivate to strive for more effective climate protection, but it also gives us reason to rethink forest management goals and practices. The primary aim is perhaps not to adapt the forest to a supposed future climate, which we can hardly project reliably. Rather, it is also about how major disturbances can be prevented by management and how post-calamity conditions can be influenced so favourably that forest recovery can succeed in the best possible way.

Understandably, there is a more lively discourse on the subject in places where the problem of post-calamity loss of forest resilience has been under investigation for a longer period of time, such as the western United States. For example, Coop (2023) highlighted the importance of cooler and mesic forest refugia for effective post-fire regeneration in this region. Coop (2023) goes so far as to question conventional wisdom in forest management, such as the focus on thinning, prescribed burning, and tree planting to restore and maintain historic forest landscapes, as these are challenged by intensifying climate change and fire risk (Loehman et al. 2018). New approaches would be guided by the RAD framework: Resist-Accept-Direct. (Schuurman et al. 2022), where acceptance includes, for example, the idea that under certain conditions it will not be possible to restore the original situation. The likely success or failure of resistance strategies or the consequences of passive management without intervention would guide decisions; some changes may even be desirable, for example, if the conversion of a conifer stand to an aspen forest reduces future flammability (Coop 2023).

North et al. (2019) note that challenges to afforestation include high seedling and sapling mortality due to water stress, competing vegetation, and reburning of young plantations, in addition to the economic costs and safety concerns of replanting large areas of standing dead trees. Maes et al. (2024) recommend drawing lessons from complex systems science, in particular feedbacks and non-linearity, when designing restoration measures. They suggest careful analysis of the typical focus on increasing resilience, which may not be sufficient to address degradation scenarios with abrupt regime shifts and hysteresis, while there is also a highly unhelpful resilience of ecosystems that requires active disruption of ecological feedbacks that maintain undesirable regimes.

A major issue is whether active or passive management is more beneficial for ecosystem recovery. Under certain circumstances and in the short term, tree planting can be more effective than natural succession (e.g. Zhang et al. 2024; Sorenson et al. 2025), but in other cases, it is more successful to rely on natural ecosystem processes (Spanos et al. 2010; Blumröder et al. 2022). Some restoration activities may also have undesirable consequences in the medium or long term. Wion et al. (2024) document the example of aerial seeding of trees after a wildfire in New Mexico. Firstly, seeding was no more effective than non-seeding in terms of vegetation cover in the long term, and secondly, seeding led to a divergence of herbaceous species, with exotic grasses showing higher cover.

A critical question and often polemic discussion concerns the management of biological legacies on disturbed sites, such as standing or downed, dead, or dying trees and snags (Franklin et al. 2000). For decades, they have been recognised as very useful in moderating microclimatic extremes and protecting disturbed soils at disaster sites, but also in providing protection from browsing herbivores, substrates, nutrients, and structural diversity for tree regeneration (e.g. Beschta et al. 2004; DellaSala et al. 2006; Leverkus et al. 2014; Martínez-Sánchez et al. 1999; Pausas et al. 2004; Moreira and Vallejo 2009; Peterson 2009; Beghin et al. 2010; Castro et al. 2011; Marañón-Jiménez et al. 2013; Marzano et al. 2013; Moya et al. 2015; Marcolin et al. 2019; Marangon et al. 2022; Blumröder et al. 2022; Zhu et al. 2024). Nevertheless, the fear of reburn, but also short-term economic interests, often steer forest management decisions towards salvage logging (compare Fig. 3). The increasing criticality of microclimatic conditions with extreme temperatures and VPD, as well as the poor recovery of salvage-logged sites after a calamity, may bring new arguments to the discussion, but this does not seem to be reflected in most of the current literature.

**Fig. 3** A manually “opened” piece of dead wood after a light ground fire in a pine plantation, which still retains moisture inside and acts as an element that would slow down the spread of the fire (Brandenburg, Germany; Photo: P.L. Ibisch)

## 7 Conclusions

The evidence of impairment and damage to forest ecosystems in the unfolding climate crisis is growing rapidly and is covering more and more aspects. A specific and apparently expanding problem is the reduction of resilience after severe disturbances or calamities. There is evidence or indication from an increasing number of biomes that recovery is not always guaranteed. It is not surprising that such problems occur mainly in forests in the area of ecotones in contact with non-forest ecosystems and where forests occur at their so-called trailing edge. The phenomenon of non-recovery of forests with a corresponding change in biome was first well documented and discussed in southwestern North America, especially after more or less extensive forest fires. In the meantime, corresponding findings have also been made in other forests with a Mediterranean climate. However, it is worrying that recovery failure even occurs in temperate and, above all, boreal forests. In particular, large forest areas are at risk of being lost in the southern taiga of Eurasia, which then has subcontinental or even global relevance – not least because carbon stores and sinks are lost.

This issue deserves significantly more attention in research, but also in ecosystem management. The loss or weakening of the recovery capacity of forest ecosystems after disasters has deliberately been linked to ecoclimatological research here, as it is becoming apparent that feedbacks can lead to critical situations developing quite suddenly and ecosystems can pass beyond threshold values into a different, then resilient, but non-forest state.

The worst-case scenario from a forest management perspective will simply have to be accepted if it is unavoidable under certain environmental conditions (see RAD framework). However, what needs to be understood quickly is that the ability of damaged forest ecosystems to recover can be further reduced by post-disaster management. If hundreds of hectares or even square kilometres are salvage-logged after forest fires or extensive drought damage, if this results in extensive soil damage or the areas are ploughed like fields – which happens on a large scale in Germany, for example – then this can lead to almost irreversible damage to the recovery capacity. It is essential for forest managers to develop a greater awareness and knowledge of micro- and mesoclimatic relationships, and to understand how landscape climate changes interact with soil and vegetation conditions, as well as global climate change. In general, we must above all learn to better understand the risks posed by more frequent compound events and how our landscapes can be made more robust and resistant in this respect. In the context of the climate crisis, cooler and wetter fire refugia deserve greater attention and a special conservation priority in their own right, regardless of their species composition or "conventionally" perceived biodiversity values. Wherever and however landscapes can be kept "greener, wetter and cooler", this is a high priority (Sheil and Ibisch 2025).

In order to better assess the loss of natural recovery capacity, but also the effectiveness of restoration measures, it is of great importance that more data be collected. There is a suspicion that forestry practitioners and afforestation projects do not always document exactly what has not worked and what mortality rates there are. There is simply not much scientific literature on this. Nevertheless, we need these results in order to learn quickly and better, especially from failures and problems.

**Funding** PYROPHOB project, funded by Waldklimafonds of the Fachagentur Nachwachsende Rohstoffe e. V.

## References

Aggestam F (2024) Crosswalking the EU Nature Restoration Regulation and the Kunming-Montreal Global Biodiversity Framework: a Forest-Centred Outlook. Sustainability 16(11): 4863

Allen CD, Macalady AK, Chenchouni H, Bachelet D, McDowell N, Vennetier M, Kitzberger T, Rigling A, Breshears DD, Hogg ET, Gonzalez P (2010) A global overview of drought and heat-induced tree mortality reveals emerging climate change risks for forests. For Ecol Manag 259(4):660–684

Allen CD, Breshears DD, McDowell NG (2015) On underestimation of global vulnerability to tree mortality and forest die-off from hotter drought in the Anthropocene. Ecosphere 6(8):1–55

Anderegg WR, Schwalm C, Biondi F, Camarero JJ, Koch G, Litvak M, Ogle K, Shaw JD, Shevliakova E, Williams AP, Wolf A et al (2015) Pervasive drought legacies in forest ecosystems and their implications for carbon cycle models. Science 349(6247):528–532

Anderson NS, Fontaine JB, Lewandrowski W, Walden L, Ruthrof KX (2025) Drought and wildfire legacies highlight vulnerability of a Mediterranean climate-type forest. Austral Ecol 50(1): e70011

Anderson-Teixeira KJ, Miller AD, Mohan JE, Hudiburg TW, Duval BD, DeLucia EH (2013) Altered dynamics of forest recovery under a changing climate. Glob Chang Biol 19(7): 2001–2021

Andrus RA, Peach LR, Cinquini AR, Mills B, Yusi JT, Buhl C, Fischer M, Goodrich BA, Hulbert JM, Holz A, Meddens AJ (2024) Canary in the forest?—Tree mortality and canopy dieback of western redcedar linked to drier and warmer summers. J Biogeogr 51(1):103–119

Baltzer JL, Day NJ, Walker XJ, Greene D, Mack MC, Alexander HD, Arseneault D, Barnes J, Bergeron Y, Boucher Y, Bourgeau-Chavez L et al (2021) Increasing fire and the decline of fire adapted black spruce in the boreal forest. Proc Natl Acad Sci 118(45):e2024872118

Baudena M, Santana VM, Baeza MJ, Bautista S, Eppinga MB, Hemerik L, Garcia Mayor A, Rodriguez F, Valdecantos A, Vallejo VR, Vasques A et al (2020) Increased aridity drives post-fire recovery of Mediterranean forests towards open shrublands. New Phytol 225(4):1500–1515

Beghin R, Lingua E, Garbarino M, Lonati M, Bovio G, Motta R, Marzano R (2010) *Pinus sylvestris* forest regeneration under different post-fire restoration practices in the northwestern Italian Alps. Ecol Eng 36(10):1365–1372

Beloiu M, Stahlmann R, Beierkuhnlein C (2020) High recovery of saplings after severe drought in temperate deciduous forests. Forests 11(5):546

Beloiu M, Stahlmann R, Beierkuhnlein C (2022) Drought impacts in forest canopy and deciduous tree saplings in Central European forests. For Ecol Manag 509:120075

Beschta RL, Rhodes JJ, Kauffman JB, Gresswell RE, Minshall GW, Karr JR, Perry DA, Hauer FR, Frissell CA (2004) Postfire management on forested public lands of the western United States. Conserv Biol 18(4):957–967

Blumröder JS, May F, Härdtle W, Ibisch PL (2021) Forestry contributed to warming of forest ecosystems in northern Germany during the extreme summers of 2018 and 2019. Ecol Solut Evid 2(3):e12087

Blumröder JS, Schmidt F, Gordon A, Grosse S, Ibisch PL (2022) Ecosystemic resilience of a temperate post-fire forest under extreme weather conditions. Front For Glob Change 5:1070958

Boehmer HJ, Galvin S (2024) Climate-induced forest mortality in the Tropical Pacific Islands: what do we really know? Pac Sci 77(2-3):139–161

Boisvenue C, Running SW (2006) Impacts of climate change on natural forest productivity–evidence since the middle of the 20th century. Glob Chang Biol 12(5):862–882

Brando PM, Balch JK, Nepstad DC, Morton DC, Putz FE, Coe MT, Silvério D, Macedo MN, Davidson EA, Nóbrega CC, Alencar A (2014) Abrupt increases in Amazonian tree mortality due to drought–fire interactions. Proc Natl Acad Sci 111(17):6347–6352

Breshears DD, Fontaine JB, Ruthrof KX, Field JP, Feng X, Burger JR, Law DJ, Kala J, Hardy GE (2021) Underappreciated plant vulnerabilities to heat waves. New Phytol 231(1):32–39

Brown PT, Hanley H, Mahesh A, Reed C, Strenfel SJ, Davis SJ, Kochanski AK, Clements CB (2023) Climate warming increases extreme daily wildfire growth risk in California. Nature 621(7980):760–766

Burke KD, Williams JW, Chandler MA, Haywood AM, Lunt DJ, Otto-Bliesner BL (2018) Pliocene and Eocene provide best analogs for near-future climates. Proc Natl Acad Sci 115(52): 13288–13293

Burrell A, Kukavskaya E, Baxter R, Sun Q, Barrett K (2021) Post-fire recruitment failure as a driver of forest to non-forest ecosystem shifts in boreal regions. In: Ecosystem collapse and climate change. Springer International Publishing, Cham, pp 69–100

Burrell AL, Sun Q, Baxter R, Kukavskaya EA, Zhila S, Shestakova T, Rogers BM, Kaduk J, Barrett K (2022) Climate change, fire return intervals and the growing risk of permanent forest loss in boreal Eurasia. Sci Total Environ 831:154885

Canelles Q, Aquilué N, James PM, Lawler J, Brotons L (2021) Global review on interactions between insect pests and other forest disturbances. Landsc Ecol 36:945–972

Castro J, Allen CD, Molina-Morales M, Marañón-Jiménez S, Sánchez-Miranda Á, Zamora R (2011) Salvage logging versus the use of burnt wood as a nurse object to promote post-fire tree seedling establishment. Restor Ecol 19(4):537–544

Catry FX, Moreira F, Tujeira R, Silva JS (2013) Post-fire survival and regeneration of Eucalyptus globulus in forest plantations in Portugal. For Ecol Manag 310:194–203

Cawson JG, Collins L, Parks SA, Nolan RH, Penman TD (2024) Atmospheric dryness removes barriers to the development of large forest fires. Agric For Meteorol 350:109990

Ceccherini G, Duveiller G, Grassi G, Lemoine G, Avitabile V, Pilli R, Cescatti A (2020) Abrupt increase in harvested forest area over Europe after 2015. Nature 583(7814):72–77

Cerioni M, Brabec M, Bače R, Bāders E, Bončina A, Brůna J, Chećko E, Cordonnier T, de Koning JH, Diaci J, Dobrowolska D (2024) Recovery and resilience of European temperate forests after large and severe disturbances. Glob Chang Biol 30(2):e17159

Chakraborty D, Ciceu A, Ballian D, Benito Garzón M, Bolte A, Bozic G, Buchacher R, Čepl J, Cremer E, Ducousso A, Gaviria J et al (2024) Assisted tree migration can preserve the European forest carbon sink under climate change. Nat Clim Chang 14(8):845–852

Cliquet A, Aragão A, Meertens M, Schoukens H, Decleer K (2024) The negotiation process of the EU Nature Restoration Law Proposal: bringing nature back in Europe against the backdrop of political turmoil? Restor Ecol 32(5):e14158

Coop JD (2023) Postfire futures in southwestern forests: climate and landscape influences on trajectories of recovery and conversion. Ecol Appl 33(1):e2725

Coop JD, DeLory TJ, Downing WM, Haire SL, Krawchuk MA, Miller C, Parisien MA, Walker RB (2019) Contributions of fire refugia to resilient ponderosa pine and dry mixed-conifer forest landscapes. Ecosphere 10(7):e02809

Coop JD, Parks SA, Stevens-Rumann CS, Crausbay SD, Higuera PE, Hurteau MD, Tepley A, Whitman E, Assal T, Collins BM, Davis KT (2020) Wildfire-driven forest conversion in western North American landscapes. Bioscience 70(8):659–673

Crockett JL, Hurteau MD (2024) Ability of seedlings to survive heat and drought portends future demographic challenges for five southwestern US conifers. Tree Physiol 44(1):tpad136

Cruz Ó, Riveiro SF, García-Duro J, Casal M, Reyes O (2024) European Atlantic deciduous forests are more resilient to fires than Pinus and Eucalyptus plantations. For Ecol Manag 561:121849

de Dios VR, Hedo J, Camprubí ÀC, Thapa P, Del Castillo EM, de Aragón JM, Bonet JA, Balaguer-Romano R, Díaz-Sierra R, Yebra M, Boer MM (2021) Climate change induced declines in fuel moisture may turn currently fire-free Pyrenean mountain forests into fire-prone ecosystems. Sci Total Environ 797:149104

del Campo AD, Segura-Orenga G, Ceacero CJ, González-Sanchis M, Molina AJ, Reyna S, Hermoso J (2020) Reforesting drylands under novel climates with extreme drought filters: the importance of trait-based species selection. For Ecol Manag 467:118156

Del Campo AD, Segura-Orenga G, Bautista I, Ceacero CJ, González-Sanchis M, Molina AJ, Hermoso J (2021) Assessing reforestation failure at the project scale: the margin for technical improvement under harsh conditions. A case study in a Mediterranean Dryland. Sci Total Environ 796:148952

DellaSala DA, Karr JR, Schoennagel T, Perry D, Noss RF, Lindenmayer D, Beschta R, Hutto RL, Swanson ME, Evans J (2006) Post-fire logging debate ignores many issues. Science 314(5796):51–52

Dirmeyer PA, Balsamo G, Blyth EM, Morrison R, Cooper HM (2021) Land-atmosphere interactions exacerbated the drought and heatwave over northern Europe during summer 2018. AGU Adv 2(2):e2020AV000283

Donato DC, Harvey BJ, Turner MG (2016) Regeneration of montane forests 24 years after the 1988 Yellowstone fires: a fire-catalyzed shift in lower treelines? Ecosphere 7(8):e01410

Doughty CE, Metcalfe DB, Girardin CA, Amézquita FF, Cabrera DG, Huasco WH, Silva-Espejo JE, Araujo-Murakami A, Da Costa MC, Rocha W, Feldpausch TR (2015) Drought impact on forest carbon dynamics and fluxes in Amazonia. Nature 519(7541):78–82

Duivenvoorden E, Wagner B, Nitschke CR, Kasel S (2024) Short-interval, high-severity wildfires cause declines in soil seed bank diversity in montane forests of south-eastern Australia. For Ecol Manag 553:121627

Erb KH, Haberl H, Le Noë J, Tappeiner U, Tasser E, Gingrich S (2022) Changes in perspective needed to forge "no-regret" forest-based climate change mitigation strategies. GCB Bioenergy 14(3):246–257

Ermitão T, Gouveia CM, Bastos A, Russo AC (2024) Recovery following recurrent fires across Mediterranean ecosystems. Glob Chang Biol 30(12):e70013

Fairman TA, Bennett LT, Nitschke CR (2019) Short-interval wildfires increase likelihood of resprouting failure in fire-tolerant trees. J Environ Manag 231:59–65

Falk DA, Watts AC, Thode AE (2019) Scaling ecological resilience. Front Ecol Evol 7:275

Forzieri G, Girardello M, Ceccherini G, Spinoni J, Feyen L, Hartmann H, Beck PS, Camps-Valls G, Chirici G, Mauri A, Cescatti A (2021) Emergent vulnerability to climate-driven disturbances in European forests. Nat Commun 12(1):1081

Franklin JF, Lindenmayer D, MacMahon JA, McKee A, Magnuson J, Perry DA et al (2000) Threads of continuity. Conserv Biol Pract 1(1):8–16

Friedlingstein P, O'Sullivan M, Jones MW, Andrew RM, Hauck J, Landschützer P, Le Quéré C, Li H, Luijkx IT, Olsen A, Peters GP et al (2025) Global carbon budget 2024. Earth Syst Sci Data 17(3):965–1039

Gamelin BL, Feinstein J, Wang J, Bessac J, Yan E, Kotamarthi VR (2022) Projected US drought extremes through the twenty-first century with vapor pressure deficit. Sci Rep 12(1):8615

García-García A, Cuesta-Valero FJ, Miralles DG, Mahecha MD, Quaas J, Reichstein M, Zscheischler J, Peng J (2023) Soil heat extremes can outpace air temperature extremes. Nat Clim Chang 13(11):1237–1241

Gazol A, Camarero JJ, Sangüesa-Barreda G, Vicente-Serrano SM (2018) Post-drought resilience after forest die-off: shifts in regeneration, composition, growth and productivity. Front Plant Sci 9:1546

Gibson RK, Hislop S (2022) Signs of resilience in resprouting *Eucalyptus* forests, but areas of concern: 1 year of post-fire recovery from Australia's Black Summer of 2019–2020. Int J Wildland Fire 31(5):545–557

Goessling HF, Rackow T, Jung T (2025) Recent global temperature surge intensified by record-low planetary albedo. Science 387(6729):68–73

González-De Vega S, De las Heras J, Moya D (2016) Resilience of Mediterranean terrestrial ecosystems and fire severity in semiarid areas: responses of Aleppo pine forests in the short, mid and long term. Sci Total Environ 573:1171–1177

Grossiord C, Buckley TN, Cernusak LA, Novick KA, Poulter B, Siegwolf RT, Sperry JS, McDowell NG (2020) Plant responses to rising vapor pressure deficit. New Phytol 226(6): 1550–1566

Guillermic M, Misra S, Eagle R, Tripati A (2021) Atmospheric CO2 estimates for the Miocene to Pleistocene based on foraminiferal δ11B at Ocean Drilling Program Sites 806 and 807 in the Western Equatorial Pacific. Clim Past Discuss 2021:1–40

Hartmann H, Bastos A, Das AJ, Esquivel-Muelbert A, Hammond WM, Martínez-Vilalta J, McDowell NG, Powers JS, Pugh TA, Ruthrof KX, Allen CD (2022) Climate change risks to global forest health: emergence of unexpected events of elevated tree mortality worldwide. Annu Rev Plant Biol 73(1):673–702

Heinken T, Blumröder JS, Adhikari Y, Balthasar C, Binder A, Birkhofer K, Bischoff A, Brosinsky A, Bubner B, Clerc D, Djoudi EA et al (2024) PYROPHOB-A post-fire ecosystem research project to inform management for resilient forest development. Tuexenia 44:33–65

Hering D, Schürings C, Wenskus F, Blackstock K, Borja A, Birk S, Bullock C, Carvalho L, Dagher-Kharrat MB, Lakner S, Lovrić N et al (2023) Securing success for the nature restoration law. Science 382(6676):1248–1250

Hernández-Duarte A, Saavedra F, González E, Miranda A, Francois JP, Somos-Valenzuela M, Sibold J (2024) Effects of drought and fire severity interaction on short-term post-fire recovery of the Mediterranean forest of South America. Fire 7(12):428

Huang M, Zhai P (2025) Protracted vegetation recovery after compound drought and hot extreme compared to general drought. Environ Res Lett 20(2):024001

Ibáñez I, Acharya K, Juno E, Karounos C, Lee BR, McCollum C, Schaffer-Morrison S, Tourville J (2019) Forest resilience under global environmental change: do we have the information we need? A systematic review. PLoS One 14(9):e0222207

Jain P, Barber QE, Taylor SW, Whitman E, Castellanos Acuna D, Boulanger Y, Chavardès RD, Chen J, Englefield P, Flannigan M, Girardin MP (2024) Drivers and impacts of the record-breaking 2023 wildfire season in Canada. Nat Commun 15(1):6764

Jandl R, Spathelf P, Bolte A, Prescott CE (2019) Forest adaptation to climate change—is non-management an option? Ann For Sci 76(2):48

Johnstone JF, Allen CD, Franklin JF, Frelich LE, Harvey BJ, Higuera PE, Mack MC, Meentemeyer RK, Metz MR, Perry GL, Schoennagel T (2016) Changing disturbance regimes, ecological memory, and forest resilience. Front Ecol Environ 14(7):369–378

Jones MW, Abatzoglou JT, Veraverbeke S, Andela N, Lasslop G, Forkel M, Smith AJ, Burton C, Betts RA, van der Werf GR, Sitch S (2022) Global and regional trends and drivers of fire under climate change. Rev Geophys 60(3):e2020RG000726

Jones MW, Veraverbeke S, Andela N, Doerr SH, Kolden C, Mataveli G, Pettinari ML, Le Quéré C, Rosan TM, van der Werf GR, van Wees D et al (2024) Global rise in forest fire emissions linked to climate change in the extratropics. Science 386(6719):eadl5889

Jouy F, Schüle M, Adhikari Y, Binder A, Clerc D, Gerwin W, Heinken T, Raab T, Repmann F, Rönnefarth S, Schirrmacher M et al (2025) Factors impacting the variability of post-fire forest regeneration in central European pine plantations. Restor Ecol 33:e70017

Kirchmeier-Young MC, Malinina E, Barber QE, Garcia Perdomo K, Curasi SR, Liang Y, Jain P, Gillett NP, Parisien MA, Cannon AJ, Lima AR (2024) Human driven climate change increased the likelihood of the 2023 record area burned in Canada. NPJ Clim Atmos Sci 7(1):316

Klesse S, Peters RL, Alfaro-Sánchez R, Badeau V, Baittinger C, Battipaglia G, Bert D, Biondi F, Bosela M, Budeanu M, Čada V (2024) No future growth enhancement expected at the northern edge for European beech due to continued water limitation. Glob Chang Biol 30(10):e17546

Knutzen F, Averbeck P, Barrasso C, Bouwer LM, Gardiner B, Grünzweig JM, Hänel S, Haustein K, Johannessen MR, Kollet S, Müller MM et al (2025) Impacts on and damage to European forests from the 2018–2022 heat and drought events. Nat Hazards Earth Syst Sci 25(1):77–117

Kurbanov E, Tarasova L, Yakhyayev A, Vorobev O, Gozalov S, Lezhnin S, Wang J, Sha J, Dergunov D, Yastrebova A (2024) Detecting trends in post-fire forest recovery in Middle Volga from 2000 to 2023. Forests 15(11):1919

Landesmann JB, Tiribelli F, Paritsis J, Veblen TT, Kitzberger T (2021) Increased fire severity triggers positive feedbacks of greater vegetation flammability and favors plant community-type conversions. J Veg Sci 32(1):e12936

Lawal S, Koch FH, Scheller RM, Costanza J (2025) Forest demographic changes across Texas associated with hot drought. Ecol Indic 171:113117

Le Breton TD, Lyons MB, Nolan RH, Penman T, Williamson GJ, Ooi MK (2022) Megafire-induced interval squeeze threatens vegetation at landscape scales. Front Ecol Environ 20(5): 327–334

Lemus-Canovas M, Insua-Costa D, Trigo RM, Miralles DG (2024) Record-shattering 2023 Spring heatwave in western Mediterranean amplified by long-term drought. NPJ Clim Atmos Sci 7(1): 25

Leuschner C (2020) Drought response of European beech (*Fagus sylvatica* L.)—a review. Perspect Plant Ecol Evol Syst 47:125576

Leuschner C, Meinzer FC (2024) Drought resistance and drought adaptation of Douglas-fir (*Pseudotsuga menziesii*) – a review. Perspect Plant Ecol Evol Syst 65:125829
Leuschner C, Weithmann G, Bat-Enerel B, Weigel R (2023) The future of European beech in northern Germany—climate change vulnerability and adaptation potential. Forests 14(7):1448
Leverkus AB, Lorite J, Navarro FB, Sánchez-Cañete EP, Castro J (2014) Post-fire salvage logging alters species composition and reduces cover, richness, and diversity in Mediterranean plant communities. J Environ Manag 133:323–331
Lloret F, Jaime LA, Margalef-Marrase J, Pérez-Navarro MA, Batllori E (2022) Short-term forest resilience after drought-induced die-off in southwestern European forests. Sci Total Environ 806:150940
Loehman R, Flatley W, Holsinger L, Thode A (2018) Can land management buffer impacts of climate changes and altered fire regimes on ecosystems of the southwestern United States? Forests 9(4):192
Maes SL, Perring MP, Cohen R, Akinnifesi FK, Bargués-Tobella A, Bastin JF, Bauters M, Bernardino PN, Brancalion PH, Bullock JM, Ellison D et al (2024) Explore before you restore: incorporating complex systems thinking in ecosystem restoration. J Appl Ecol 61(5):922–939
Marangon D, Marchi N, Lingua E (2022) Windthrown elements: a key point improving microsite amelioration and browsing protection to transplanted seedlings. For Ecol Manag 508:120050
Marañón-Jiménez S, Castro J, Querejeta JI, Fernández-Ondoño E, Allen CD (2013) Post-fire wood management alters water stress, growth, and performance of pine regeneration in a Mediterranean ecosystem. For Ecol Manag 308:231–239
Marcolin E, Marzano R, Vitali A, Garbarino M, Lingua E (2019) Post-fire management impact on natural forest regeneration through altered microsite conditions. Forests 10(11):1014
Maringer J, Wohlgemuth T, Hacket-Pain A, Ascoli D, Berretti R, Conedera M (2020) Drivers of persistent post-fire recruitment in European beech forests. Sci Total Environ 699:134006
Martín-Alcón S, Coll L (2016) Unraveling the relative importance of factors driving post-fire regeneration trajectories in non-serotinous *Pinus nigra* forests. For Ecol Manag 361:13–22
Martinez del Castillo E, Zang CS, Buras A, Hacket-Pain A, Esper J, Serrano-Notivoli R, Hartl C, Weigel R, Klesse S, Resco de Dios V, Scharnweber T (2022) Climate-change-driven growth decline of European beech forests. Commun Biol 5(1):163
Martínez-Sánchez JJ, Ferrandis P, de las Heras J, Herranz JM (1999) Effect of burnt wood removal on the natural regeneration of *Pinus halepensis* after fire in a pine forest in Tus valley (SE Spain). For Ecol Manag 123(1):1–10
Marzano R, Garbarino M, Marcolin E, Pividori M, Lingua E (2013) Deadwood anisotropic facilitation on seedling establishment after a stand-replacing wildfire in Aosta Valley (NW Italy). Ecol Eng 51:117–122
Mata-Guel EO, Soh MC, Butler CW, Morris RJ, Razgour O, Peh KS (2023) Impacts of anthropogenic climate change on tropical montane forests: an appraisal of the evidence. Biol Rev 98(4): 1200–1224
McDowell NG, Allen CD, Anderson-Teixeira K, Aukema BH, Bond-Lamberty B, Chini L, Clark JS, Dietze M, Grossiord C, Hanbury-Brown A, Hurtt GC et al (2020) Pervasive shifts in forest dynamics in a changing world. Science 368(6494):eaaz9463
McFarland JR, Coop JD, Balik JA, Rodman KC, Parks SA, Stevens-Rumann CS (2025) Extreme fire spread events burn more severely and homogenize postfire landscapes in the southwestern United States. Glob Chang Biol 31(2):e70106
Miralles DG, Teuling AJ, Van Heerwaarden CC, Vilà-Guerau de Arellano J (2014) Mega-heatwave temperatures due to combined soil desiccation and atmospheric heat accumulation. Nat Geosci 7(5):345–349
Miralles DG, Gentine P, Seneviratne SI, Teuling AJ (2019) Land–atmospheric feedbacks during droughts and heatwaves: state of the science and current challenges. Ann N Y Acad Sci 1436(1): 19–35

Moreira F, Vallejo R (2009) What to do after fire? Post-fire restoration. In: Birot Y (ed) Living with wildfires: what science can tell us, EFI discussion paper 15. European Forest Institute, Joensuu, pp 53–58

Moya D, de las Heras J, López-Serrano FR, Ferrandis P (2015) Post-fire seedling recruitment and morpho-ecophysiological responses to induced drought and salvage logging in *Pinus halepensis* Mill. stands. Forests 6(6):1858–1877

Nikinmaa L, Lindner M, Cantarello E, Jump AS, Seidl R, Winkel G, Muys B (2020) Reviewing the use of resilience concepts in forest sciences. Curr For Rep 6:61–80

North MP, Stevens JT, Greene DF, Coppoletta M, Knapp EE, Latimer AM, Restaino CM, Tompkins RE, Welch KR, York RA, Young DJ et al (2019) Tamm review: reforestation for resilience in dry western US forests. For Ecol Manag 432:209–224

Novick KA, Ficklin DL, Grossiord C, Konings AG, Martínez-Vilalta J, Sadok W, Trugman AT, Williams AP, Wright AJ, Abatzoglou JT, Dannenberg MP (2024) The impacts of rising vapour pressure deficit in natural and managed ecosystems. Plant Cell Environ 47(9):3561–3589

Paneghel M, Torné G, Morin X, Alday JG, Coll L (2024) Increasing temperature threatens post-fire auto-successional dynamics of a Mediterranean obligate seeder. J Ecol 112(12):2929–2943

Pardos M, Del Río M, Pretzsch H, Jactel H, Bielak K, Bravo F, Brazaitis G, Defossez E, Engel M, Godvod K, Jacobs K et al (2021) The greater resilience of mixed forests to drought mainly depends on their composition: analysis along a climate gradient across Europe. For Ecol Manag 481:118687

Parks SA, Dobrowski SZ, Shaw JD, Miller C (2019) Living on the edge: trailing edge forests at risk of fire-facilitated conversion to non-forest. Ecosphere 10(3):e02651

Pausas JG, Bladé C, Valdecantos A, Seva JP, Fuentes D, Alloza JA, Vilagrosa A, Bautista S, Cortina J, Vallejo R (2004) Pines and oaks in the restoration of Mediterranean landscapes of Spain: new perspectives for an old practice—a review. Plant Ecol 171:209–220

Pearson PN, Palmer MR (2000) Atmospheric carbon dioxide concentrations over the past 60 million years. Nature 406(6797):695–699

Peterson DL (2009) Effects of timber harvest following wildfire in western North America. US Department of Agriculture, Forest Service, Pacific Northwest Research Station, Portland

Petritan AM, Petritan IC, Hevia A, Walentowski H, Bouriaud O, Sánchez-Salguero R (2021) Climate warming predispose sessile oak forests to drought-induced tree mortality regardless of management legacies. For Ecol Manag 491:119097

Qin Y, Xiao X, Wigneron JP, Ciais P, Canadell JG, Brandt M, Li X, Fan L, Wu X, Tang H, Dubayah R et al (2022) Large loss and rapid recovery of vegetation cover and aboveground biomass over forest areas in Australia during 2019–2020. Remote Sens Environ 278:113087

Richardson D, Black AS, Irving D, Matear RJ, Monselesan DP, Risbey JS, Squire DT, Tozer CR (2022) Global increase in wildfire potential from compound fire weather and drought. NPJ Clim Atmos Sci 5(1):23

Ridder NN, Ukkola AM, Pitman AJ, Perkins-Kirkpatrick SE (2022) Increased occurrence of high impact compound events under climate change. NPJ Clim Atmos Sci 5(1):3

Rifai SW, De Kauwe MG, Gallagher RV, Cernusak LA, Meir P, Pitman AJ (2024) Burn severity and post-fire weather are key to predicting time-to-recover from Australian forest fires. Earths Future 12(4):e2023EF003780

Rodman KC, Veblen TT, Battaglia MA, Chambers ME, Fornwalt PJ, Holden ZA, Kolb TE, Ouzts JR, Rother MT (2020) A changing climate is snuffing out post-fire recovery in montane forests. Glob Ecol Biogeogr 29(11):2039–2051

Rodrigues M, Camprubí ÀC, Balaguer-Romano R, Megía CJ, Castañares F, Ruffault J, Fernandes PM, de Dios VR (2023) Drivers and implications of the extreme 2022 wildfire season in Southwest Europe. Sci Total Environ 859:160320

Rodrigues M, de Dios VR, Sil Â, Camprubí AC, Fernandes PM (2024) VPD-based models of dead fine fuel moisture provide best estimates in a global dataset. Agric For Meteorol 346:109868

Rohde R (2025) Global temperature report for 2024. https://berkeleyearth.org/global-temperature-report-for-2024/

Romanou A, Hegerl G, Seneviratne S, Abis B, Bastos A, Conversi A, Landolfi A, Kim H, Lerner P, Mekus J, Otto-Bliesner B (2024) Extreme events contributing to tipping elements and tipping points. Surv Geophys 46:1–46

Rönnefarth S, Adhikari Y, Jouy F, Ibisch PL (in prep.) From ashes to extremes: dynamics of land surface temperature and vapor pressure deficit in post-fire pine forests

Ruggirello MJ, Bustamante G, Rodriguez P, Cruz-Alonso V, Soler R (2023) Post-fire forest recovery at high latitudes: tree regeneration dominated by fire-adapted, early-seral species increases with latitude. Ann For Sci 80(1):47

Salesa D, Baeza MJ, Pérez-Ferrándiz E, Santana VM (2022) Longer summer seasons after fire induce permanent drought legacy effects on Mediterranean plant communities dominated by obligate seeders. Sci Total Environ 822:153655

Sangüesa-Barreda G, Camarero JJ, Oliva J, Montes F, Gazol A (2015) Past logging, drought and pathogens interact and contribute to forest dieback. Agric For Meteorol 208:85–94

Schillerberg TA, Tian D (2024) Global assessment of compound climate extremes and exposures of population, agriculture, and forest lands under two climate scenarios. Earths Future 12(9): e2024EF004845

Schuldt B, Buras A, Arend M, Vitasse Y, Beierkuhnlein C, Damm A, Gharun M, Grams TE, Hauck M, Hajek P, Hartmann H (2020) A first assessment of the impact of the extreme 2018 summer drought on Central European forests. Basic Appl Ecol 45:86–103

Schüle M, Domes G, Schwanitz C, Heinken T (2023) Early natural tree regeneration after wildfire in a Central European Scots pine forest: forest management, fire severity and distance matters. For Ecol Manag 539:120999

Schumacher DL, Keune J, Dirmeyer P, Miralles DG (2022) Drought self-propagation in drylands due to land–atmosphere feedbacks. Nat Geosci 15(4):262–268

Schuurman GW, Cole DN, Cravens AE, Covington S, Crausbay SD, Hoffman CH, Lawrence DJ, Magness DR, Morton JM, Nelson EA, O'Malley R (2022) Navigating ecological transformation: resist–accept–direct as a path to a new resource management paradigm. Bioscience 72(1): 16–29

Seidl R, Thom D, Kautz M, Martin-Benito D, Peltoniemi M, Vacchiano G, Wild J, Ascoli D, Petr M, Honkaniemi J, Lexer MJ (2017) Forest disturbances under climate change. Nat Clim Chang 7(6):395–402

Seidl R, Potterf M, Müller J, Turner MG, Rammer W (2024) Patterns of early post-disturbance reorganization in Central European forests. Proc R Soc B 291(2031):20240625

Sheil D, Ibisch PL (2025) Reversing desiccation: cooler, moister, greener. eLetter to Seo KW, Ryu D, Jeon T, Youm K, Kim JS, Oh EH, Chen J, Famiglietti JS, Wilson CR (2025) Abrupt sea level rise and Earth's gradual pole shift reveal permanent hydrological regime changes in the 21st century. Science 387(6741):1408–1413. https://doi.org/10.1126/science.adq6529

Shrestha S, Williams CA, Rogers BM, Rogan J, Kulakowski D (2024) Divergent biophysical responses of western United States forests to wildfire driven by eco-climatic gradients. Biogeosciences 21(9):2207–2226

Shvetsov EG, Kukavskaya EA, Buryak LV, Barrett K (2019) Assessment of post-fire vegetation recovery in Southern Siberia using remote sensing observations. Environ Res Lett 4(5):055001

Sorenson QM, Young DJ, Latimer AM (2025) Tree planting outcomes after severe wildfire depend on climate, competition, and priority. For Ecol Manag 575:122346

Spanos I, Raftoyannis Y, Platis P, Xanthopoulou E (2010) Post-fire management and recovery of a pine forest in Greece. Web Ecol 10(1):27–31

Stevens-Rumann CS, Kemp KB, Higuera PE, Harvey BJ, Rother MT, Donato DC, Morgan P, Veblen TT (2018) Evidence for declining forest resilience to wildfires under climate change. Ecol Lett 21(2):243–252

Strickland MK, Jenkins MA, Ma Z, Murray BD (2024) How has the concept of resilience been applied in research across forest regions? Front Ecol Environ 22(3):1–9

Sun Q, Burrell A, Barrett K, Kukavskaya E, Buryak L, Kaduk J, Baxter R (2021) Climate variability may delay post-fire recovery of boreal forest in southern Siberia, Russia. Remote Sens 13(12): 2247

Tepley AJ, Thompson JR, Epstein HE, Anderson-Teixeira KJ (2017) Vulnerability to forest loss through altered postfire recovery dynamics in a warming climate in the Klamath Mountains. Glob Chang Biol 23(10):4117–4132

Treydte K, Liu L, Padrón RS, Martínez-Sancho E, Babst F, Frank DC, Gessler A, Kahmen A, Poulter B, Seneviratne SI, Stegehuis AI (2024) Recent human-induced atmospheric drying across Europe unprecedented in the last 400 years. Nat Geosci 17(1):58–65

Tripathy KP, Mishra AK (2023) How unusual is the 2022 European compound drought and heatwave event? Geophys Res Lett 50(15):e2023GL105453

Turner MG, Braziunas KH, Hansen WD, Harvey BJ (2019) Short-interval severe fire erodes the resilience of subalpine lodgepole pine forests. Proc Natl Acad Sci 116(23):11319–11328

Vallet L, Schwartz M, Ciais P, Van Wees D, De Truchis A, Mouillot F (2023) High-resolution data reveal a surge of biomass loss from temperate and Atlantic pine forests, contextualizing the 2022 fire season distinctiveness in France. Biogeosciences 20(18):3803–3825

Vancutsem C, Achard F, Pekel JF, Vieilledent G, Carboni S, Simonetti D, Gallego J, Aragao LE, Nasi R (2021) Long-term (1990–2019) monitoring of forest cover changes in the humid tropics. Sci Adv 7(10):eabe1603

Viana-Soto A, Aguado I, Salas J, García M (2020) Identifying post-fire recovery trajectories and driving factors using landsat time series in fire-prone mediterranean pine forests. Remote Sens 12(9):1499

Volkova L, Paul KI, Roxburgh SH, Weston CJ (2025) Recovery of south-eastern Australian temperate forest carbon is influenced by post-fire drought as well as fire severity. For Ecol Manag 585:122666

Walden L, Fontaine JB, Ruthrof KX, Matusick G, Harper RJ (2023) Drought then wildfire reveals a compound disturbance in a resprouting forest. Ecol Appl 33(2):e2775

Wang C, Li Z, Chen Y, Ouyang L, Zhao H, Zhu J, Wang J, Zhao Y (2024) Characteristic changes in compound drought and heatwave events under climate change. Atmos Res 305:107440

Wang W, Wang X, Flannigan MD, Guindon L, Swystun T, Castellanos-Acuna D, Wu W, Wang G (2025) Canadian forests are more conducive to high-severity fires in recent decades. Science 387(6729):91–97

Wegler M, Kuenzer C (2024) Potential of earth observation to assess the impact of climate change and extreme weather events in temperate forests—a review. Remote Sens 16(12):2224

Windisch MG, Humpenöder F, Merfort L, Bauer N, Luderer G, Dietrich JP, Heinke J, Müller C, Abrahao G, Lotze-Campen H, Popp A (2025) Hedging our bet on forest permanence for the economic viability of climate targets. Nat Commun 16(1):2460

Wion AP, Stevens JT, Beeley K, Oertel R, Margolis EQ, Allen CD (2024) Multidecadal vegetation transformations of a New Mexico ponderosa pine landscape after severe fires and aerial seeding. Ecol Appl 34(6):e3008

Xoplaki E, Ellsäßer F, Grieger J, Nissen KM, Pinto JG, Augenstein M, Chen TC, Feldmann H, Friederichs P, Gliksman D, Goulier L et al (2025) Compound events in Germany in 2018: drivers and case studies. Nat Hazards Earth Syst Sci 25(2):541–564

Xu H, Chen HW, Chen D, Wang Y, Yue X, He B, Guo L, Yuan W, Zhong Z, Huang L, Zheng F (2024) Global patterns and drivers of post-fire vegetation productivity recovery. Nat Geosci 17(9):874–881

Young DJ, Werner CM, Welch KR, Young TP, Safford HD, Latimer AM (2019) Post-fire forest regeneration shows limited climate tracking and potential for drought-induced type conversion. Ecology 100(2):e02571

Zahura FT, Bisht G, Li Z, McKnight S, Chen X (2024) Impact of topography and climate on post-fire vegetation recovery across different burn severity and land cover types through random forest. Eco Inform 82:102757

Zhang Y, Peng C, Li W, Tian L, Zhu Q, Chen H, Fang X, Zhang G, Liu G, Mu X, Li Z (2016) Multiple afforestation programs accelerate the greenness in the "Three North" region of China from 1982 to 2013. Ecol Indic 61:404–412

Zhang X, Zhou T, Zhang W, Ren L, Jiang J, Hu S, Zuo M, Zhang L, Man W (2023) Increased impact of heat domes on 2021-like heat extremes in North America under global warming. Nat Commun 14(1):1690

Zhang X, Yin G, Ma Y, Fan J, Zhou J (2024) Comparison between artificial restoration and natural recovery in vegetation regrowth following high-frequency fire disturbances in the Hengduan Mountains, Southwest China. Ecol Indic 167:112692

Zhang M, Yuan X, Zeng Z, Pan M, Wu P, Xiao J, Keenan TF (2025) A pronounced decline in northern vegetation resistance to flash droughts from 2001 to 2022. Nat Commun 16(1):2984

Zhou G, Peng C, Li Y, Liu S, Zhang Q, Tang X, Liu J, Yan J, Zhang D, Chu G (2013) A climate change-induced threat to the ecological resilience of a subtropical monsoon evergreen broad-leaved forest in Southern China. Glob Chang Biol 19(4):1197–1210

Zhou S, Wu S, Gao J, Liu L, Li D, Yan R, Wang J (2024) Increased stress from compound drought and heat events on vegetation. Sci Total Environ 949:175113

Zhu Q, Liu Y, Wu Y (2024) Timescale effects on the driving mechanisms of post-fire recruitment in Larix gmelinii population. For Ecol Manag 554:121691

Zscheischler J, Westra S, Van Den Hurk BJ, Seneviratne SI, Ward PJ, Pitman A, AghaKouchak A, Bresch DN, Leonard M, Wahl T, Zhang X et al (2018) Future climate risk from compound events. Nat Clim Chang 8(6):469–477

# Four Decades of Forest Development in the Context of Nitrogen Eutrophication and Climate Change

**Sabine Braun, Simon Tresch, Sven-Eric Hopf, and Christian Schindler**

## Contents



Communicated by Christoph Leuschner


S. Braun (✉), S. Tresch, and S.-E. Hopf
Institute for Applied Plant Biology, Witterswil, Switzerland
e-mail: sabine.braun@iap.ch; simon.tresch@iap.ch; sven.hopf@iap.ch

C. Schindler
University of Basel, Swiss Tropical and Public Health Institute, Basel, Switzerland
e-mail: christian.schindler@unibas.ch







Progress in Botany (2026) 86: 211–267, https://doi.org/10.1007/124_2025_94,
Published online: 08 October 2025

**Abstract** Long-term forest monitoring allows to document developments which would otherwise remain undetected. Observed changes need to be interpreted by disentangling the various influencing factors. In addition to robust measurements, the selection of sites and the coverage of important ecological gradients are crucial for epidemiological data analysis. The significance of monitoring has been increasing in recent years, as the environment has undergone rapid changes and climate change has become the predominant factor influencing tree health. Four decades of monitoring European beech (*Fagus sylvatica*), Norway spruce (*Picea abies*), and oak species (*Quercus robur, Q. petraea, Q. pubescens*) in 247 sites of the intercantonal long-term forest monitoring in Switzerland show that direct or indirect effects of air pollution, such as nitrogen (N) deposition, causing eutrophication and soil acidification, continue to exert a significant influence on forest health. Evidence of soil acidification was provided through the measurement of soil solution samples on a monthly basis at 47 sites over a 25-year period, encompassing both topsoil and subsoil layers. The ratio between base cations and aluminium (BC/Al ratio), which is an indicator for soil acidification, revealed an increasing trend in soil acidification. N deposition, as the main driver of soil acidification in Switzerland, is causing nutrient losses in the soil and tree nutrient imbalances. The mycorrhiza communities are affected as well, impacting drought and pathogen resistance, as evidenced by the repeated summer drought years (2018–2023). High N deposition is linked to substantial growth reduction in beech and Norway spruce, further exacerbated by drought. Additionally, ozone was found to reduce growth and to increase water uptake.

The hot summer 2018 was a tipping point for the vitality of European beech in Switzerland with the occurrence of acute cavitation and subsequent dieback of branches and whole trees, especially in lower elevations. In beech, the proportion of strongly damaged trees increased during the 2019–2024 period compared to 1984–2018 by a factor of 6.9 and the mortality by a factor of 4.2. Our data show that the cumulative impact of the consecutive drought years from 2015 to 2022 is a significant factor contributing to the exacerbated damage observed after 2018 in comparison to previous drought years, such as 2003.

The observed vitality changes in Norway spruce in the last decades have accelerated in recent time. During the drought years 2018, 2019, 2020, and 2022, mortality increased by a factor of 9.9 compared to the years 1985–2017. For Norway spruce, we found evidence of increased mortality under high N deposition. The oaks performed better, even though pubescent oak was also affected by the drought, especially on dry sites. In pubescent oak, the number of strongly damaged trees increased by a factor of 16.6 during the period 2019–2024, and mortality increased by a factor of 15.5. Stem volume increment per ground area has recently decreased in all observed tree species. This is due to changes in forest structure as a result of mortality and wind damage, and, in the case of beech, also due to a decrease in growth at the tree level.

The rapid changes caused by climate change, in combination with N eutrophication effects and the appearance of new parasites, emphasize the importance of intensive forest monitoring with cause-related data analysis in the future.

**Keywords** Drought, Fagus sylvatica, Nitrogen deposition, Picea abies, Quercus sp., Tree nutrition

# 1 Introduction

Human activity has greatly changed the environmental conditions forests are exposed to. While in the 1980s, the research focus was on acid deposition and the consequences of acidification (Reuss et al. 1987; Ulrich 1991), the focus later changed to eutrophication driven by increased deposition of nitrogen (N) (Nihlgård 1985; Aber et al. 1998). In former times, N was a macronutrient, limiting forest growth in large parts of Europe. Thus, the increased growth observed in European forests in the 1980s as described by Spiecker et al. (1996) has been attributed to increases in temperature, $CO_2$ and N deposition (Pretzsch et al. 2014). The visible positive relationship between tree growth and N deposition was used in carbon sequestration models, and there has been much debate about how much carbon can be sequestered by additional nitrogen deposition (Sutton et al. 2008; De Vries and Posch 2011).

Anthropogenic climate change has led to global warming, which has occurred at a faster rate over land surfaces with an average increase by 1.5°C between 2006 and 2015 compared to 1850–1900 (Shukla et al. 2019). In Switzerland, the average temperature increase has been 2.5°C for the last decade (2011–2020) compared to 1871–1900 (MeteoSwiss 2021). The projected heat waves are expected to occur more frequently and to last longer (IPCC 2021), leading to hotter droughts (Allen et al. 2010), as summer precipitation is expected to decrease in parts of Central Europe (IPCC 2021). The European heatwave in 2003 has highlighted the urgency of better understanding and predicting the response of ecosystems to extreme drought events due to large impacts on human health, crop yields, and forest health (Buras et al. 2020; García-Herrera et al. 2010). The drought event in 2018 was even more severe with unprecedented hot droughts in many parts of Central and Northern Europe (Buras et al. 2020). It was one of the most severe and long-lasting droughts ever recorded, including in Switzerland, and it caused massive damage to the Swiss forests (Schuldt et al. 2020).

In the 1980s, a forest monitoring network was established in Switzerland by Swiss cantons (hereafter called intercantonal forest monitoring program) to gain information on the state and development of the forest and to detect the causes of the observed changes (Braun et al. 2020). The monitoring plots were selected such as to cover gradients in stand age, altitude, drought, soil base saturation, N deposition, and ozone concentration in the atmosphere. With sufficient repetitions, data can be analyzed with epidemiological methods (Braun et al. 2017), allowing quantitative relationships to be drawn on ecological effects, including interactions between different driving factors.

## 2 Materials and Methods

The monitoring plots cover important ecological gradients (Fig. 1) and include sufficient replicates for N deposition, drought, soil acidity, tree age, and altitude (Braun et al. 2017). In total, 9846 European beech (*Fagus sylvatica*), 9131 Norway spruce (*Picea abies*), and 2738 oak trees (*Quercus robur, Q. petraea, Q. pubescens*) were under observation in 108, 86, and 53 plots, respectively, starting in 1984 (beech and Norway spruce) and 2009 (oaks) (Table 1). The time interval of field measurements is shown in Table 2. Information on the analysis of tree mortality, crown defoliation, and discoloration including model selection and data preparation can be found in Braun et al. (2021) for beech and in Tresch et al. (2023) for Norway spruce. Details on growth measurements are given in Braun et al. (2022), on methods of foliar analysis in Braun et al. (2020) and of soil analysis in Braun et al. (2020). The use of gradient studies in epidemiological research and modeling methods has been addressed in an ICP Vegetation workshop (Braun et al. 2017). Equal spatial distribution is not a prerequisite for such studies, as long as there are no problems with spatial autocorrelation. However, a good coverage of ecological gradients with sufficient replicates, the inclusion of important covariates, and the analysis of residuals against the predictors are essential.

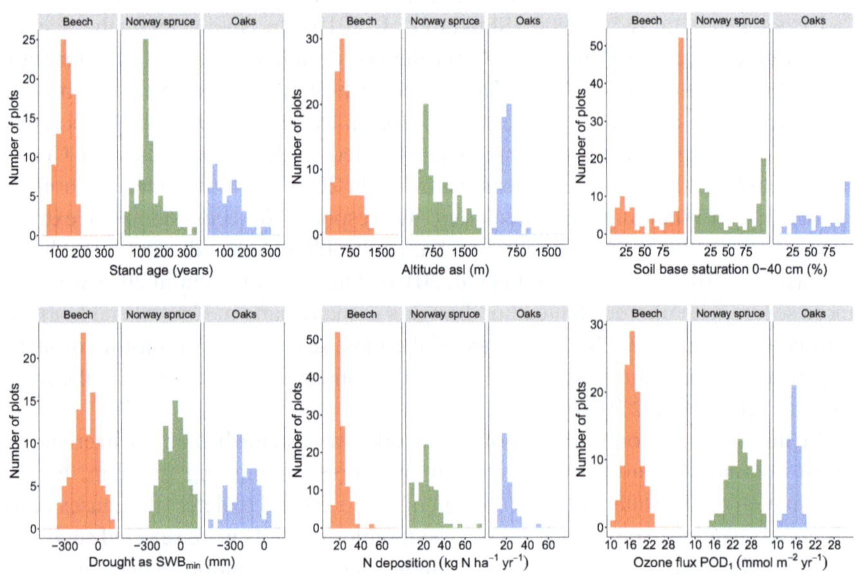


**Fig. 1** Gradients covered with the monitoring plots of the intercantonal forest observation program. Soil base saturation: saturation of the soil cation exchangers with base cations, averaged over a depth of 0–40 cm. $SWB_{min}$: minimum site water balance according to Spellmann et al. (2007) and Braun (2015) for the years 1981–2020. Nitrogen deposition: Total nitrogen deposition with a 1 ha spatial resolution modeled by Rihm and Künzle (2019), reference year 2020. Ozone flux: ozone uptake (CLRTAP and ICP Vegetation 2017) calculated using the $DO_3SE$ model (Büker et al. 2012) for the years 1991–2019

**Table 1** Overview over the intercantonal forest monitoring program in Switzerland since 1984. The monitoring of European beech (*Fagus sylvatica*) and Norway spruce (*Picea abies*) has started in 1984. The three oak species (*Quercus robur, Q. petraea, Q. pubescens*) were added to the monitoring network in 2009. The table lists the total numbers of plots and individual trees ever assessed, including removed and dead trees. The age has been calculated in reference to the year 2020

| | | European beech (*Fagus sylvatica*) | Norway spruce (*Picea abies*) | Oak species (*Quercus* spp.) |
|---|---|---|---|---|
| Plots | | 108 | 86 | 53 |
| Trees | | 9846 | 9131 | 2738 |
| Starting year | | 1984 | 1984 | 2009 |
| Age (year 2020) | Min | 62 | 46 | 34 |
| | Mean | 130 | 143 | 127 |
| | Max | 196 | 328 | 300 |
| Altitude (m) | Min | 260 | 290 | 270 |
| | Mean | 617 | 878 | 521 |
| | Max | 1220 | 1870 | 1010 |
| Tree height (m) | Min | 21 | 18 | 10 |
| | Mean | 33 | 32 | 26 |
| | Max | 45 | 44 | 36 |
| BHD (cm) | Min | 29 | 25 | 15 |
| | Mean | 52 | 52 | 48 |
| | Max | 74 | 72 | 94 |

**Table 2** Assessment intervals of field measurements in the intercantonal forest observation monitoring program

| Assessment interval | |
|---|---|
| Crown condition | 1 year |
| Mortality, stem breakage, and uprooting | 1 year |
| Pathogens and symptoms | 1 year |
| Foliar nutrients and shoot phenology | 4 years |
| Stem increment | 4 years |
| Stand density | 12 years |
| Soil chemistry | 12 years |
| Soil physics | Once |

# 3 Observed Developments

## *3.1 Crown Transparency*

The visual assessment of crown transparency is a fast measure for tree vitality which is widely used in forest monitoring (ICP Forests 2016). The assessment in the plots is similar to the ICP-Forest method, but differs in the evaluation, as the data assessment started 11 years earlier. The same method has been used since then, which is important for the continuity of the data. Quality control is carried out annually at several locations by an experienced field team with many years of experience. It is

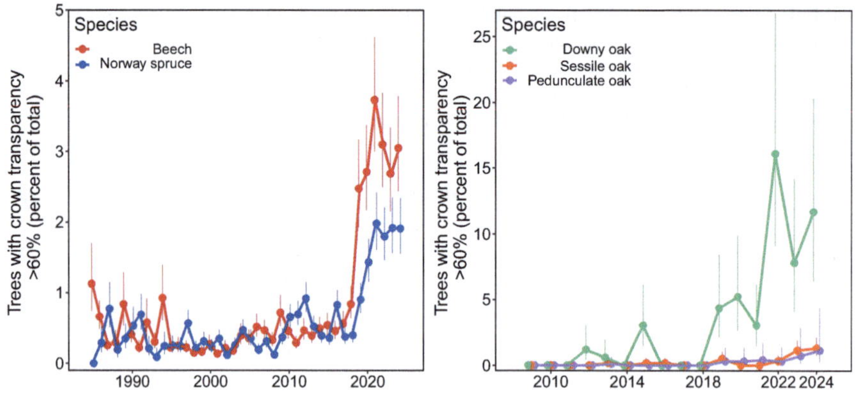


**Fig. 2** Percentage of trees with crown transparency $> 60\%$. Left: European beech and Norway spruce, right: oaks. Points are weighted annual means, and the error bars are standard error estimates from a mixed logistic regression with site as random variable

obvious that the number of observations is essential for an estimation of crown transparency.

A crown transparency of >25% is taken as a threshold for "damaged" and a crown transparency of >60% as a threshold for "strongly damaged" trees. After the drought of 2018, the proportion of "strongly damaged" trees increased in beech, pubescent oak, and Norway spruce, but not in sessile oak and pedunculate oak (Fig. 2). The proportion of strongly damaged trees remains elevated even six years after 2018, despite the wet years of 2021 and 2024, suggesting slow recovery. The proportion of "damaged" trees increased only slightly after the multiple drought years since 2018 (data not shown). In the case of deciduous trees, severe damage is mainly caused by dead branches, which probably results from cavitation. In Norway spruce, the loss of needles manifests in a decrease in the number of needle years (Fig. 15). In addition, the decline in shoot growth also points to a low vitality of the monitored spruce trees (Fig. 13).

## 3.2 *Mortality*

The mortality of beech, Norway spruce, and pubescent oak strongly increased after the dry year 2018 (Fig. 3). For beech, the increase in mortality was lower than the increase in the proportion of strongly damaged trees, but it remained high for at least six consecutive years. Although beech mortality was also elevated in the 1980s and 1990s, the episode with increased forest damage is on a much higher level since 2018. In Norway spruce, mortality has increased after dry years (2003, 2018/2019, 2022) and as a consequence of storm damage (2000, after the gale "Lothar" in the end of 1999), when bark beetle populations increased. However, the mortality rates of Norway spruce reached in the years after 2018 (cumulated mortality rate 2018–2023: 15.9%) are clearly at a non-sustainable level and have led to the abandonment

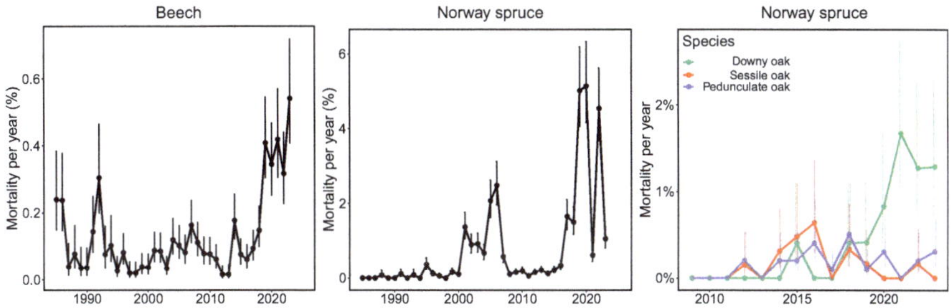


**Fig. 3** Mortality of beech, Norway spruce, and oaks. Points are estimated mean annual mortality rates with error bars as 95% confidence intervals from mixed-effect generalized linear regression models

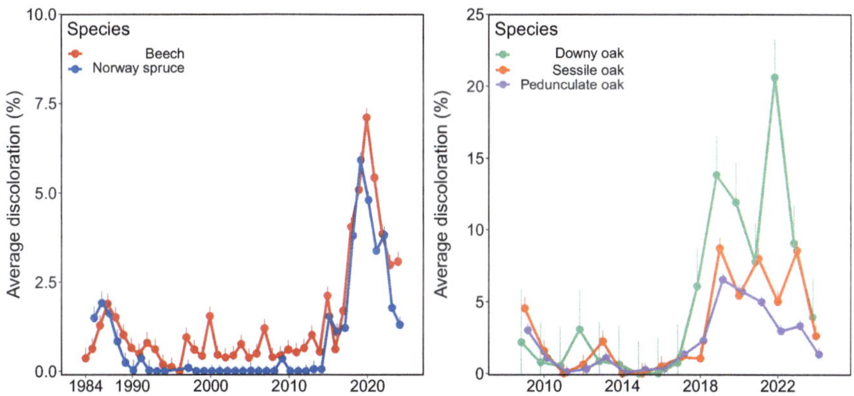


**Fig. 4** Average discoloration assessed from the ground with scores between 0 (no discoloration) and 100% (completely yellow or brown). Left: beech and Norway spruce, right: oaks. Points are estimates from a mixed regression with site as random variable, and error bars give 95% confidence intervals

of a number of plots at lower elevations. Mortality has increased in pubescent oak, but not in the other oak species, after the dry summer of 2018, with a maximum reached in 2024.

## 3.3 *Discolorations*

The discoloration of leaves and needles has been assessed annually from the ground together with crown transparency according to the methodology of ICP Forests (2022). In addition, leaf phenology together with discoloration has been assessed in the shoots harvested for nutrient analysis every four years. Both measures of discoloration (Figs. 4, 5, 6) have strongly increased over time. Figure 4 shows the development of the sum of yellow and brown discolorations assessed from the ground. We observed that the discoloration increase after 2018 is mainly caused by the leaves turning yellow, comparable to the symptoms observed in the harvested shoots. Premature leaf fall was rarely observed, as the assessments took place between mid-July and mid-August. The dominant phenomenon observed in the harvested shoots was photobleaching of the sun-exposed parts of the leaves (Braun et al. 2021). It has strongly increased since 2007 in beech (Fig. 5) and since 2011 in the oaks (Fig. 6). Intercostal chloroses were observed nearly exclusively in beech. They are a typical symptom of Mg deficiency and increased from

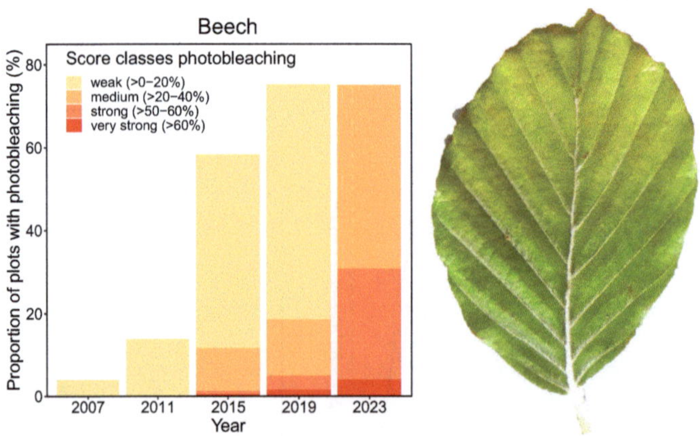


**Fig. 5** Development of photobleaching in harvested shoots of beech since 2007 and beech leaf with strong photobleaching from site Sagno TI 2023

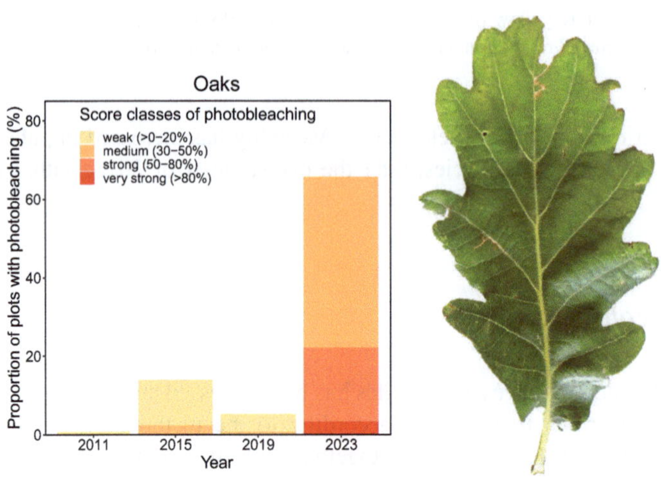


**Fig. 6** Temporal development of photobleaching in harvested oak shoots since 2007 and oak leaf with strong photobleaching from site Therwil BL 2023

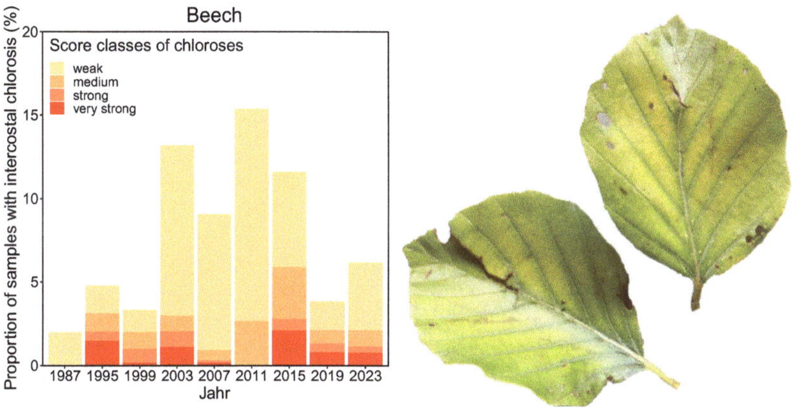


**Fig. 7** Development of intercostal chloroses in beech. Right: Beech leaf sample from Liesberg BL 2023 with a foliar Mg concentration of 0.18 mg $g^{-1}$ d.m.

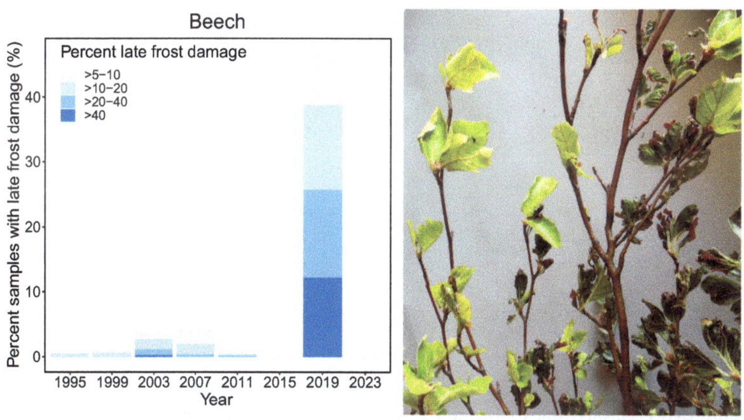


**Fig. 8** Development of late-frost damages in beech (left) and typical late-frost damage in the foliage (brown and curled leaf margins, production of new leaves)

1987 to 2015 (Fig. 7), the latter being the year with the lowest recorded foliar Mg concentrations (Fig. 9). The year 2019 was extraordinary due to severe late-frost damages in beech (Fig. 8), with almost 50% of the plots affected. Typically, the leaves have a brown curled margin, and a second leaf flush is often produced.

## 3.4 *Tree Nutrition*

Since 1984, branches from the top of the crowns of 6–8 trees per species and site have been harvested every four years for the analysis of foliar nutrient concentrations, measurement of shoot growth and beech fructification and the assessment of

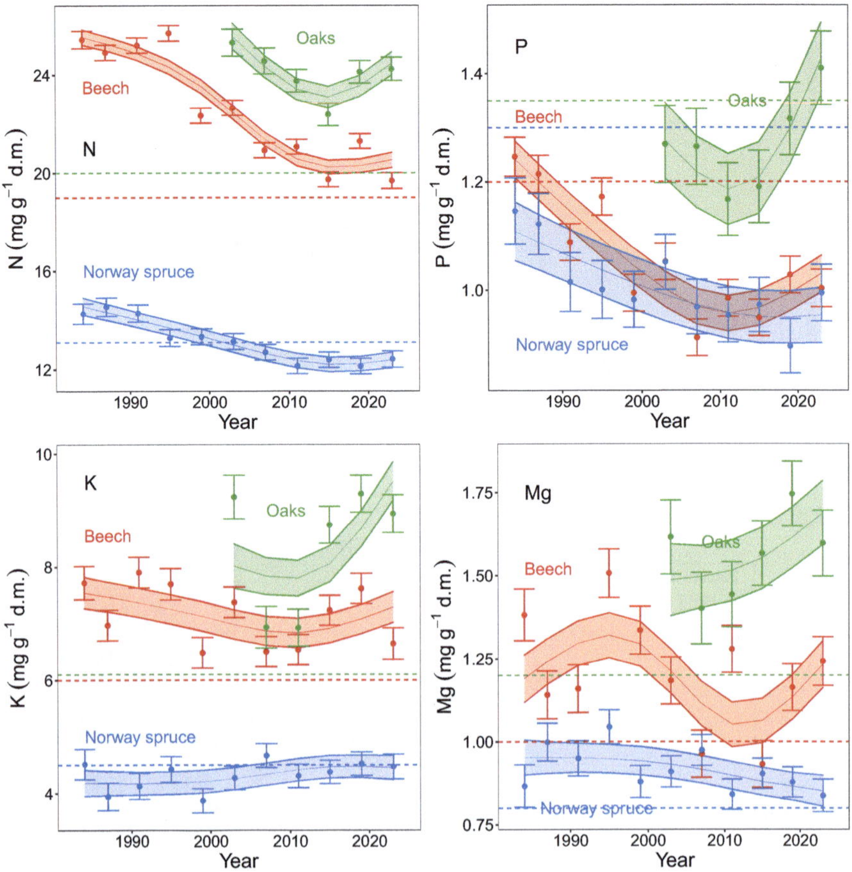


**Fig. 9** Development of foliar concentrations of nitrogen (N), phosphorus (P), potassium (K), and magnesium (Mg). Dashed lines indicate the thresholds for normal nutrition according to Göttlein (2016). Points are estimated annual mean concentrations from mixed-effect linear regression models, with error bars representing 95% confidence intervals. Data were smoothed using local natural spline regression fitting with $\alpha = 3$, results being displayed as lines (point estimates) and shaded areas (95% confidence intervals)

discolorations. Figure 9 shows the development of foliar nutrient concentrations in beech, Norway spruce, and oaks in the period 1984–2023 in comparison to the thresholds of normal nutrition according to Göttlein (2016). Foliar N concentrations have decreased in all investigated species since 1984. In beech, they have shifted from a very high level to near normal values in recent time. In Norway spruce, N concentrations showed an average decrease below the threshold for normal nutrition, whereas the ratios of N to P and K still clearly remain above the limit for harmonic nutrition (Fig. 9). Foliar P concentrations have strongly decreased both in beech and in Norway spruce and are far below the threshold of normal nutrition. In beech, the decrease has leveled off since 2007, but the values measured are still in the range of

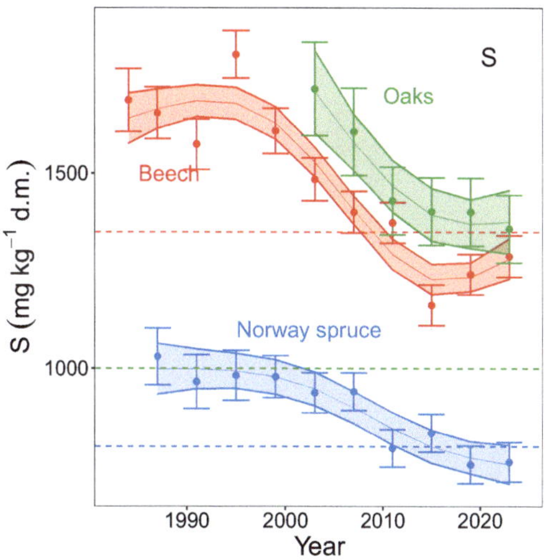


**Fig. 10** Development of foliar sulfur concentrations, analyzed by TUM Munich (A. Göttlein). Dashed lines indicate the thresholds for normal nutrition according to Göttlein (2016). Points are estimated annual mean concentrations from mixed-effect linear regression models, with error bars representing 95% confidence intervals. Data were smoothed using local natural spline regression fitting with $\alpha = 3$, results being displayed as lines (point estimates) and shaded areas (95% confidence intervals)

strong deficiency. K concentrations did not show a clear trend, but the concentrations in Norway spruce are close to the deficiency level. Foliar Mg concentrations in beech have recovered since 2015, when the average foliar concentrations were below the threshold for normal nutrition. This was also visible in the proportion of trees with intercostal chlorosis (Fig. 7) which was highest in the years with low foliar Mg concentration. Foliar sulfur concentrations were analyzed by TUM Munich in a cooperation project with A. Göttlein in a subset of mixed samples. In all species, sulfur concentrations have dropped strongly during the last 30 years (Fig. 10). On average, the current concentrations are below the thresholds for normal nutrition in beech and Norway spruce.

The N:P ratio is clearly above harmonic ratios in all three species (Fig. 11). In beech and Norway spruce, it reached a maximum around the year 2000. There is a clear declining trend in the N:K ratio, reaching in recent years the threshold for normal nutrition for the oaks.

## 3.5 *Fructification in Beech*

Fruit production in European beech differs by orders of magnitude between mast and non-mast years and occurs more frequently after warm and dry summers (Müller-Haubold et al. 2015). Fructification in beech can be quantified backward in harvested

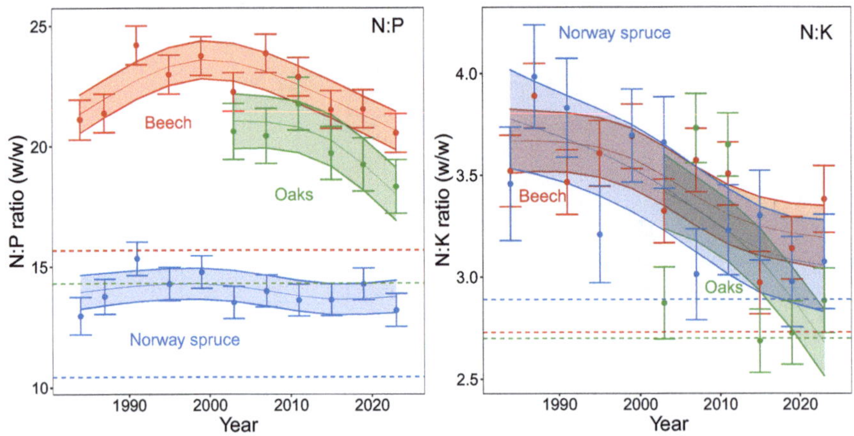


**Fig. 11** Development of foliar ratios of N:P and N:K. Dashed lines indicate the thresholds for normal nutrition according to Göttlein (2016). Points are estimated annual mean ratios from mixed-effect linear regression models, with error bars representing 95% confidence intervals. Lines and shades are predicted by natural spline regression fitting with $\alpha = 3$

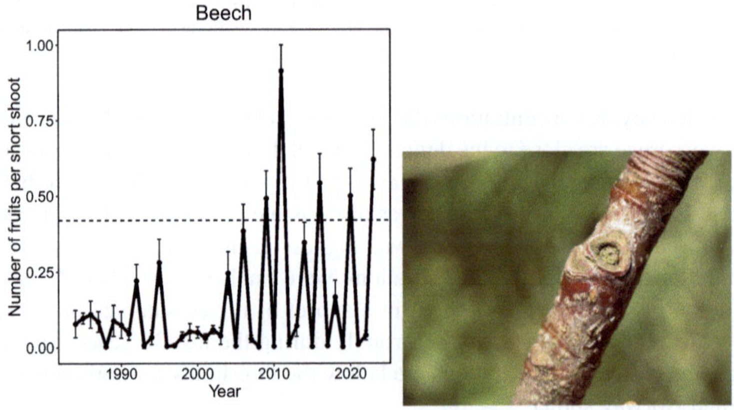


**Fig. 12** Development of beech fructification in the period 1984–2023 (left). Dashed line: Mast threshold (150 fruits m$^{-2}$). Right: Scar left after fruit production

shoots, because beech fruits leave a typical scar on the branches (Fig. 12 right), and the age of the corresponding branch can be determined by counting annual branch increments backward. Fructification has increased strongly after the 2003 hot drought (Fig. 12). The threshold for mast years used in forestry of 150 fruits/m$^2$ corresponds to 0.42 fruits per short shoot as assessed by litter fall measurements. The average of all plots has exceeded this threshold in the years 2009, 2011, 2016, 2020, and 2023. In addition, the frequency of years with high fruit production has increased. Since 2009, this has been the case every 2–3 years, which is a clear increase compared to reported mast year intervals of 3–7 years in central Europe before 1990 (Paar et al. 2011).

## 3.6 Vitality Indicators in Shoots

**Shoot Growth**

The development of shoot growth over the last five decades differs between the five species (Fig. 13). The data shown are corrected for age effects. In beech, the variation of shoot growth is high and no clear time trend can be recognized, even when removing the effect of fructification which is a strong predictor for shoot growth of beech. In Norway spruce, there is a strong decrease of shoot growth over time, suggesting a low vitality of the trees in more recent years. Drought is the main factor explaining the decrease. In the oaks, the shoot growth of sessile oak differs from pedunculate and pubescent oak. While the time trend of shoot growth is similar in *Q. petraea* and *Q. pubescens*, downy oak shows a stronger relation with drought (data not shown).

Mean leaf area of beech was reduced in 2011 and 2023 compared to the long-term mean (Fig. 14). There was no such pattern in the oaks, but there were clear species differences, with leaf size increasing from *Q. pubescens* over *Q. robur* to *Q. petraea* in all years.

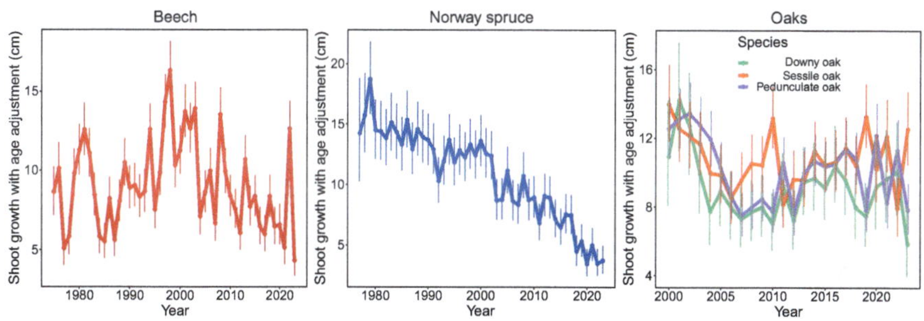


**Fig. 13** Shoot growth of beech, Norway spruce, and oaks with age correction. Points are estimated annual means of growth from mixed-effect linear regression models, with bars representing 95% confidence intervals

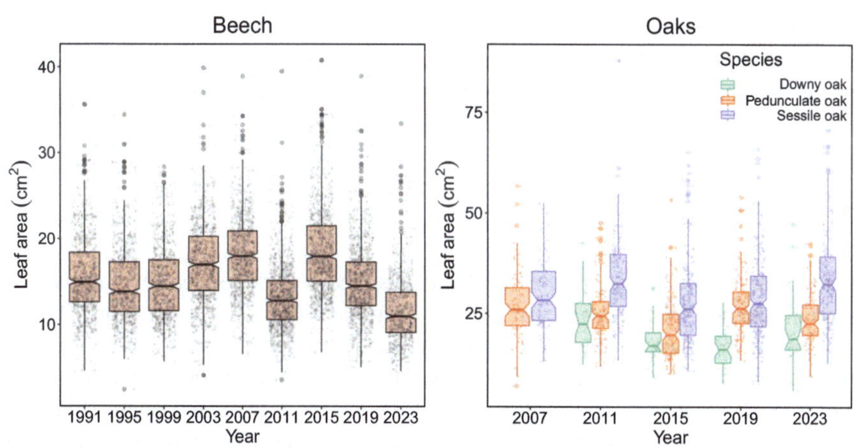


**Fig. 14** Boxplots of leaf area in beech (left) and oaks (right)

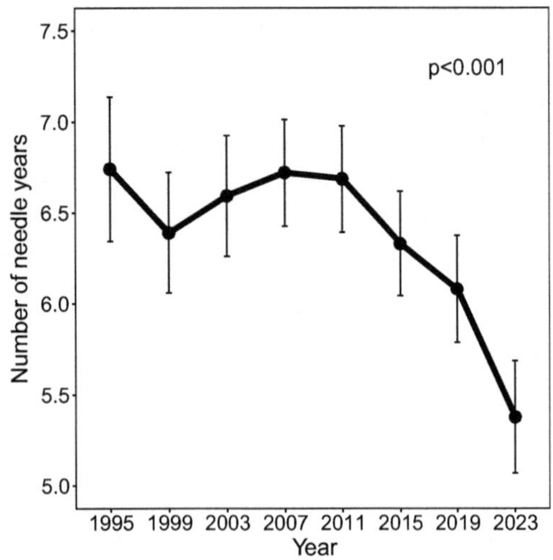


**Fig. 15** Development of the number of needle years in Norway spruce. Points are estimated annual means of the number of needle years from mixed-effect linear regression models, with bars representing 95% confidence intervals

**Needle Years**

The number of needle years in Norway spruce has clearly decreased since 2011 (Fig. 15). This is a further indication of reduced vitality of Norway spruce and is in line with the increased proportion of trees with a high crown transparency and low shoot growth.

## *3.7 Stem Growth*

The volume increment on a stand-level basis has decreased since 1987 (Fig. 16) in all species. In the beech stands, this decrease is also reflected by a decrease in the growth of the single trees (Fig. 16 right). The decrease in Norway spruce is the result of stand disturbance such as mortality due to bark beetle attack, uprooting or stem breakage, while the volume increment of the individual trees has even increased. In common plots with beech, Norway spruce and oak showed higher volume increments than beech (by 27% and 6%, respectively (data not shown)).

## *3.8 Pathogens*

During the time of forest decline in the 1980s, beech bark necroses were observed more frequently. *Neonectria coccinea* has been identified several times on different monitoring plots in symptomatic beech trees, often together with the beech scale (*Cryptococcus fagisuga*). In the 1990s, beech bark necroses declined rapidly. After the 2018 drought year, the frequency of bark necroses increased again (Fig. 17), while the beech scale is still rare. In the oaks, stem bleeding symptoms have also

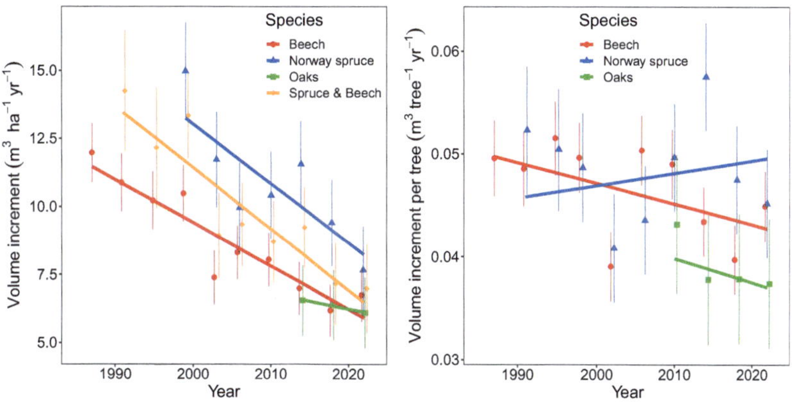


**Fig. 16** Left: Volume increment in stands of beech, Norway spruce, oak and mixed beech/Norway spruce. Right: Development of volume increment per tree. Points are estimated annual mean volume increments from mixed-effect linear regression models, with bars representing 95% confidence intervals

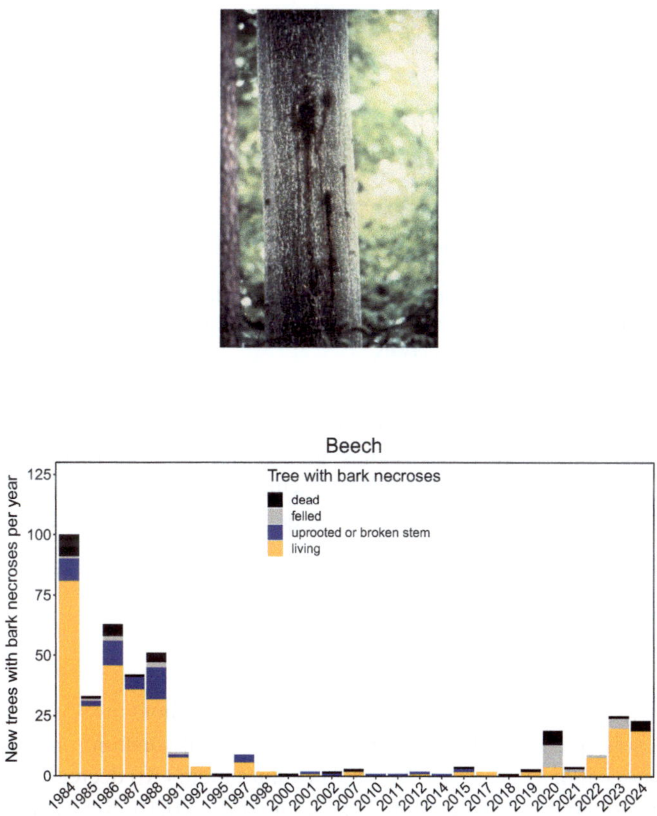


**Fig. 17** Development of bark necroses in beech, with indication of the fate of the trees. Foto: Bark necroses in combination with a heavy infestation by the beech scale (*Cryptococcus fagisuga*) in 1984

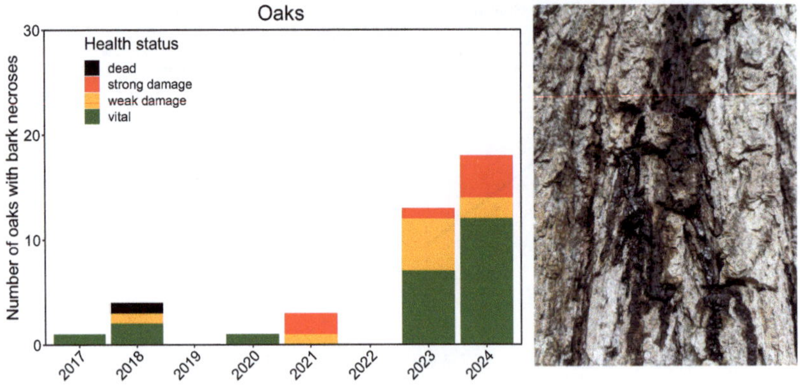


**Fig. 18** Development of stem bleeds in oak. Foto: Stem bleeding symptoms in an oak tree in Sissach BL 2023, caused by AOD bacteria

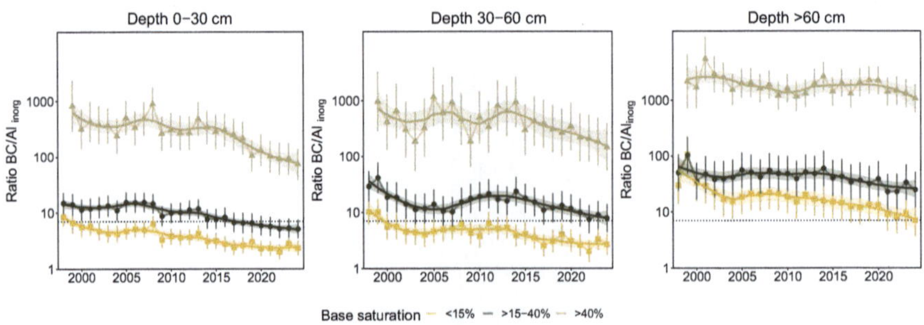


**Fig. 19** Development of the acidity indicator BC/Al in different classes of soil base saturation. Points are estimated annual means of BC/Al ratio from mixed-effect linear regression models, with bars representing 95% confidence intervals

increased (Fig. 18). In some trees, the pathogens causing acute oak decline (AOD) and currently endangering oak stands in the UK have been identified (Forest Research 2022; Brown et al. 2015). Oaks with symptoms are currently monitored to follow the spread in Switzerland and the development of infected trees.

## 3.9 *Soil Acidification*

During the monitoring period, soil acidity has increased both in the soil solution and in the soil solid phase. In 47 monitoring plots with soil solution samplers, the ratio between base cations (BC = Ca, Mg, K) and aluminium (Al) decreased over time in all soil base saturation classes (Fig. 19), Braun et al. (2020). Repeated sampling of the soil solid phase of all plots in the years 2005 and 2016 revealed a decrease in pH by 0.22 units and of base saturation by 2.9% in the lime-free horizons. Dendrochemical analyses of aluminium in tree rings also suggest substantial soil acidification during the last decades in the monitoring plots (Hopf et al. 2023).

## *3.10 Soil Drought*

Soil water content and tension (matric potential) were measured in 44 plots at depths of 20, 40, and 60 cm. To quantify soil drought, we calculated the proportion of daily averages during which available soil water fell below 40% (Bréda et al. 1995) or a soil water tension threshold of pF 3.5 ( – 0.35 MPa) was exceeded. Both thresholds lead to a similar assessment of drought intensity, but the latter value was less sensitive to technical failure. The exceedance of pF 3.5 was used for site comparisons, the validation of the hydrological model used (Wasim-ETH, Schulla (2019)), and the analysis of the monthly distribution of soil drought.

The water potential probes used after 2016 allow pF values of >4.2 to be measured which is beyond the permanent wilting point. In eight plots, pF values of >5 were reached. Seven of these eight plots were oak plots. These results suggest that some trees can take up water beyond the permanent wilting point from the corresponding soil depths, regardless of the fact that oaks also extract water from deep soil. This finding is supported by Carrasco-Molina et al. (2024) who report predawn water potentials of Mediterranean oaks which are higher than pF 3.5.

### 3.10.1 Long-Term Variation

There is no clear time trend in the exceedance of the drought threshold of pF 3.5 (Fig. 20), but the dry years 2011, 2015, 2018, 2022, and 2023 can be identified. Figure 20 also shows a comparison with the corresponding output from Wasim-ETH for the subset of plots with soil water measurements. The time trends are similar for the modeled and measured data, with a lower level of exceedance for the measured data.

### 3.10.2 Variation Within the Year

The annual average time of the exceedance of pF 3.5 was used to classify plots into wet and dry plots. Figure 21 shows the predicted percentage of days with pF >3.5 for each of the 12 months, for a dry site (where pF exceeds 3.5 on 30% of all days) and a wet site (where pF exceeds 3.5 on 2% of all days). In dry plots, the soil is drying out almost simultaneously in all depths and, on average, stays at levels of >60% for about three months, whereas, in wet plots, the proportion of days with pF >3.5 does not exceed 20% in any of the 12 months and is very low in most of the months. The proportion of coniferous trees in a plot was correlated with the monthly

**Fig. 20** Development of the annual exceedance of pF 3.5 between 2010 and 2023 (daily averages of the whole year), with shaded areas representing 95% confidence intervals

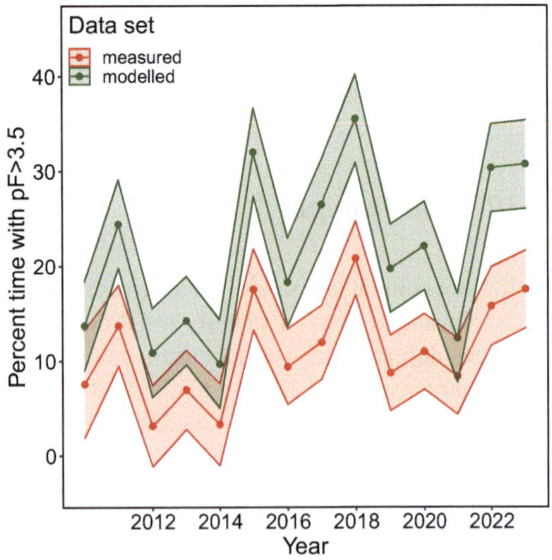


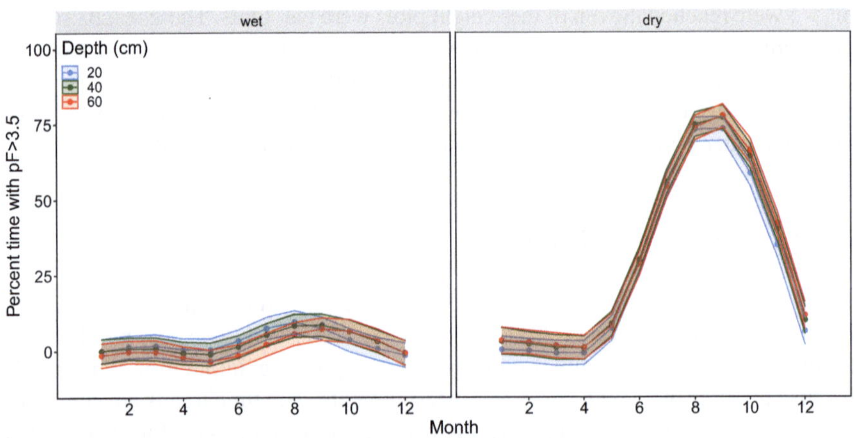


**Fig. 21** Exceedance of the proportion of time with a relative proportion of pF > 3.5 during the year in wet (average exceedance of pF > 3.5 2%) and dry plots (average exceedance of pF > 3.5 30%), with shaded areas representing 95% confidence intervals. 15399 monthly averages from 39 plots and 21 years

distribution of threshold exceedance. For sites with a high coniferous proportion, the drought level reached is smaller, and the drought peak is attained later at 60 cm compared to 20 cm. If the deciduous proportion is high, the time interval of maximum drought lasts longer, especially at 60 cm depth (data not shown). These results are consistent with the hypothesis that the water uptake of Norway spruce occurs at smaller soil depths than in the other species (see also Sect. 3.10.3).

#### 3.10.3 Depth of Water Uptake

Based on the $\delta^2H$ signature in the xylem water of 1214 trees in 182 plots from three years, Allen et al. (2019) developed a seasonal origin index (SOI) which indicates the proportion of summer water available to trees. Beech and oak xylem show a predominance of winter precipitation, while the proportion of summer precipitation is higher in Norway spruce (Allen et al. 2019). The results suggest that Norway spruce is taking up water from smaller soil depths. They also highlight the importance of winter precipitation for trees. More details of these results are given in Fig. 48.

## 4 Proposed Causalities

### *4.1 Nitrogen Deposition*

#### 4.1.1 Eutrophication

In Switzerland, N deposition is high, mainly due to livestock farming (Rihm et al. 2013; Schweizerischer Bundesrat 2022). In 87% of the studied forests, the critical loads for nitrogen deposition are exceeded (Rihm and Künzle 2019). Total N deposition in the forest monitoring plots ranges between 8 and 97 kg N $ha^{-1}$ $yr^{-1}$. Negative effects of excess nitrogen on ecosystems have been suggested as early as in the 1980s (Nihlgård 1985; Aber et al. 1989). Several stages of nitrogen saturation have been proposed by Aber et al. (1989) and Emmett (2007), starting with increased net production and followed by increased nitrification. A later stage is characterized by increased nitrate leaching, decreased fine root mass, and decreased net production. N deposition loads of $> 15$ kg N $ha^{-1}$ $yr^{-1}$ are regarded as harmful for forest ecosystems (Braun et al. 2022). Critical loads for N deposition have been convened in international working groups for various ecosystems within the framework of the UNECE Air Convention (Bobbink et al. 2022).

Indicators for nitrogen saturation are low C:N ratios in the forest floor mass, increased N leaching (Aber et al. 2003; Dise and Wright 1995), or an increase in nitrophilic species (for a review, see Bobbink et al. (2022)). The comparison of nitrate concentrations and leaching losses with the soil solution of 44 monitoring plots with the corresponding critical thresholds for nitrogen leaching (UNECE; CLRTAP (2017)) indicates a high level of nitrogen saturation. Although excess N leaching has decreased since the start of the soil solution measurements in 2000 (Braun and Tresch 2021), the measurements of the last 10 years still show an exceedance of the threshold in 43% of the annual leaching rates per plot (Fig. 22). Eutrophication of the ground flora is frequently observed, e.g., by high cover of blackberries, which can be related to N deposition both within the stand (not shown) and in disturbed parts of the plots (Fig. 23). In disturbed parts, the relation between N

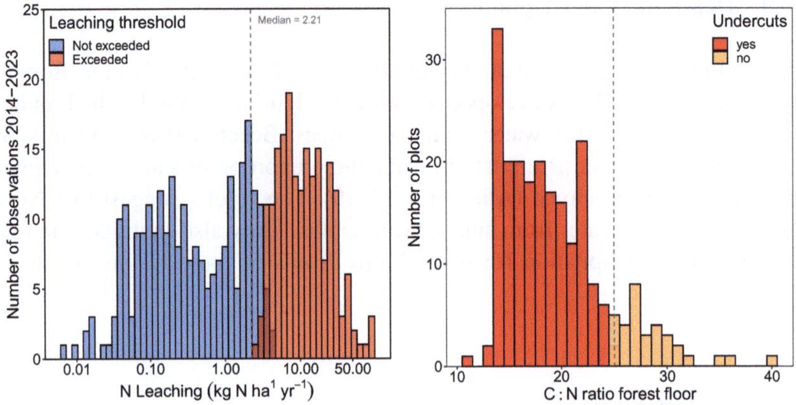


**Fig. 22** Left: Frequency distribution of annual leaching rates during the latest 10 years (2014–2023). Leaching threshold: altitude-adjusted leaching rate of 2–4 kg N ha$^{-1}$ yr$^{-1}$ (CLRTAP 2004). Right: Frequency distribution of C:N ratios in the forest floor grouped according to the "nitrogen saturation threshold" of 25

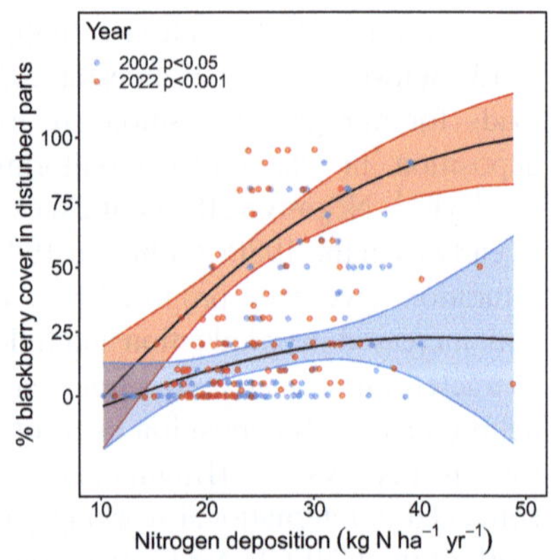


**Fig. 23** Observed and modeled blackberry cover in disturbed parts of the plots in relation to N deposition, with shaded areas representing 95% confidence intervals: survey years 2002 and 2022

deposition and blackberry cover strengthened from 2002 to 2022, resulting in almost 100% blackberry cover at deposition loads of $>40$ kg N ha$^{-1}$ yr $^{-1}$. A C:N ratio of $<25$ is regarded as indication of nitrogen saturation (Gundersen et al. 2006). This threshold is undercut in the majority of forest monitoring plots (Fig. 22). Taken together, all presented indicators suggest widespread eutrophication of Swiss forests.

### 4.1.2 Ectomycorrhiza and Foliar Nutrient Concentrations

Ectomycorrhizal fungi play an important role for the water and nutrient uptake of forest trees (Agerer 2001; Smith and Read 2008; Pickles and Simard 2017). Water uptake is especially important in times of stress (Lehto and Zwiazek 2011). It has been shown that increased N deposition reduces the diversity of these symbiotic fungi (van der Linde et al. 2018; Suz et al. 2021) and also the mycorrhization rate of the root tips (de Witte et al. 2017; Tresch et al. 2025). de Witte et al. (2017) further showed that the mycorrhizal species *Cenococcum geophilum*, which was more frequent when foliar P concentrations were high, disappeared at N deposition rates of $> 25$ kg N ha$^{-1}$ yr$^{-1}$. This fungus is known to be rather drought resistant (Jany et al. 2003; di Pietro et al. 2007). Its disappearance may be one explanation of the increased sensitivity of beech to drought under high N deposition (Pickles and Simard 2017).

The sensitivity of ectomycorrhizae to excess N is probably the reason for negative relations observed between N deposition and foliar P and K concentrations (Suz et al. 2021; Wallander 2000). A decrease of foliar concentrations of P, K, or Mg has been shown both in N addition experiments (Balsberg-Påhlsson 1992; Flückiger 1999) and in gradient studies. A relation between the N:P ratio and total N deposition has been demonstrated for the ICP forest plots Du et al. (2021). This study in the ICP forest plots suggests an effect threshold of 15 kg N ha$^{-1}$ yr$^{-1}$ (Roth et al. 2022). The negative effect of N deposition on foliar P and K concentrations is also visible in our monitoring plots for beech (Braun et al. 2020), Norway spruce, and oaks (Fig. 24). For *Fagus sylvatica*, the relation between N deposition and foliar concentrations of N and P changed over time: While N was slightly positive (and P uncorrelated) with N deposition in 1984, the relation of N deposition with foliar N turned slightly negative, and that with foliar P strongly negative, in 2023 (Fig. 25). These results suggest a change in the eutrophication status of the forest ecosystems during the last 40 years.

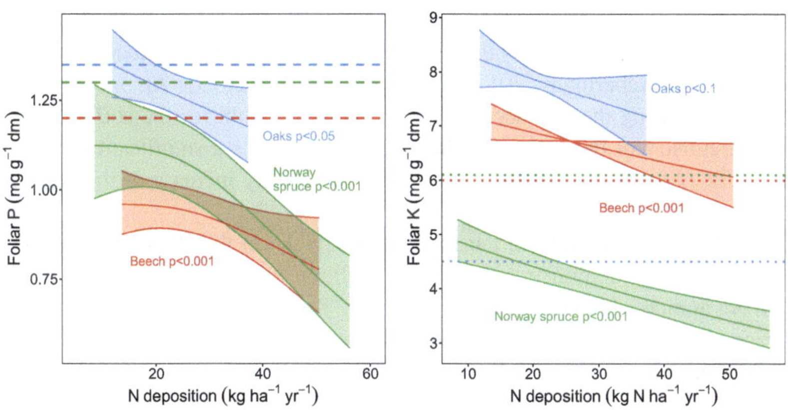


**Fig. 24** Relation between modeled N deposition and foliar concentrations of P and K in beech, Norway spruce, and oaks, with shaded areas representing 95% confidence intervals (for statistics, see Braun et al. (2020)). Dataset 1984–2023

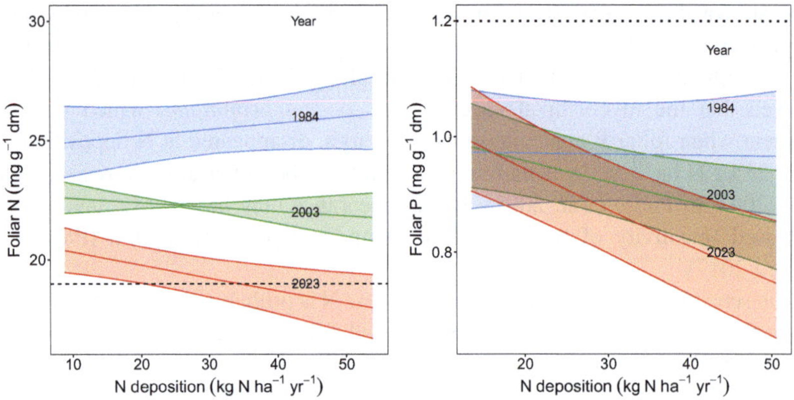


**Fig. 25** Change of the relation between modeled N deposition and foliar concentrations of N (left) and P (right) over time, with shaded areas representing 95% confidence intervals. Interactions with time significant ($p < 0.001$) in both cases. Dataset 1984–2023

### 4.1.3 Nitrogen Deposition and Soil Acidification

Increased N deposition leads to increased nitrate leaching (Dise and Wright 1995; Gundersen et al. 1998), which is also visible in the dataset of our monitoring plots (Fig. 26). Nitrate leaching leads to increased loss of base cations from the soil and thus to soil acidification, when the loss of cations exceeds the supply by mineral weathering (Ulrich 1986; CLRTAP 2017). In the 1980s, forest soils in Europe were mostly acidified by sulphur deposition. The subsequent reduction in sulphur deposition has led to a relative increase in the significance of N deposition for soil acidification. In Switzerland, N deposition is still high enough to cause acidification (Augustin and Achermann 2012; Rihm and Künzle 2019). An increase of soil acidification during the last decades has been shown both for the soil solution and for the chemistry of the soil solid phase. N deposition is a significant predictor for both the BC/Al ratio in the soil solution and the acidity increase of the solid phase (Fig. 27).

The effect of N addition in the form of $NH_4NO_3$ on the BC/Al ratio in the soil solution is clearly demonstrated in an N addition experiment (Braun and Tresch 2021). An annual addition of 20 kg N ha$^{-1}$ yr$^{-1}$ already led to a significantly faster decrease in the BC/Al ratio in the soil solution after 10 years than in control plots receiving only the ambient N deposition of 35 kg N ha$^{-1}$ yr$^{-1}$.

### 4.1.4 Nitrogen Deposition and Growth

In the past, N has been a limiting nutrient in most ecosystems (Tamm 1991). Therefore, the first response to N addition is growth stimulation until another

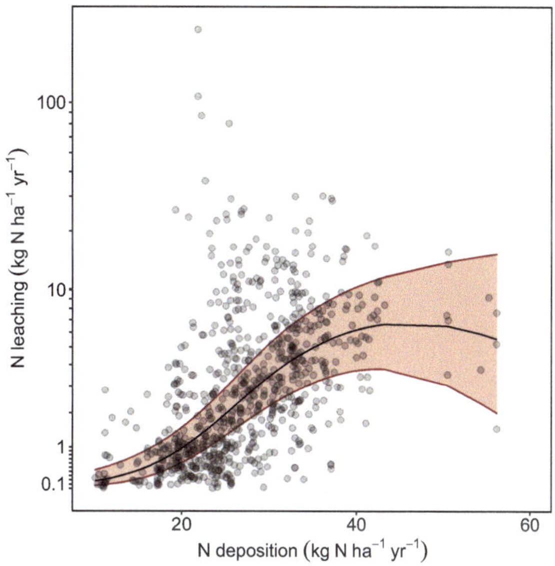


**Fig. 26** Relation between N deposition and N leaching, with shaded area representing 95% confidence intervals and points representing partial residuals from a multivariate mixed linear regression model including tree mortality over three previous years, drought and soil field capacity as covariates, after Braun et al. (2020), including 764 site years

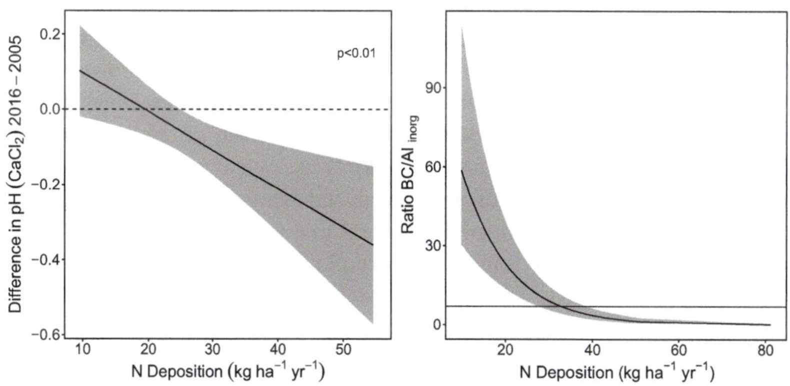


**Fig. 27** Relation between N deposition and the difference in pH ($CaCl_2$) between 2016 and 2005 in 467 lime-free samples from 124 plots (left) and with the decrease in the BC/Al ratio observed in 43 soil solution plots (right, see also Fig. 19; model including depth as factor variable). Shaded areas represent 95% confidence intervals

nutrient or water is limiting a further growth increase (Emmett 2007; Aber et al. 1998). The growth stimulation observed in European forests in the 1980s and 1990s was partially attributed to enhanced N deposition (EFI 2002; Kahle et al. 2008). However, during the last decade, reports of tree growth decreases have been published from various Central European regions (Meining et al. 2008; Gschwantner 2006; Charru et al. 2010; Aertsen et al. 2014) and also for the European level (Nabuurs et al. 2013).

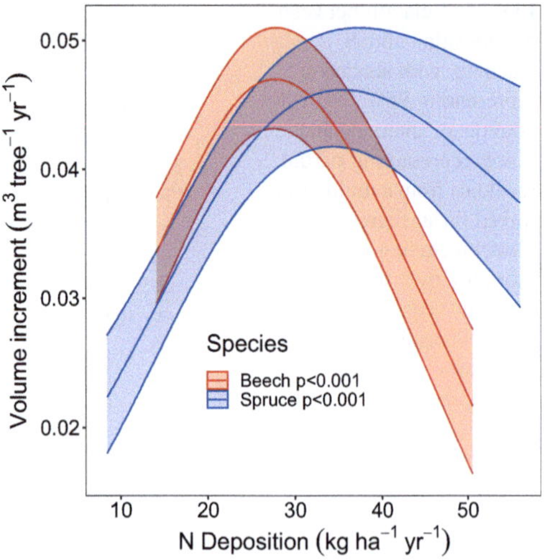


**Fig. 28** Volume growth of beech and Norway spruce in relation to N deposition (with shaded areas representing 95% confidence intervals)

In our dataset, the relation between N deposition and growth showed an essential unimodal pattern in both beech and Norway spruce, with peaks at 28 and 32 kg N ha$^{-1}$ yr$^{-1}$, respectively. At higher N deposition rates, growth decreased strongly. Similarly, Etzold et al. (2014) and Etzold et al. (2020) found that the growth increase caused by N leveled off above 20–25 kg N ha$^{-1}$ yr$^{-1}$ in Swiss forests. This is in accordance with results from other forest growth studies in Europe and the United States. Kint et al. (2012) observed that maximum growth of oak occurred around 25 kg N ha$^{-1}$ yr$^{-1}$ and of beech around 20 kg N ha$^{-1}$ yr$^{-1}$ in Belgium. A nonlinear relationship between N deposition and growth and an increased mortality at higher N deposition rates has been reported for several tree species in the United States by Thomas et al. (2010). In the forest inventory plots of Thomas et al. (2010), total N deposition ranged from 6 to 55 kg N ha$^{-1}$ yr$^{-1}$, which is less than the N deposition rates in our study. The species with a decreasing or a unimodal relationship between N deposition and growth (25 out of 94 species) showed maximum growth between 4.5 and 21 kg N ha$^{-1}$ yr$^{-1}$. The N response may change over time, when a shift in the status of N saturation occurs as proposed by Aber et al. (1989) and Emmett (2002). A transient growth stimulation by N deposition has been reported in Norway by Nellemann and Thomsen (2001): In plots with an (wet) N deposition of $>15$ kg N ha$^{-1}$ yr$^{-1}$, growth was stimulated until 1980 and then reduced, while growth stayed constant in plots with lower N deposition. Furthermore, Kint et al. (2012) observed that the growth response of beech and oak to N deposition switched from positive to negative in Belgium throughout the observation period. Without limitation by P or other nutrients, the growth increase driven by N addition is much less restricted (Gentilesca et al. 2013).

Detailed regression analyses of basal area growth in the Swiss monitoring plots suggest that a combination of N deposition and drought is playing an important role in explaining the observed growth decrease, along with the tree age increase and the increase in fructification as covariates (Braun et al. 2017). This is in line with the conclusions of Kint et al. (2012). The comparison between tree-level and stand-level volume growth trajectories shows that changes in stand structure explain part of the growth decrease in beech, and most of the changes in Norway spruce. In the latter species, increased bark beetle attack as a consequence of drought and of the interaction between N deposition and drought (Tresch et al. 2023) was the most important factor driving the decrease in stand density. For a more comprehensive discussion of the interactions between N deposition and drought, see also Sect. 4.5.

## 4.2 *Effects of Soil Acidification*

Anthropogenic soil acidification is a consequence of N and S deposition in excess of the buffer capacity of soils as determined by the natural weathering rate. Geology and the related weathering rate thus determine the buffering capacity of different soils against acidity (CLRTAP 2017). The acidifying effect of anthropogenic S and N deposition on ecosystems is the topic of the International Cooperative Programme on Modelling and Mapping of Critical Levels and Loads and Air Pollution Effects, Risks and Trends (ICP Modelling and Mapping) (CLRTAP 2017) in the frame of the UNECE Air convention. Damage thresholds have been established for the ratio between base cations and aluminium in the soil solution (BC/Al; initially 1.0) (Sverdrup and Warfvinge 1993) and for pH (4.0), later also for base saturation. Effects of high soil acidity have been shown for tree growth (Sverdrup and Warfvinge 1993), root morphology (Puhe 1994), and the susceptibility to wind throw (Mayer et al. 2004). In our monitoring plots, an increase of uprooting was observed on acidified soils (Braun et al. 2003), suggesting an effect threshold of 40% base saturation in the soil for increased uprooting. This was one of the reasons for adjusting the BC/Al threshold for the calculation of critical loads of acidity to 7–10 (CLRTAP 2017). A possible explanation for increased uprooting on acid soils could be a decreased rooting depth in soils with low base saturation (Braun et al. 2005; Puhe 1994). While the study of Braun et al. (2003) focused on the effects of the single event of the extreme gale "Lothar," a comprehensive data analysis on the forest monitoring plots over the time period 1985–2022 showed that the responses to "Lothar" differed somewhat from those in other years, but that there was still a relation between uprooting of beech and foliar Ca concentrations (Braun et al. 2023). In comparison to the other years, the "Lothar" year 1999 was characterized by exceptionally high wind speeds also in the lowlands (up to 47 m $s^{-1}$; WSL and BUWAL (2001)) and very wet soils before the gale. The "Lothar" year also differed in foliar K concentrations: In Norway spruce, 1999 was the year with the lowest K concentrations in the entire time series, on average 14% below the threshold of normal nutrition according to Göttlein (2016). In beech, the average K concentration was among the three lowest values out of 10 harvests within 40 years, but still within the range of normal nutrition (Braun et al. 2020).

Soil base saturation was also a significant predictor of the number of needle years in Norway spruce, with an effect threshold of 15% (Fig. 29 left). This is similar to

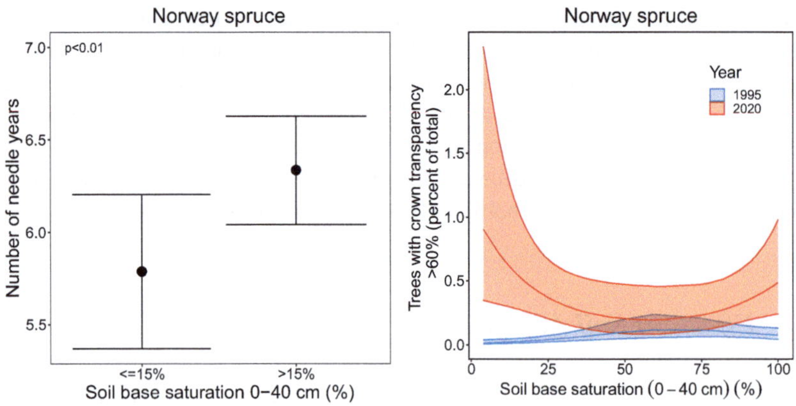


**Fig. 29** Relation between soil base saturation and vitality of Norway spruce. Left: Number of needle years in relation to soil base saturation. Bars indicate 95% confidence intervals. Multivariate analysis with N deposition, drought, and altitude. Data from seven harvests between 1999 and 2023, 78 plots, 426 plot years. Right: Interaction between time and base saturation on the proportion of Norway spruce with >60% crown transparency, with shaded areas representing 95% confidence intervals (interaction $p < 0.001$). Dataset, see Table 1

observations by Becher (1986) who observed increased damage in Norway spruce when manganese concentrations in the needles were higher, which is an indicator for soil acidification. In the monitoring plots, we also found a highly significant interaction between base saturation and time (1995 vs. 2020) on the proportion of Norway spruce with a crown transparency of >60%: In later years, the proportion of strongly damaged Norway spruce was increased on soils with low base saturation (Fig. 29, right). These two results suggest that soil acidity does affect the vitality of Norway spruce and that increasing soil acidity may have contributed to the weakening of Norway spruce vitality.

## *4.3 Foliar Nutrient Concentrations*

The nutrition of trees is both an explaining factor for the evolution of several tree vitality indicators and a factor for which the development has to be explained. The reason for the decrease of foliar N and P in both beech and Norway spruce is not so clear. As foliar S concentrations have decreased strongly (Fig. 10), and S is also an important macronutrient (Göttlein and Mellert 2019), it was tested if the decrease in foliar S concentrations contributed to the decrease in foliar N and P concentrations by regressing these two nutrients against S and time. The time coefficient was decreased substantially (from 22% for P in beech to 65% for N in Norway spruce), when S was added to the regression, but this did not explain the decrease completely (Table 3), suggesting that the S decrease is only one of the reasons for the decrease of N and P.

**Table 3** Time coefficients in the regression models for foliar N and P concentrations, with and without adjustment for foliar S concentrations, for beech and Norway spruce

| | | Beech | Norway spruce |
|---|---|---|---|
| N | With S | – 0.212 | – 0.112 |
| | Without S | – 0.486 | – 0.322 |
| P | With S | – 0.291 | – 0.100 |
| | Without S | – 0.372 | – 0.215 |

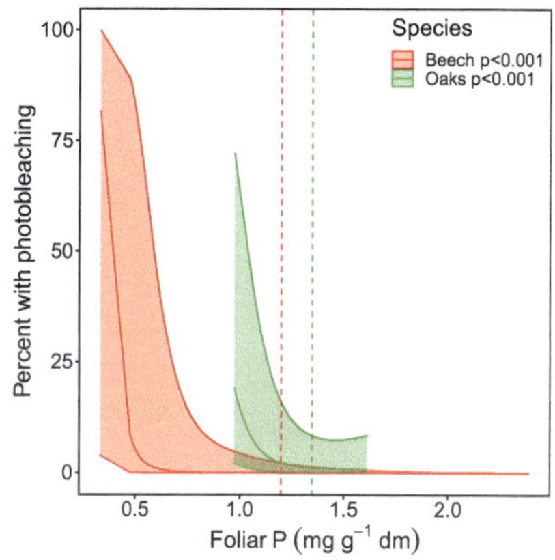


**Fig. 30** Relation between photobleaching and foliar P concentrations of beech and oak (with shaded areas representing 95% confidence intervals). Dashed line: threshold for normal nutrition according to Göttlein (2016)

However, the negative relation of foliar P and K with N deposition (Fig. 24) also suggests an important role of N deposition. The foliar N:P ratio showed a small peak around the year 2000 in beech, while it stayed constant in Norway spruce during the same period. However, the ratio was above the normal nutrition thresholds in all three species, and the foliar P concentrations currently observed must be considered very low, as they are related to the discolorations observed in the field with mature trees and on harvested branches (Fig. 30). Part of the ceasing decrease in beech foliar P concentrations can also be explained by the small leaves; the amount of P per leaf was lowest in 2023 (data not shown). In beech, photobleaching increased strongly at foliar P concentrations below 0.6 mg $g^{-1}$, suggesting that this symptom can be regarded as a P deficiency symptom. In oaks, a similar relationship between foliar P concentration and photobleaching was observed in 2023, the first year with clear photobleaching symptoms, although the P supply of oaks is better than in beech. The threshold of photobleaching is higher in the oaks, starting at about 0.7–0.8 mg $g^{-1}$. Photobleaching in beech was also related to high temperatures, especially at low foliar P concentrations. Also in oak, a high June vapor pressure deficit (VPD) was positively correlated with photobleaching, particularly in combination with low foliar P (Fig. 44). These interactions of low foliar P with high temperature or high VPD are probably also the reason for the observed increase in discoloration in the monitoring plots (Fig. 4). In beech, the discoloration observed from the ground was directly related to foliar P (Table 4).

**Table 4** Overview over significant relations with vitality indicators of beech, Norway spruce, and oaks with foliar P and K concentrations. Positive correlations are indicated with the sign +, and negative ones with the sign -. The number of signs indicates the level of significance (+ or - $p < 0.05$, ++ or – $p < 0.01$, +++ or — $p < 0.001$). The asterisk (*) indicates an interaction with N deposition. For the size of the dataset, see Table 1

| | | P | K | Mg |
|---|---|---|---|---|
| Beech | Crown transparency >60% | - - - | - - - | |
| | Mortality | - | - - - | |
| | Active vessels | | + + + | |
| | Discoloration | - - | | |
| | Photobleaching | - - - | | |
| | Uprooting | | - - - | |
| | Stem increment | | + + + | |
| | Shoot growth | + + + | | |
| Norway spruce | Crown transparency >60% | - - - | | |
| | Mortality | - - - * | - - - | |
| | Uprooting | | - - - | |
| | Discoloration | - - | | |
| | Stem increment | + + + | + + + | |
| | Shoot growth | + + + | + + + | |
| Oaks | Crown transparency >60% | | - | |
| | Discoloration | | - - | |
| | Photobleaching | - - - | | |
| | Active vessels | | + + | |
| | Stem increment | | + + + | + + + |
| | Shoot growth | | + + + | |

Both foliar P and K were important predictors in various analyses of forest vitality (Table 4). Low P and K nutrition is related with an increased proportion of beech trees with strong crown transparency, and with increased mortality of both beech (Braun et al. 2021) and Norway spruce (Tresch et al. 2023). For the evaluation of stem increment patterns and the proportion of active vessels in the xylem, a multi-nutrient ratio of P and K (Diagnosis and Recommendation Integrated System, DRIS index) was a better predictor than foliar nutrient concentrations. For calculating the DRIS index, all possible ratios of macronutrients are related to threshold values and then summed up. The method has been developed in agriculture (Beaufils 1973) and applied to forest trees by Talkner et al. (2011). The reference nutrient ratios were taken from Göttlein (2016). The stem increment of all three species decreased with an unbalanced DRIS index of K (Fig. 31), and the proportion of active vessels in the branch xylem was also decreased at unbalanced K nutrition in beech and oak (Figs. 35, 43). In the case of stem increment, strongly negative index values of K were associated with an increased drought effect in beech (Fig. 32 left) and with a stronger growth reduction at high N deposition in Norway spruce (Fig. 32 right). Unbalanced P nutrition was a predictor of reduced stem increment of spruce, and low foliar Mg concentration a predictor of reduced stem increment of oak.

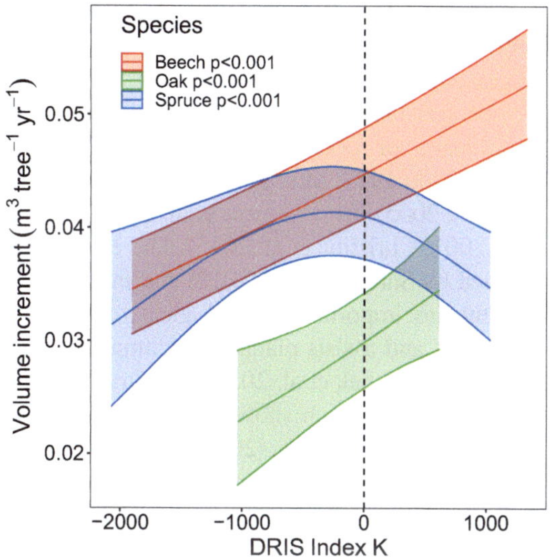


**Fig. 31** Relation between stem increment and DRIS index of K (with shaded areas representing 95% confidence intervals). Values below 0 (dashed line) mean unbalanced K nutrition

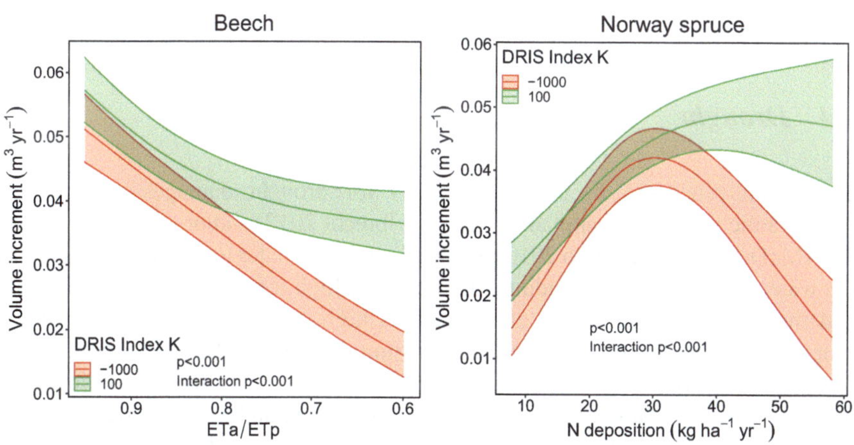


**Fig. 32** Interactions of volume increment with K nutrition as DRIS index value (with shaded areas representing 95% confidence intervals). Left: Interaction with drought in beech, right: interaction with N deposition in Norway spruce

The importance of P and K for forest vitality can be explained by these elements' roles in plant physiology. P is essential in all energy transfer reactions. An unbalanced nutrition will thus impair photosynthesis (Reich et al. 2009) as well as biomass production (Güsewell 2004; Vitousek et al. 2010). As strong P deficiency caused discolorations, P may also be involved in drought-induced processes described as "drought starvation" and the subsequent attack by weakness parasites. P has been reported to be important for the resistance against drought. An increased cavitation sensitivity at low P nutrition has been found by Harvey and van den Driessche (1997). P deficiency may also affect the control of stomata (Singh et al. 2000) or decrease frost resistance (Pfeiffer et al. 2013; Jönsson et al. 2004), while the role of P for the resistance against parasites is less clear (Veresoglou et al. 2014). The significance of potassium (K) lies in the resistance against parasite attacks (Chaboussou 1973; Mrazek 1983; review by Amtmann et al. 2008), oxidative stress (Cakmak 2005), frost (Senser et al. 1990; Jönsson et al. 2004), and drought (Egilla et al. 2001; Christina et al. 2015). The essential role of K under dry climates is also outlined by Sardans et al. (2012) and more generally by Johnson et al. (2022): The element has important physiological functions in stomatal control under drought conditions and assists plants to acclimate under water stress conditions (Aksu and Altay 2020; Pathak et al. 2020). The importance of N and K nutrition for cavitation vulnerability has been shown by Harvey and van den Driessche (1999) and more recently by Oddo et al. (2014), who demonstrated that sufficient K nutrition is improving the hydraulic conductance of trees. Unbalanced K nutrition, along with increased N deposition, is associated with increased drought sensitivity of beech stem increment (Fig. 32). The observed effects of K nutrition on various processes observed in the monitoring plots suggest that drought-related physiological processes are important.

## *4.4 Drought*

Two main processes discussed as causes for drought damages of trees are (i) hydraulic failure (Choat et al. 2012; Engelbrecht 2012) and/or (ii) carbon starvation (McDowell and Sevanto 2010). Hydraulic failure relates to the phenomenon that the water filaments in the conduits rupture under high tension, when the soil is dry and the water demand of the atmosphere is high. Death occurs when a certain proportion of the vessels has lost their function due to embolism (Urli et al. 2013; Choat et al. 2018). Death by carbon starvation is a process usually occurring in understory trees because of the lack of light (McDowell and Sevanto 2010), but it is supposed to occur in adult overstory trees when stomata are closed for a long time. The understanding of this process is mainly based on indirect evidence (McDowell and Sevanto 2010). Carbon starvation can be seen as a process facilitating the attack by weakness parasites (McDowell et al. (2011), Sect. 4.4.4).

The striking increase in strongly defoliated beech trees after the dry summer of 2018, with the appearance of many dead branches, suggests that cavitation became more important after 2018. This conclusion is also supported by the relationship between the proportion of active vessels in branch xylem and drought of the previous year, which was observed on harvested shoots from the top crown in 2019. It is a matter of discussion, whether cavitated vessels can be refilled at favorable xylem pressures (Brodersen et al. 2018; Love and Sperry 2018). In line with our findings, Arend et al. (2022) concluded that beech xylem was not refilled after 2018. The significant relation of the cavitated vessels observed in 2023 to the drought in 2018 (see Fig. 35) also suggests that cavitation is long-lasting in these populations. In 2018, soil water was refilled late in the year, but the cavitated vessels did not even refill in spring during budbreak, when a refill would be expected under normal conditions, as shown for beech (Cochard et al. 2001). The increase in discolorations suggests that carbon starvation may also be an issue, with photobleaching as a mechanism where energy is lacking for the repair of oxidized chlorophyll (Powles 1984).

In Norway spruce, the physiological processes differ somewhat, as Norway spruce has tracheids instead of tracheae, whose diameters are smaller and which are not continuous. They can also suffer from cavitation, but the mortality thresholds are different to angiosperm trees (Choat et al. 2018). In the Hölstein experiment in Switzerland, mature Norway spruce trees died during the summer 2018 as a consequence of cavitation (Arend et al. 2021). In the monitoring plots, however, most dying Norway spruce trees were killed by a bark beetle attack, although acute cavitation cannot be excluded in a few cases.

For the data analysis of drought effects on vitality, drought has to be quantified. Braun (2015) compared a variety of indices. The best indicators were the ratio between actual and potential evapotranspiration (ETa/ETp) and the site water balance according to Spellmann et al. (2007). From the latter, the minimum value reached during the summer was used ($SWB_{min}$). Evapotranspiration, which is used for both indicators, was calculated using Wasim-ETH (Schulla 2019). The mortality of Norway spruce was even closer related with average VPD of July than with ETa/ETp (see Fig. 33).

### 4.4.1 Mortality or Strong Crown Damages

The proportion of trees with strong crown damages ($>60\%$ crown transparency) or of dead trees was related to the drought of two to six preceding years according to a nonlinear multilag model. The result is illustrated in Fig. 33. In beech, the drought of six preceding years was needed to explain the strong crown transparency, in Norway spruce, it was three years, while in the oaks, the drought of the current and the preceding year was needed. In beech, the results for mortality are similar to those for strong crown damages. The maximum effects of drought on the mortality of Norway spruce occurred with a time delay of two and three years. In this case, the best drought indicator was average July VPD.

In order to compare the modeled risk predictions shown in Fig. 33 with observations, the drought effect estimates cumulated over 6 years (i.e., including the effect of the actual year and of the 5 preceding years) have been plotted together with observed data of strong crown transparency and mortality for beech (Fig. 34). The lag model explains the increase in strong crown transparency of beech after 2018 very well, suggesting that the cumulation of dry years (2015, 2018, 2019, 2020) was the main reason for the surprisingly high damages after 2018, which were much higher than after the drought year 2003. Likewise, the increase in mortality after

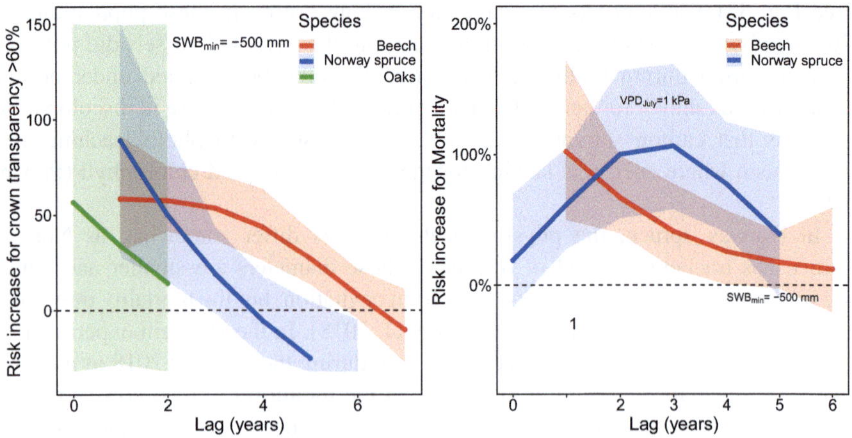


**Fig. 33** Lagged effects of drought on crown transparency $> 60\%$ (left) and on mortality (right) of beech, Norway spruce, and the oaks. Results of a nonlinear multilag model (DLNM, Gasparrini (2011)). Lines represent predictions of drought effects in the current year (0), the previous year (1), and the years before (2–6) for a fixed drought level of $SWB_{min} = -500$ mm, corresponding to a VPD of 1 kPa in Norway spruce. The predictions indicate the risk increase in % compared to a fixed drought level of 0 $SWB_{min}$ and VPD 0, respectively. The shaded areas represent the 95% confidence intervals. For mortality of Norway spruce, the best drought indicator was average July VPD

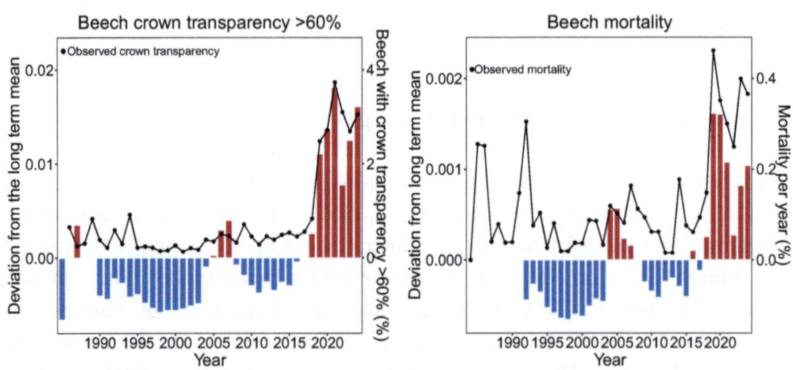


**Fig. 34** Illustration of the time lags between estimated drought effects and crown transparency $> 60\%$ (left) and between estimated drought effects and mortality (right) of beech. Estimated effects were derived from the regression model underlying Fig. 33. Years with risk estimates higher than the long-term average are shown in red, and the others in blue. Black line and dots: observed proportion of trees with crown transparency $> 60\%$ or mortality (right scale)

2018 has been predicted rather well. However, the mortality peaks in the 1980s and 1990s seem to be unrelated with drought.

One explanation for these lag effects is the persisting cavitation after drought. In 2023, the quantification of active vessels published in Braun et al. (2021) was repeated. In the dataset of 2023, the only drought predictor for the proportion of

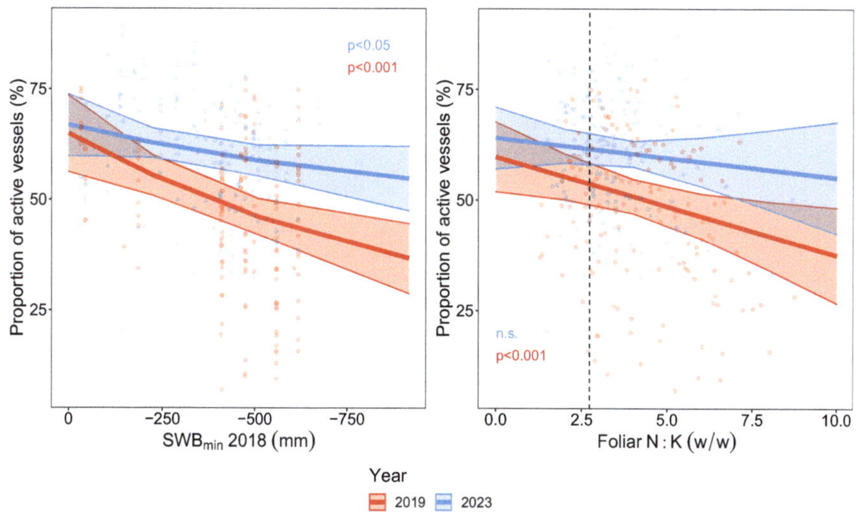


**Fig. 35** Active vessels stained in beech branches (red = 2019, blue = 2023) in relation to drought in 2018 (left) and to foliar N:K ratio (right). Shaded areas represent 95% confidence intervals

active vessels was the count from 2018 (Fig. 35), suggesting that the drought effect of 2018 was still decisive five years later. The relation with the foliar N:K ratio was less tight but still visible. It is obvious that a branch with reduced hydraulic conductivity is much less able to refill the water in the leaves when the evaporative demand of the air is high, and that the threshold proportion of cavitated vessels for mortality ($\Psi^{88}$) (Choat et al. 2018) is reached faster. In oaks, the proportion of cavitated vessels was related with N deposition (Fig. 43) and K nutrition (Fig. 43) (for an explanation of the DRIS index, refer to Sect. 4.3).

### 4.4.2 The Role of Site Conditions for Oak Tree Health

The bad health status of *Q. pubescens* was surprising, as this oak species is usually judged as very drought resistant. It was hypothesized that this could be due to the drier site conditions usually colonized by this oak species. To test this hypothesis, the moisture axis of the ecogram of the local ground vegetation communities was used (Frehner et al. 2005) as described by Braun et al. (2023). For the data analysis, the moisture axis was combined with site water balance as a drought indicator. The proportion of oaks with crown damage > 60% is higher on dry sites (Fig. 36). This relationship explains some of the decline in health status, but not all of it. Furthermore, these results suggest that *Q. robur* is more damaged than *Q. petraea*.

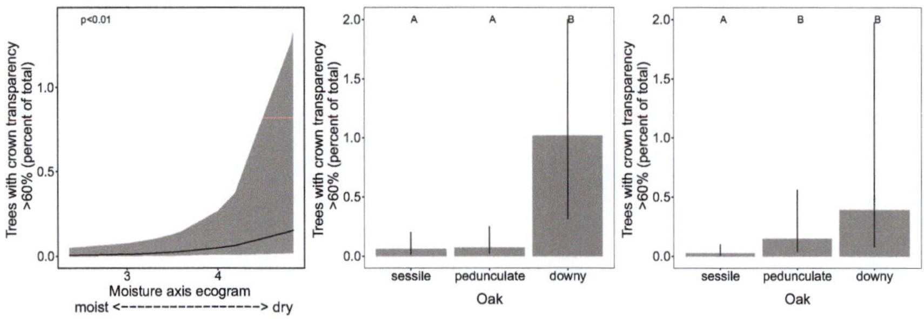


**Fig. 36** Left: Proportion of oaks with crown transparency >60% in relation to the moisture axis of the ecogram (with shaded area representing 95% confidence intervals). Middle: uncorrected species differences, right: predictions from a regression model including the ecogram moisture axis adjusting for differences in average moisture exposure between the three species

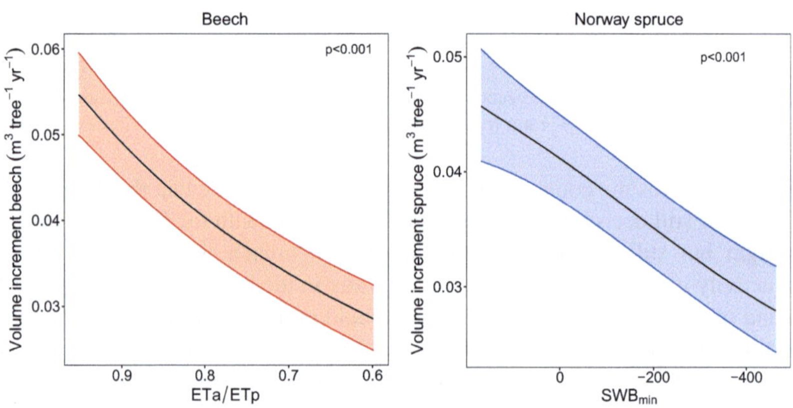


**Fig. 37** Volume increment in relation to drought (with shaded areas representing 95% confidence intervals). Left: beech with drought indicator ETa/ETp, right: Norway spruce with drought indicator $SWB_{min}$. The drought axes are reversed to show increasing drought from left to right. Dataset as described in Braun et al. (2022) and Table 1

### 4.4.3 Tree Growth

#### Drought Effect on Tree Growth

Drought is a very important predictor for stem increment in both beech and Norway spruce (Fig. 37), although the best drought indicators differ between the two species. There was an interaction between drought and N deposition on tree growth in beech and Norway spruce, showing that strong drought and high N deposition reduce growth much more than drought and low N deposition alone (see also Sect. 4.5).

#### Drought Damages and Tree Growth

We found that beech trees with a high crown transparency in the years 2019–2022, i.e., drought-affected trees, have shown a lower volume increment since 1987 (Fig. 38). In addition, the growth decrease observed in general (Fig. 16) was stronger

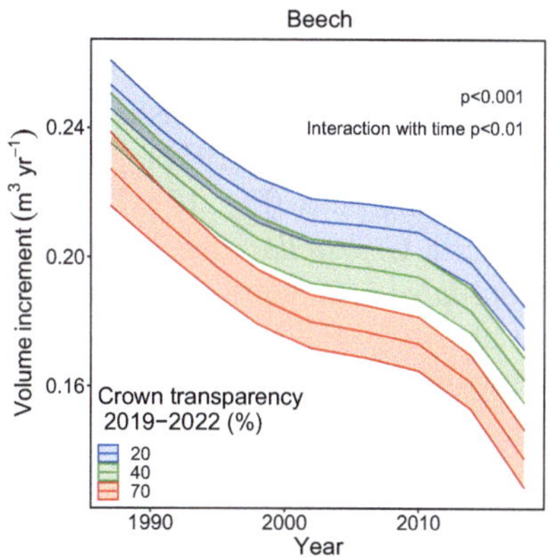


**Fig. 38** Estimated annual average volume increment of beech during the entire monitoring period in relation to crown transparency in the years 2019–2022 (shaded areas representing 95% confidence intervals). Estimates were adjusted for differences in age, position within the stand, initial diameter, and crown size

in the damaged trees, as suggested by a significant interaction with time. The relation between crown transparency and growth can, however, be observed also for the trees with lower crown transparency, e.g., 40%. This result is in line with the results from Klesse et al. (2022) and Neycken et al. (2022), suggesting that drought damages and growth are causally related. This relation is further supported by the relation between visible drought necroses and height growth of beech and oaks observed in an N addition experiment (data not shown).

These surprisingly clear results may be interpreted in several ways. Growth occurs only, when the internal tree water deficit is low (Zweifel et al. 2021). This can be an explanation for the relation between drought damages and current year's shoot growth in the N addition experiment, as drought damages would be related with a high internal tree water deficit. However, in mature forest trees, the low vitality may make the trees more susceptible to drought damages. Interestingly, old wind-thrown beech trees in Belgium also showed a growth decrease during decades before the event (Verschuren et al. 2023). The authors argue that death by windthrow in old-grown beech forests does not hit trees randomly, but predominantly occurs in trees which are already in growth decline. Similar results on mortality were reported by Gillner et al. (2013).

### 4.4.4 Drought and Pathogens

It is well known that weakened trees are less resistant against parasites (Manion 1981). Carbon starvation can be seen as a facilitating process for the attack by weakness parasites (McDowell et al. 2011). Often, drought is followed by insect outbreaks (Anderegg et al. 2015). The most common example is the spruce bark beetle (*Ips typographus*) (e.g., Seidl et al. (2016); Müller et al. (2022)), which is killing Norway spruce in large areas of drought-affected forests. Fungal outbreaks after drought are less frequent, but also known. Weakness parasites usually lead to death with some delay. *Armillaria* sp. is a well-known weakness parasite attacking drought-stressed trees (Kubiak et al. 2017; Holuša et al. 2018). An increase of beech bark disease caused by *Neonectria coccinea* after drought has been shown for European beech by Jarcuska et al. (2013) and already after the dry years 1947–1949 by Schindler (1951). After 2018, related symptoms have increased in the monitoring plots (Fig. 17). The acute oak decline (AOD), which is caused by the bacteria *Gibbsiella quercinecans*, *Brenneria goodwinii*, *Rahnella victoriana*, and *Lonsdalea britannica* (Crampton et al. 2020), is also promoted by drought (Brown et al. 2018). For the discussion on the role of N deposition in combination with drought, see Sect. 4.5.

## *4.5 Interaction Between N Deposition and Drought*

The drought sensitivity of trees may be increased, when N deposition is high. Such an interaction was observed for several parameters in the monitoring plots. The stem increment of both beech and Norway spruce increased slightly with increasing N deposition up to a threshold of 30 kg N ha$^{-1}$ yr$^{-1}$ and decreased at higher deposition loads. In the presence of drought, however, the decrease started at clearly lower N deposition loads (Braun et al. 2022) (Fig. 39). When drought occurred in combination with high N deposition, uprooting was also increased (Braun et al. 2023). In Norway spruce, this interaction was also observed, when analyzing mortality (Tresch et al. 2023). When these effects are quantified and related to the average volume increment per hectare in the forest plots, it is obvious that, in the case of Norway spruce, the loss by mortality and stem breakage under the combined action of drought and high N deposition exceeds the volume increment by far, suggesting that sustainability in growth is completely lost under these conditions (Fig. 40).

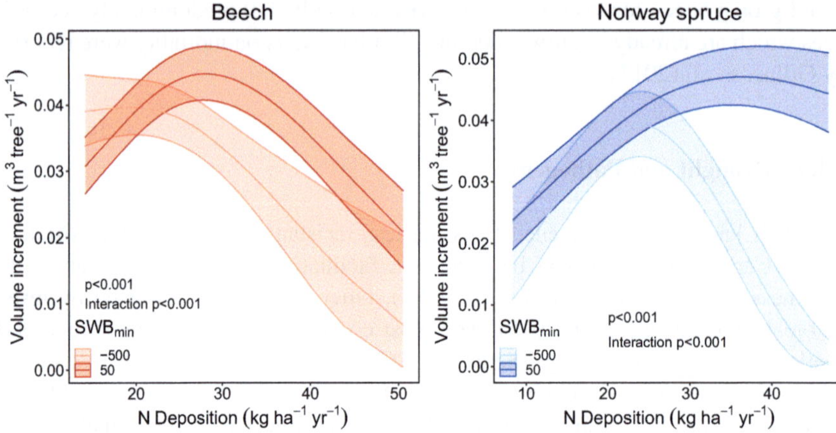


**Fig. 39** Interaction between N deposition and drought on the growth of beech (left) and Norway spruce (right), with shaded areas representing 95% confidence intervals

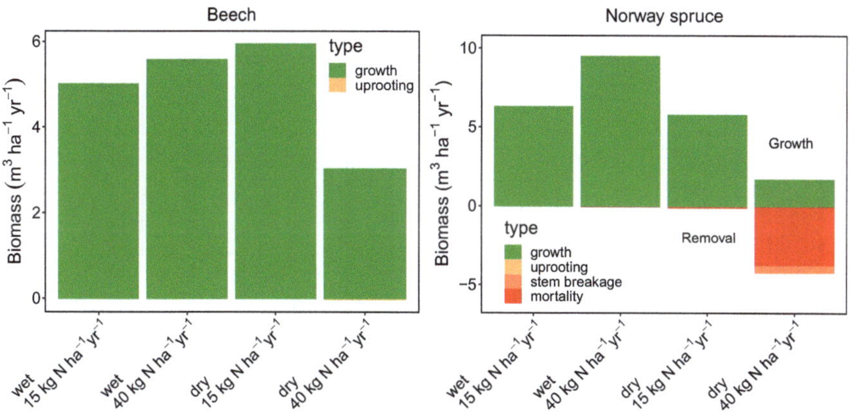


**Fig. 40** Biomass estimate at two drought and two N deposition levels. Estimates of growth are taken from Braun et al. (2022) and Fig. 28, those of uprooting and stem breakage from Braun et al. (2023), and the one of spruce mortality from Tresch et al. (2023). Drought levels used for the calculation: $SWB_{min}$ -500 vs. 0 mm. The increased uprooting in beech at the combination of high N deposition levels and high drought is too small to be seen. For the size of the dataset, see Table 2

In the case of beech, the loss of volume increment in the combination of high N deposition and drought is substantial, while the observed increase in uprooting in the combination of high N deposition and drought does not cause a remarkable change in the biomass accumulation. Figure 40 combines the estimates of growth and disturbances to provide estimates of change in biomass per hectare; the figure can be used to estimate total carbon sequestration.

Visible symptoms of drought in dry years have been observed to increase in both beech and the oaks in an N addition experiment (Fig. 41). In mature trees, the proportion of active vessels in oak branch xylem decreased at high N deposition (Figs. 42, 43).

Interactions between N deposition and drought have been documented in a number of studies. They are usually attributed to an increased transpiring leaf area as a result of increased N availability, without, however, showing direct evidence for this assumption (Harvey and van den Driessche 1997, 1999; Hacke et al. 2010; Plavcová and Hacke 2012; Plavcová et al. 2013). The results of our study speak against this explanation, as average leaf area was not influenced by high N deposition. Experimental evidence of an interaction between N addition and drought on beech was published by Dziedek et al. (2016): In a one-year pot experiment, the combination of drought and N (50 kg N ha$^{-1}$ yr$^{-1}$) affected stem diameter, necrotic leaf biomass, and root biomass.

Sap flow measurements in an N addition experiment with strong K deficiency suggested that N increases night-time water loss and thus the ability of trees to retain water (Braun and Flückiger 2013). Low concentrations of foliar P and K at high N deposition rates are also observed in mature forest trees. The observed relationships of K and P deficiency with crown transparency, mortality or cavitation rate of both

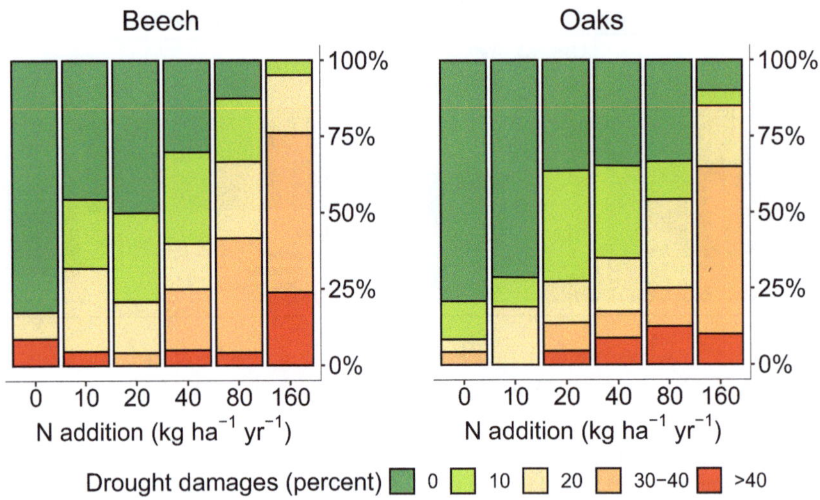


**Fig. 41** Foliar drought symptoms in an N addition experiment. Relation with N addition rates significant at $p < 0.001$ in both species

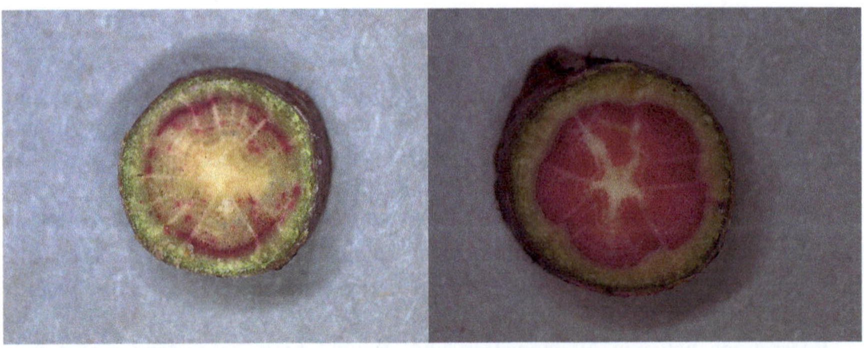

**Fig. 42** Active vessels stained in oak branches (Braun et al. 2021). Left: branch with intact vessels, right: cavitated branch

beech and Norway spruce are also a hint that foliar nutrient concentrations play an important role in increased drought sensitivity. An additional explanation could be N effects on ectomycorrhizal fungi, which are essential in the uptake of both nutrients and water in forest trees (Sect. 4.1.2).

The interactions between N deposition and drought on biomass accumulation shown in Fig. 40 suggest that the C sequestration of stressed forests is decreased, as it has been outlined by Nabuurs et al. (2013) and Wolf and Paul-Limoges (2023). Furthermore, these results also suggest a largely negative N effect on C sequestration, which is in contradiction to many assumptions made in climate models. Sutton et al. (2008) came to the conclusion that one kg N may sequester 50–75 kg C, while experimental data with $^{15}$N propose 41–88 kg C (de Vries et al. 2009). These are still

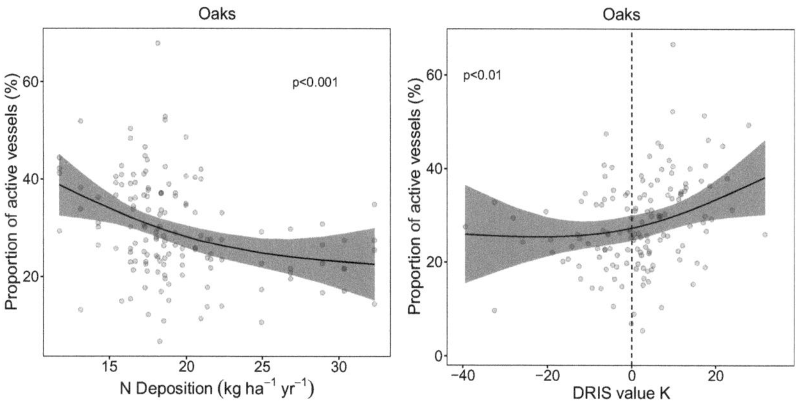


**Fig. 43** Active vessels in oaks in relation to total N deposition (left) and to K nutrition expressed as DRIS index (right), with shaded areas representing 95% confidence intervals (142 trees in 34 plots, analysis in 2023)

fairly high positive numbers, while our data from the monitoring plots suggest a positive relation only at N deposition loads of $<20$ kg N ha$^{-1}$ yr$^{-1}$ (Fig. 37) (Tresch et al. 2023). The combination of drought and high N deposition may even result in a decrease in stand-level biomass as shown in Fig. 40 for Norway spruce, suggesting a decline in vitality and growth sustainability.

## *4.6 Temperature, Vapor Pressure Deficit (VPD), and Irradiation*

High temperatures lead to a high evaporative demand of the atmosphere and thus to increasing atmospheric drought. How much heat is tolerated by trees is discussed by Gauthey et al. (2024). They examined in three oak species $T_{crit}$, the temperature which may lead to severe physiological dysfunctions and local tree mortality. $T_{crit}$ ranged between 51.4 and 59.5°C, but as canopy temperature may be up to 14° higher than air temperature, depending on the efficiency of evaporative cooling, the thermal safety margin decreased to low values during the season. High VPD causes stomatal closure (Jarvis 1976; Emberson et al. 2000), which reduces leaf cooling, and it may cause carbon starvation (McDowell and Sevanto 2010).

In the examined monitoring plots, heat damage cannot be detected directly in the field, in contrast to damages by drought, which is discussed in Sect. 4.4. There are, however, also direct effects of temperature, mainly visible in the regression models for discoloration and wind damages (Table 5) or late-frost damage of beech (Fig. 8). In beech, photobleaching is strongly increased at high June temperatures (Fig. 44). In oak, photobleaching was increased at high VPD in June. These relations with temperature or high VPD may explain the strong increase of photobleaching that has recently been observed in both beech and oaks. In Norway spruce, lagged drought

**Table 5** Overview over significant relations of vitality indicators of beech and Norway spruce with temperature parameters. Positive correlations are indicated with the sign **+**, negative ones with the sign **-**. The number of signs indicates the level of significance (**+** or **-** $p < 0.05$, **++** or **- -** $p < 0.01$, **+++** or **- - -** $p < 0.001$)

| | | T July | T spring | T vegetation period | T winter | T < −2.0 | VPD |
|---|---|---|---|---|---|---|---|
| | | | | | Dec-Jan | after budbreak | |
| Beech | Photobleaching | +++ | | | | | |
| | Late-frost damage | | | | +++ | +++ | +++ |
| | Uprooting | | | | + | | |
| | Leaf area | | +++ | | | - - - | |
| | Stem breakage | | | +++ | | | |
| | Stem increment | | | | | - - - | |
| Norway spruce | Mortality | | | | | | +++ |
| | Discoloration | | ++ | | - - | | |
| | Stem breakage | +++ | | | | | |
| | Stem increment | | | | | - - - | |
| Oaks | Discoloration | | +++ | | | | |
| | Photobleaching | | | | | | ++ |
| | Stem increment | | | | | - - - | |

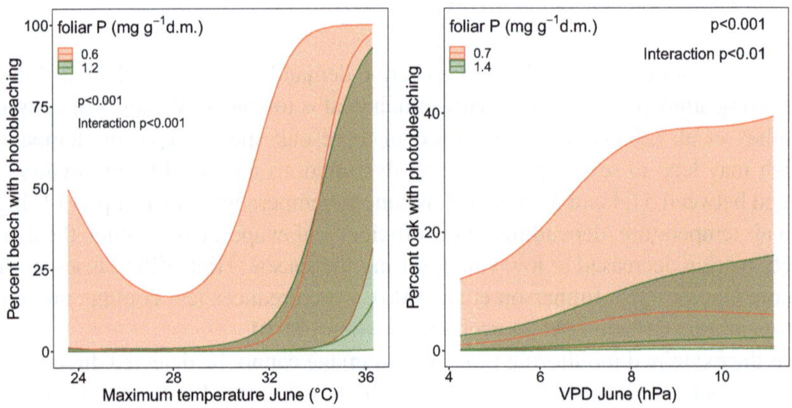


**Fig. 44** Interaction between climate variables and foliar P concentrations on photobleaching. Left: beech, interaction with maximum temperature in June (with shaded areas representing 95% confidence intervals), right: oaks, interaction with average VPD in June

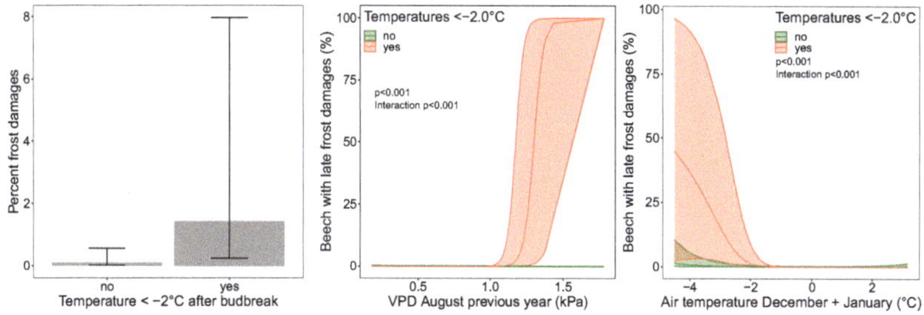


**Fig. 45** Left: Relation between late-frost damage in the harvested shoots and the occurrence of temperatures $< -2.0°C$ after beech budbreak. Middle: Interaction between temperature after budbreak and VPD of the previous August. Right: Interaction between temperature after budbreak and winter temperature. Error bars and shaded areas represent 95% confidence intervals

effects on mortality were better explained by the average VPD of July than by the ratio between actual and potential evapotranspiration (Fig. 33). The discoloration of the Norway spruce crowns observed in the field increases with increasing spring temperature. Stem breakage was also positively related with temperature, either averaged over the vegetation period (beech) or of July (Norway spruce).

Irradiation was a highly significant predictor for the discoloration of beech crowns (not shown). It was, however, the irradiation of the previous year that was influential, and not of the current year.

The late-frost damages observed in the harvested shoots (Fig. 8) were related to the occurrence of temperatures $< -2°C$ after budbreak had occurred in beech (Fig. 45). Lower or higher threshold temperatures ($-1.9$ to $-2.2$) provided a lower fit, and it was essential to use observed budbreak dates as delivered by the phenological networks of MeteoSwiss (Defila and Clot 2005). The temperature sum proposed by Vitasse et al. (2019) as an indicator for late-frost damages was a much weaker predictor of frost damage in our study. The effect of low temperatures after budbreak was modified by VPD in August of the previous year and air temperature during December and January (Fig. 45 center and right). Late-frost damages were strongly increased, when average VPD of the previous August was >1.2 kPa or when the average air temperature in December and January was $< -2.5°C$. With these two covariates, the variation over time was much better explained than with late-frost temperatures alone. The temperature indicator derived for the visible late-frost damages in beech shoots was also a good predictor of stem increment: In all tree species, stem increment was lower when the temperature dropped to $< -2°C$ after beech budbreak (Fig. 46).

## 4.7 *Ozone*

Ozone ($O_3$) is a highly phytotoxic air pollutant (Fuhrer et al. 1997; Matyssek et al. 2010). Apart from negative growth departures (Karlsson et al. 2007; Wittig et al.

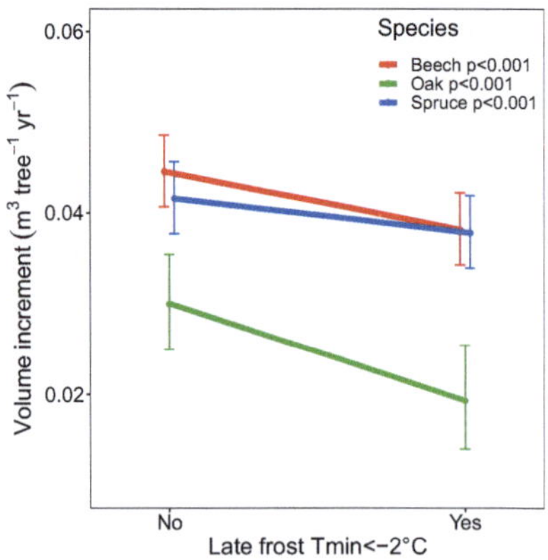


**Fig. 46** Occurrence of temperatures $< -2.0°C$ after beech budbreak and stem increment of beech, oak, and Norway spruce (with bars representing 95% confidence intervals)

2009) and decreasing carbon sequestration (Büker et al. 2012), harmful effects of ozone on the vegetation also include visible injury of leaves (Günthardt-Goerg et al. 1993; Van der Heyden et al. 2001; Paoletti et al. 2009; Hicks 1978), decreasing foliar chlorophyll content (Dalstein and Vas 2005), inhibited carbon allocation to the roots (Coleman et al. 1995; Samuelson and Kelly 1996; Lux et al. 1997), premature leaf senescence (Pääkkönen et al. 1997; Gielen et al. 2006), a changed susceptibility to abiotic and biotic stress factors (Braun and Flückiger 1989; Wellburn and Wellburn 1994; Manning and von Tiedemann 1995; Karnosky et al. 2002, 2007), and sluggish or impaired response of stomata to water deficits (Pearson and Mansfield 1993; Maier-Maercker and Koch 1991; Paoletti 2005). The United Nations Economic Commission for Europe (UNECE) has defined critical levels for external ozone concentration and stomatal ozone uptake to protect from negative impacts (CLRTAP 2017).

The critical level for tropospheric ozone is based on the dose-response relationships with biomass data from fumigation experiments with young trees in chambers (CLRTAP and ICP Vegetation 2017). A set of fumigation experiments, mostly in open-top chambers (OTC), was compiled, and the biomass response at a specific ozone flux was related to the biomass of the controls. Ozone flux, i.e., the stomatal uptake of ozone during a season, proved to be a better predictor than the accumulated ozone over a threshold of 40 ppb (AOT40, Karlsson et al. (2007)). Dose-response relationships were established for beech, birch, and Norway spruce (Mills et al. 2011), and critical levels were set to 5.2 (4% biomass reduction for beech and birch) and 9.2 mmol $m^{-2}$ $yr^{-1}$ (2% biomass reduction for Norway spruce) (LRTAP 2017).

The stem growth data from the monitoring plots were used for an epidemiological analysis of ozone effects. Volume increment per tree was clearly negatively related with annual ozone flux (Braun et al. 2022). Whereas the dose-response relationship from epidemiology very well agrees with the dose-response relationship from the chamber experiments used to set the critical level in beech, the observed response in Norway spruce is stronger than that expected from the experiments (Braun et al. 2022). Visible injuries are not an issue in the monitoring plots, as the harvest of

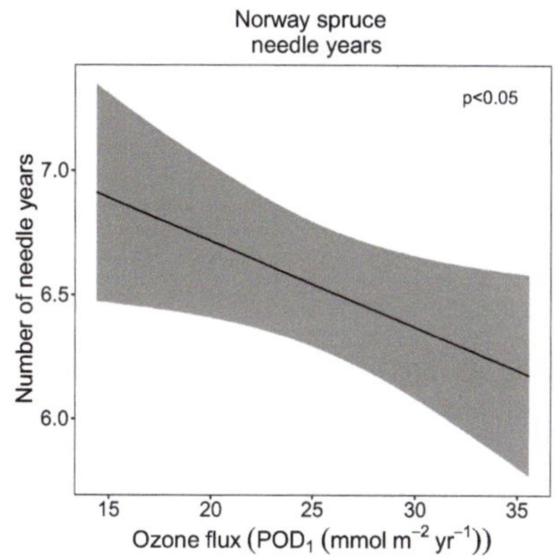


**Fig. 47** Relation between ozone flux and the number of needle years of Norway spruce (with shaded area representing 95% confidence intervals)

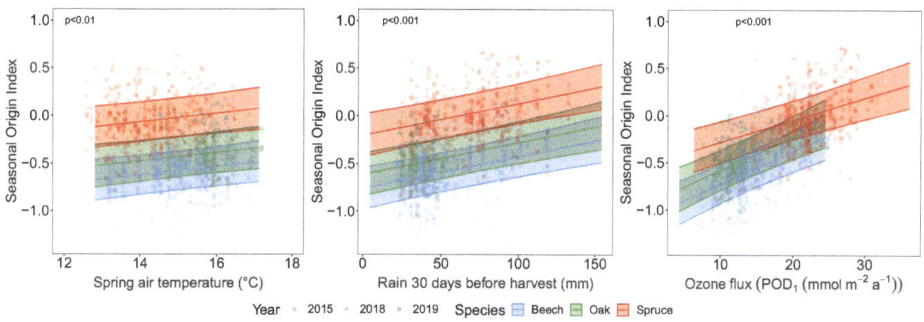


**Fig. 48** Relation between the seasonal origin index according to Allen et al. (2019) in xylem water and three significant predictors: spring air temperature, precipitaiton 30 days before shoot harvest and ozone flux (with shaded areas representing 95% confidence intervals).

branches at the end of July is too early in the season, although they were occasionally observed. However, a relation between ozone flux of the previous year and the number of needle years was observed in Norway spruce (Fig. 47).

Based on $\delta^2$H signatures in the xylem water of 1214 trees in 182 plots during three years, Allen et al. (2019) developed a seasonal origin index (SOI), which indicates the proportion of summer water that is available to trees. This SOI was correlated with spring air temperature, precipitation 30 days before harvest, and ozone flux of the corresponding year (Fig. 48). The higher value at higher ozone flux values suggests a larger proportion of summer precipitation being available. A larger proportion of summer precipitation available could be interpreted as a higher amount of cumulated evapotranspiration before sampling. Ozone can affect the water consumption of trees either through stomatal closure (reduced water loss) or through a sluggish or impaired response of stomata to drought (increased water loss) (Mills et al. 2013). Increased cumulated evapotranspiration suggests that the latter mechanism had been dominant in the monitoring plots. These results are supported by the analysis of stream waters in the United States by McLaughlin et al. (2007) who also

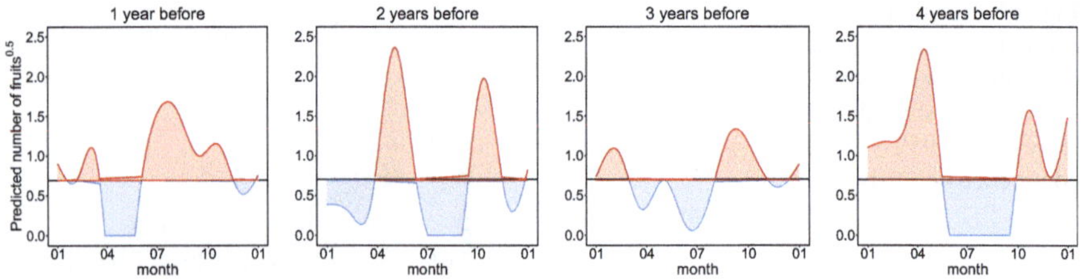


**Fig. 49** Relations of fructification in beech with temperature over one, two, three, and four years before the event. Positive relations are marked in red, and negative relations in blue

found that stream water amounts decreased during high-ozone episodes, pointing at higher transpiration.

## 4.8 *Fructification*

### 4.8.1 Climate Effect on Beech Fructification

The relation between beech fructification and weather conditions in the previous years has been analyzed by several authors (Drobyshev et al. 2014; Hacket-Pain et al. 2015). These authors state that low summer temperatures with high precipitation two years before masting and a warm and dry year directly before the masting year are associated with fructification in beech (Nussbaumer et al. 2018). A polynomial model, which includes time before the event, was applied to the data of the monitoring plots, including polynomial terms of the products of temperature and VPD with time as well. This analysis suggests that the weather of the four preceding years affects the fructification (Fig. 49). In the red parts of the graph, temperatures above average promote fructification, and in the blue parts, temperatures above average decrease fructification. The model predicted the time course of fructification very well. Thus, increased temperatures do not linearly translate to an increased fructification, but the data analysis shows that the increase in fructification observed after the year 2000 is due to climate change. Recent analyses of European datasets even report a decrease in fructification in beech, which may also be consistent with a high-temperature effect (Foest et al. 2024).

### 4.8.2 Fructification and Stem Growth

In beech, fructification was negatively related to stem increment, both in a dataset that included only the average of fruit numbers within the four-year increment period (Fig. 50 left) and in a dendrochronological study with annual rings (Fig. 50 center). Together with the increase in fructification, this negative relationship explained about 25% of the decrease in stem increment observed in beech (Braun et al.

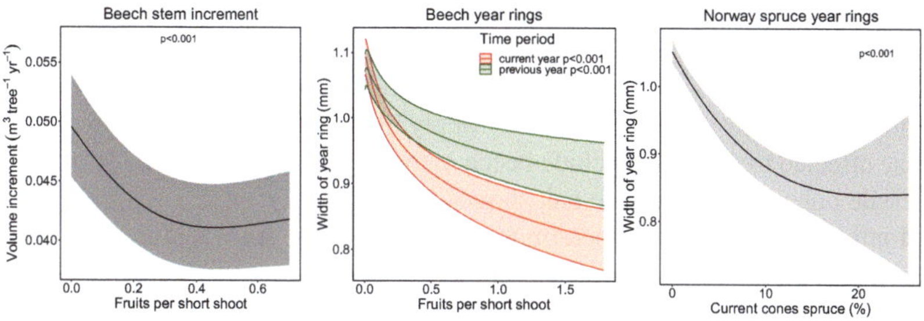


**Fig. 50** Relation between fructification and stem growth (with shaded areas representing 95% confidence intervals). Left: Four-year data of beech, all trees. Center: Annual dendrochronological data from 27 beech sites and 37 years, right: Annual dendrochronological data from 31 Norway spruce sites and 25 years

2017). A data analysis with dendrochronological data showed that tree ring width was affected during the current year of fructification and in the previous year (Fig. 50 center). In both analyses, the effect was largest at moderate fructification levels (0–0.4 fruits per short shoot). Similar observations were also published by Nussbaumer et al. (2021) who also found a reduction of BAI during mast years, and they interpreted this reduction as resource switching between vegetative and generative growth. We observed a similar relationship for Norway spruce (Fig. 50 right), where a fructification with 10% current cones was associated with a growth reduction by 16% compared to no fructification (0% cones).

### 4.8.3 Beech Fructification and Foliar Nutrient Concentrations

The relations between fructification and foliar nutrient concentrations were outlined by Braun et al. (2020). Foliar nutrient concentrations increased, when fructification was high. For fruits of the previous year, concentrations increased for Ca and K and decreased for N. The reason for this is still unclear, as leaf area was not related to fruit production.

## 5 Conclusion

Long-term monitoring is essential for detecting dynamic change in tree and forest health and for understanding the underlying processes in forests. Our long-term data suggest that today climate change is an important driver, but N deposition is intensifying its impact due to interactions with drought. The most important changes that happened during four decades of observation can be summarized as follows:

- Foliar P concentrations of beech and Norway spruce have decreased to levels of strong deficiency.

- Stem increment on a hectare basis has decreased in all tree species and also on a tree basis in beech.
- Fructification of beech has increased strongly both in frequency and in extent.
- Mortality and crown dieback have strongly increased after the dry summer 2018. The affected tree species include beech, pubescent oak, and Norway spruce, while sessile oak and pedunculate oak show a higher vitality.
- Monthly soil solution measurements show that soil acidification is still progressing, driven by N deposition.
- Photobleaching has increased strongly in both beech and the oaks. It can be regarded as a symptom of P deficiency.
- Stem bleeding symptoms in beech and the oaks have been rising since 2018. In the oaks, the pathogens of the acute oak decline (AOD) have been detected.

The most important findings concerning likely causal relations were:

- A series of subsequent dry years, together with an already reduced tree vitality, was the main reason for the strong drought damages observed after 2018.
- The drought damages observed after 2018 are partly associated with cavitation (driven by water deficits) and photobleaching (probably caused by drought starvation).
- Pubescent oak is more damaged than sessile and pedunculate oak, which is only partly explained by the drier sites of its occurrence.
- The effect of drought is stronger under high N deposition, as shown for stem increment of beech and Norway spruce, the mortality of Norway spruce, and xylem cavitation in the oaks.
- The status of P and K nutrition is related to a number of vitality indicators, such as stem increment, strong crown damages, discoloration, and mortality.
- The relation of tree health to K nutrition has increased over time, suggesting an increasing significance of drought-related physiological processes.
- The foliar concentrations of P and K are reduced at high N deposition loads.
- The species diversity of mycorrhizal fungi and their success in colonizing root tips are reduced at high N deposition loads.
- Ozone is explaining 11% of the observed growth reductions in the Swiss forests. It has increased tree water consumption.
- After reduction of sulphur emissions, N deposition is the most important driver of soil acidification.
- On base poor soils, the risk of uprooting is increased.
- The sensitivity of strong crown transparency of Norway spruce to soil acidification has increased over time.
- The increasing soil acidity may have contributed to the weakening of Norway spruce vitality.
- Foliar concentrations of S have decreased due to the decrease in S emissions. This reduction explains part of the decrease in foliar concentrations of P and N.

**Acknowledgements** We thank all funders for their support and all forest owners for their patience in supplying us with monitoring plots.

**Declarations**


- Funding: The forest monitoring was funded by the forestry departments of the cantons of AG, BL, BS, GR, TG, SO, and ZH as well as the environmental agencies of the cantons of Central Switzerland (OW, NW, LU, UR, SZ, ZG). Part of the soil solution measurement and of the data analysis were supported by the Federal Office for the Environment.


- Data availability: Data will be available on request.
- Code availability: The codes of the statistical evaluation will be available on request.
- Author contribution: SB, ST, and SeH were responsible for the collection of the monitoring data, SB and ST for the data analysis, and CS for statistical support in epidemiology.

## References


Aber J, McDowell W, Nadelhoffer K, Magill A, Berntson G, Kamakea M, McNulty W, Currie W, Rustad L, Fernandez I (1998) Nitrogen saturation in temperate forest ecosystems. BioScience 48(11):921–934. https://doi.org/10.2307/1313296

Aber JD, Goodale CL, Ollinger SV, Smith ML, Magill AH, Martin ME, Hallett RA, Stoddard JL (2003) Is nitrogen deposition altering the nitrogen status of northeastern forests? BioScience 53(4):375–389

Aber JD, Nadelhoffer KJ, Steudler P, and Melillo JM (1989, jun) Nitrogen saturation in northern forest ecosystems. BioScience 39(6):378–386. https://doi.org/10.2307/1311067

Aertsen W, Janssen E, Kint V, Bontemps JD, J. Van Orshoven, Muys B (2014) Long-term growth changes of common beech (Fagus sylvatica L.) are less pronounced on highly productive sites. Forest Ecol Manag 312:252–259

Agerer R (2001) Exploration types of ectomycorrhizae. Mycorrhiza 11(2):107–114. https://doi.org/10.1007/s005720100108

Aksu G, Altay H (2020) The effects of potassium applications on drought stress in sugar beet. Sugar Tech 22(6):1092–1102. https://doi.org/10.1007/s12355-020-00851-w

Allen MF, Allen EB, Lansing JL, Pregitzer KS, Hendrick RL, Ruess RW, Collins SL (2010) Responses to chronic N fertilization of ectomycorrhizal pinon but not arbuscular mycorrhizal juniper in a pinon-juniper woodland. J Arid Environ 74(10):1170–1176. https://doi.org/10.1016/j.jaridenv.2010.05.001

Allen ST, Kirchner JW, Braun S, Siegwolf RT, Goldsmith GR (2019) Seasonal origins of soil water used by trees. Hydrol Earth Syst Sci 23(2):1199–1210. https://doi.org/10.5194/hess-23-1199-2019

Amtmann A, Troufflard S, Armengaud P (2008) The effect of potassium nutrition on pest and disease resistance in plants. Physiologia Plantarum 133(4):682–691. https://doi.org/10.1111/j.1399-3054.2008.01075.x

Anderegg WRL, Schwalm C, Biondi F, Camarero JJ, Koch G, Litvak M, Ogle K, Shaw JD, Shevliakova E, Williams AP, Wolf A, Ziaco E, Pacala S (2015) Pervasive drought legacies in forest ecosystems and their implications for carbon cycle models. Science 349(6247):528–532

Arend M, Link RM, Patthey R, Hoch G, Schuldt B, Kahmen A (2021, apr) Rapid hydraulic collapse as cause of drought-induced mortality in conifers. Proc Natl Acad Sci 118(16):e2025251118. https://doi.org/10.1073/pnas.2025251118

Arend M, Link RM, Zahnd C, Hoch G, Schuldt B, Kahmen A (2022, mar) Lack of hydraulic recovery as a cause of post-drought foliage reduction and canopy decline in European beech. New Phytologist. https://doi.org/10.1111/nph.18065

Augustin S, Achermann B (2012, sep) Deposition von Luftschadstoffen in der Schweiz: Entwicklung, aktueller Stand und Bewertung. Schweizerische Zeitschrift fur Forstwesen 163(9):323–330. https://doi.org/10.3188/szf.2012.0323

Balsberg-Påhlsson AMB (1992, jan) Influence of nitrogen fertilization on minerals, carbohydrates, amino acids and phenolic compounds in beech (Fagus sylvatica L.) leaves. Tree Physiology 10(1):93–100. https://doi.org/10.1093/treephys/10.1.93

Beaufils ER (1973) Diagnosis and recommendation integrated system (DRIS). Soil Sci Bull Natal Univ, Pietermaritzburg, South Africa, 132 pp

Becher G (1986) Ergebnisse und methodisch-theoretische Überlegungen zur immissionsökologischen Waldzustandserfassung (IWE) - dargestellt am Beispiel Hamburgs. Forstarchiv 57:167–174

Bobbink R, Loran C, Tomassen H (2022) Review and revision of empirical critical loads of nitrogen for Europe. Texte Umweltbundesamt 110:1–358

Braun S (2015) Die Eignung verschiedener Trockenheitsindices zur Abschätzung des Trockenheitsrisikos für den Wald und ihre standortspezifische Entwicklung bis 2050/2100. Schlussbericht des Projektes im Forschungsprogramm "Wald und Klimawandel". Schönenbuch/Bern: Institut für Angewandte Pflanzenbiologie/BAFU

Braun S, Achermann B, De Marco A, Pleijel H, Karlsson PE, Rihm B, Schindler C, Paoletti E (2017, dec) Epidemiological analysis of ozone and nitrogen impacts on vegetation – Critical evaluation and recommendations. Sci Total Environ 603–604:785–792. https://doi.org/10.1016/j.scitotenv.2017.02.225

Braun S, Alonso R, Ashwood F, Chuman T, de Witte LC, Dirnböck T, Karlsson PE, Manninen S, Perring MP, Tömmervik H, Tresch S, Ukomaanaho L, Vanguelova EI (2022) Effects of nitrogen deposition on woodland, forest and other wooded land (EUNIS class T, formerly G). Bobbink et al, Review and revision of empirical critical loads and dose-response relationships. In: Proceedings of an Expert Workshop, Berne Oct 25–28, 2021, Chapter 9. UBA Berlin

Braun S, Cantaluppi L, Flückiger W (2005, oct) Fine roots in stands of Fagus sylvatica and Picea abies along a gradient of soil acidification. Environ Pollut 137(3):574–579. https://doi.org/10.1016/j.envpol.2005.01.042

Braun S, Flückiger W (1989) Effect of ambient ozone and acid mist on aphid development. Environ Pollut 56(3):177–187

Braun S, Flückiger W (2013) Wie geht es unserem Wald. Schönenbuch: Institut für Angewandte Pflanzenbiologie

Braun S, Frehner M, Rihm B, Augustin S (2023, jan) Feuchteachse von Ökogrammen: Quantifizierung und Abschätzung zukünftiger Veränderungen. Schweizerische Zeitschrift fur Forstwesen 174(1):24–31. https://doi.org/10.3188/szf.2023.0024

Braun S, Hopf SE, Tresch S, Remund J, Schindler C (2021, oct) 37 Years of forest monitoring in Switzerland: Drought effects on fagus sylvatica. Front Forests Global Change 4(October):1–10. https://doi.org/10.3389/ffgc.2021.765782

Braun S, Rihm B, Schindler C (2022, mar) Epidemiological estimate of growth reduction by Ozone in Fagus sylvatica L. and Picea abies Karst.: Sensitivity analysis and comparison with experimental results. Plants 11(6):777. https://doi.org/10.3390/plants11060777

Braun S, Rihm B, Tresch S, Schindler C (2023, oct) Long-term risk assessment of uprooting and stem breakage under drought conditions and at high N deposition in beech and Norway spruce. Agricult Forest Meteorol 341(August):109669. https://doi.org/10.1016/j.agrformet.2023.109669

Braun S, Schindler C, Rihm B (2017, dec) Growth trends of beech and Norway spruce in Switzerland: The role of nitrogen deposition, ozone, mineral nutrition and climate. Sci Total Environ 599–600:637–646. https://doi.org/10.1016/j.scitotenv.2017.04.230

Braun S, Schindler C, Rihm B (2020, mar) Foliar nutrient concentrations of European beech in Switzerland: Relations with nitrogen deposition, ozone, climate and soil chemistry. Front Forests Global Change 3(March):1–15. https://doi.org/10.3389/ffgc.2020.00033

Braun S, Schindler C, Volz R, Flückiger W (2003) Forest damages by the storm 'Lothar' in permanent observation plots in Switzerland: The significance of soil acidification and nitrogen deposition. Water Air Soil Pollut 142(1–4):327–340. https://doi.org/10.1023/A:1022088806060

Braun S, Tresch S (2021) Untersuchungen über die Zusammensetzung der Bodenlösung in Standorten der Interkantonalen Walddauerbeobachtung Bericht Periode 2017-2020. Technical report, Institut für Angewandte Pflanzenbiologie AG, Witterswil

Braun S, Tresch S, Augustin S (2020, jul) Soil solution in Swiss forest stands: A 20 year's time series. PLOS ONE 15(7):e0227530. https://doi.org/10.1371/journal.pone.0227530

Bréda N, Granier A, Barataud F, Moyne C (1995) Soil water dynamics in an oak stand - I. Soil moisture, water potentials and water uptake by roots. Plant Soil 172(1):17–27. https://doi.org/10.1007/BF00020856

Brodersen CR, Knipfer T, McElrone AJ (2018, jan) In vivo visualization of the final stages of xylem vessel refilling in grapevine (Vitis vinifera) stems. New Phytologist 217(1):117–126. https://doi.org/10.1111/nph.14811

Brown N, Inward DJG, Jeger M, Denman S (2015, jan) A review of Agrilus biguttatus in UK forests and its relationship with acute oak decline. Forestry 88(1):53–63. https://doi.org/10.1093/forestry/cpu039

Brown N, Vanguelova E, Parnell S, Broadmeadow S, Denman S (2018, jan) Predisposition of forests to biotic disturbance: Predicting the distribution of Acute Oak Decline using environmental factors. Forest Ecol Manag 407:145–154. https://doi.org/10.1016/j.foreco.2017.10.054

Büker P, Morissey T, Briolat A, Falk R, Simpson D, Tuovinen JP, Alonso R, Barth S, Baumgarten M, Grulke NE, Karlsson PE, King J, Lagergren F, Matyssek R, Nunn A, Ogaya R, Penuelas J, Rhea L, Schaub M, Uddling J, Werner W, Emberson LD (2012) DO3SE modelling of soil moisture to determine ozone flux to European forest trees. Atmospheric Chem Phys 12:5537–5562

Buras A, Rammig A, and Zang CS (2020) A novel approach for the identification of pointer years. Dendrochronologia 63:125746

Cakmak I (2005) The role of potassium in alleviating detrimental effects of abiotic stresses in plants. J Plant Nutrition Soil Science-Zeitschrift fur Pflanzenernahrung und Bodenkunde 168(4): 521–530

Carrasco-Molina T, Marzuoli R, Calatayud V, Simpson D, Gerosa G, Carrara A, I. Rábago, Alonso R, González-Fernández I (2024, jul) Validation and parametrization of the soil moisture index for stomatal conductance modelling and flux-based ozone risk assessment of Mediterranean plant species. Agricultural Forest Meteorol 354:110080. https://doi.org/10.1016/j.agrformet.2024.110080

Chaboussou F (1973) Le rôle du potassium et de l'équilibre cationique dans la résistance de la plante aux parasites et aux maladies. Au Service de l'Agriculture (SCPA) No 2, Document technique 16:1–26

Charru M, Seynave I, Morneau F, Bontemps JD (2010) Recent changes in forest productivity: An analysis of national forest inventory data for common beech (Fagus sylvatica L.) in north-eastern France. Forest Ecol Manag 260(5):864–874. https://doi.org/10.1016/j.foreco.2010.06.005

Choat B, Brodribb TJ, Brodersen CR, Duursma RA, López R, Medlyn BE (2018, jun) Triggers of tree mortality under drought. Nature 558(7711):531–539. https://doi.org/10.1038/s41586-018-0240-x

Choat B, Jansen S, Brodribb TJ, Cochard H, Delzon S, Bhaskar R, Bucci SJ, Feild TS, Gleason SM, Hacke UG, Jacobsen AL, Lens F, Maherali H, J. Martínez-Vilalta, Mayr S, Mencuccini M, Mitchell PJ, Nardini A, Pittermann J, Pratt RB, Sperry JS, Westoby M, Wright IJ, Zanne AE (2012) Global convergence in the vulnerability of forests to drought. Nature 491(7426): 752–755. https://doi.org/10.1038/nature11688

Christina M, Le Maire G, Battie-Laclau P, Nouvellon Y, Bouillet JP, Jourdan C, leonardo de Moraes Gonçalves J, Laclau JP (2015) Measured and modeled interactive effects of potassium deficiency and water deficit on gross primary productivity and light-use efficiency in Eucalyptus grandis plantations. Global Change Biol 21(5):2022–2039. https://doi.org/10.1111/gcb.12817

CLRTAP (2004) Manual on methodologies and criteria for modelling and mapping critical loads and levels and air pollution effects, risks and trends. Number 52. UNECE Convention on Long-range Transboundary Air Pollution

CLRTAP (2017) Mapping critical loads for ecosystems, Chapter V of Manual on methodologies and criteria for modelling and mapping critical loads and levels and air pollution effects, risks and trends. UNECE Convention on Long-range Transboundary Air Pollution. UNECE Convention on Long-range Transboundary Air Pollution

CLRTAP, ICP Vegetation (2017) Mapping Critical Levels for Vegetation. In: Mills G, Harmens H, Hayes F, Pleijel H, Büker P, González-Fernández I, Alonso R, Bender J, Bergmann E, Bermejo V, Braun S, Danielsson H, Gerosa G, Grünhage L, Karlsson PE, Marzuoli R, Simpson D (eds.) Manual on methodologies and criteria for modelling and mapping critical loads and levels of air pollution effects, risks and trends, vol 66. UNECE, Bangor UK

Cochard H, Lemoine D, Améglio T, Granier A (2001) Mechanisms of xylem recovery from winter embolism in Fagus sylvatica. Tree Physiology 21:27–33

Coleman MD, Dickson RE, Isebrands JG, Karnosky DF (1995) Carbon allocation and partitioning in aspen clones varying in sensitivity to tropospheric ozone. Tree Physiology 15:593–604

Crampton BG, Plummer SJ, Kaczmarek M, McDonald JE, Denman S (2020, sep) A multiplex real-time PCR assay enables simultaneous rapid detection and quantification of bacteria associated with acute oak decline. Plant Pathol 69(7):1301–1310. https://doi.org/10.1111/ppa.13203

Dalstein L, Vas N (2005, jan) Ozone concentrations and ozone-induced symptoms on coastal and alpine mediterranean pines in Southern France. Water Air Soil Pollut 160(1-4):181–195. https://doi.org/10.1007/s11270-005-4144-7

De Vries W, Posch M (2011) Modelling the impact of nitrogen deposition, climate change and nutrient limitations on tree carbon sequestration in Europe for the period 1900–2050. Environ Pollut 159(10):2289–2299. https://doi.org/10.1016/j.envpol.2010.11.023

de Vries W, Solberg S, Dobbertin M, Sterba H, Laubhann D, van Oijen M, Evans C, Gundersen P, Kros J, Wamelink G, Reinds G, Sutton M (2009, sep) The impact of nitrogen deposition on carbon sequestration by European forests and heathlands. Forest Ecol Manag 258(8): 1814–1823. https://doi.org/10.1016/j.foreco.2009.02.034

de Witte L, Rosenstock N, van der Linde S, Braun S (2017, dec) Nitrogen deposition changes ectomycorrhizal communities in Swiss beech forests. Sci Total Environ 605–606:1083–1096. https://doi.org/10.1016/j.scitotenv.2017.06.142

Defila C, Clot B (2005) Phytophenological trends in the Swiss Alps, 1951–2002. Meteorologische Zeitschrift 14(2):191–196

di Pietro M, Churin JL, Garbaye J (2007) Differential ability of ectomycorrhizas to survive drying. Mycorrhiza 17(6):547–550. https://doi.org/10.1007/s00572-007-0113-x

Dise NB, Wright RF (1995) Nitrogen leaching from European forests in relation to nitrogen deposition. Forest Ecol Manag 71:153–161

Drobyshev I, Niklasson M, Mazerolle MJ, Bergeron Y (2014) Reconstruction of a 253-year long mast record of European beech reveals its association with large scale temperature variability and no long-term trend in mast frequencies. Agricult Forest Meteorol 192-193. https://doi.org/10.1016/j.agrformet.2014.02.010

Du E, van Doorn M, de Vries W (2021, jun) Spatially divergent trends of nitrogen versus phosphorus limitation across European forests. Sci Total Environ 771:145391. https://doi.org/10.1016/j.scitotenv.2021.145391

Dziedek C, Härdtle W, Von Oheimb G, Fichtner A (2016) Nitrogen addition enhances drought sensitivity of young deciduous tree species. Front Plant Sci 7(JULY2016):1100. https://doi.org/10.3389/fpls.2016.01100

EFI (2002) Causes and consequences of increased forest growth in Europe. In: Conference Held in Copenhagen on Aug 28, 2002. European Forest Institute

Egilla JN, Davies FT, Drew MC (2001) Effect of potassium on drought resistance of Hibiscus rosa-sinensis cv. Leprechaun: Plant growth, leaf macro- and micronutrient content and root longevity. Plant Soil 229(2):213–224

Emberson LD, Ashmore MR, Cambridge HM, Simpson D, Tuovinen JP (2000) Modelling stomatal ozone flux across Europe. Environ Pollut 109(3):403–414

Emmett B (2002) The impact of nitrogen depostion in forest ecosystems: a review. Bangor: Centre for Ecology and Hydrology

Emmett B (2007) Nitrogen saturation of terrestrial ecosystems: some recent findings and their implications for our conceptual framework. Water Air Soil Pollut Focus 7:99–109

Engelbrecht BM (2012) Plant ecology: Forests on the brink. Nature 491(7426):675–677. https://doi.org/10.1038/nature11756

Etzold S, Ferretti M, Reinds GJ, Solberg S, Gessler A, Waldner P, Schaub M, Simpson D, Benham S, Hansen K, Ingerslev M, Jonard M, Karlsson PE, Lindroos AJ, Marchetto A, Manninger M, Meesenburg H, Merilä P, Nöjd P, Rautio P, Sanders TG, Seidling W, Skudnik M, Thimonier A, Verstraeten A, Vesterdal L, Vejpustkova M, de Vries W (2020, feb) Nitrogen deposition is the most important environmental driver of growth of pure, even-aged and managed European forests. Forest Ecol Manag 458:117762. https://doi.org/10.1016/j.foreco.2019.117762

Etzold S, Waldner P, Thimonier A, Schmitt M, Dobbertin M (2014) Tree growth in Swiss forests between 1995 and 2010 in relation to climate and stand conditions: Recent disturbances matter. Forest Ecol Manag 311:41–55

Flückiger W (1999) Fortschreitende Bodenversauerung in unseren Wäldern. Zürcher Wald 31(3): 8–12

Foest JJ, Bogdziewicz M, Pesendorfer MB, Ascoli D, Cutini A, Nussbaumer A, Verstraeten A, Beudert B, Chianucci F, Mezzavilla F, Gratzer G, Kunstler G, Meesenburg H, Wagner M, Mund M, Cools N, Vacek S, Schmidt W, Vacek Z, Hacket-Pain A (2024, may) Widespread breakdown in masting in European beech due to rising summer temperatures. Global Change Biol 30(5). https://doi.org/10.1111/gcb.17307

Forest Research (2022) Acute oak decline (AOD): Incidence and distribution

Frehner M, Wasser B, Schwitter R (2005) Nachhaltigkeit und Erfolgskontrolle im Schutzwald. In: Wegleitung für Pflegemassnahmen in Wäldern mit Schutzfunktion, Vollzug Umwelt, vol 564. S. Bern: Bundesamt für Umwelt (BAFU)

Fuhrer J, Skärby L, Ashmore MR (1997) Critical levels for ozone effects on vegetation in Europe. Environ Pollut 97(1-2):91–106

García-Herrera R, J. Díaz, Trigo RM, Luterbacher J, Fischer EM (2010, mar) A review of the European summer heat wave of 2003. Crit Rev Environ Sci Technol 40(4):267–306. https://doi.org/10.1080/10643380802238137

Gasparrini A (2011) Distributed Lag Linear and Non-Linear Models in R : The Package dlnm. J Stat Software 43(8):2–20. https://doi.org/10.18637/jss.v043.i08

Gauthey A, Kahmen A, Limousin J, Vilagrosa A, M Didion-Gency, Mas E, Milano A, Tunas A, Grossiord C (2024, aug) High heat tolerance, evaporative cooling, and stomatal decoupling regulate canopy temperature and their safety margins in three European oak species. Global Change Biol 30(8):1–15. https://doi.org/10.1111/gcb.17439

Gentilesca T, Vieno M, Perks M, Borghetti M, Mencuccini M (2013) Effects of long-term nitrogen addition and atmospheric nitrogen deposition on carbon accumulation in picea sitchensis plantations. Ecosystems 16(7):1310–1324

Gielen B, Low M, Deckmyn G, Metzger U, Franck F, Heerdt C, Matyssek R, Valcke R, Ceulemans R (2006, nov) Chronic ozone exposure affects leaf senescence of adult beech trees: a chlorophyll fluorescence approach. J Exp Botany 58(4):785–795. https://doi.org/10.1093/jxb/erl222

Gillner S, Rüger N, Roloff A, Berger U (2013, aug) Low relative growth rates predict future mortality of common beech (Fagus sylvatica L.). Forest Ecol Manag 302:372–378. https://doi.org/10.1016/j.foreco.2013.03.032

Göttlein A (2016) Ableitung von Nährelelementrelationen für die mitteleuropäischen Hauptbaumarten aus dem Wertebereich normaler Ernährung im Vergleich zu verfügbaren Literaturdaten. Allgemeine Forst- und Jagdzeitung 187:237–246

Göttlein A, Mellert KH (2019) Ableitung von Schwefel-Stickstoff-Relationen für wichtige mitteleuropäische Baumarten aus dem Wertebereich normaler Ernährung und deren Veränderung von BZE I nach BZE II. Allgemeine Forst- und Jagdzeitung 189(11-12): 221–229. https://doi.org/10.23765/afjz0002032

Gschwantner T (2006) Zuwachsänderungen nach den Daten der Österreichischen Waldinventur und ihre klimatischen Ursachen: Abdruck der im April 2004 approbierten Dissertation. Technical Report 133, BOKU, Wien

Gundersen P, Callesen I, de Vries W (1998) Nitrate leaching in forest ecosystems is related to forest floor ratios. Environ Pollut 102(1):403–407. https://doi.org/10.1016/S0269-7491(98)80060-2

Gundersen P, Schmidt IK, Raulund-Rasmussen K (2006) Leaching of nitrate from temperate forests - effects of air pollution and forest management. Environ Rev 14:1–57

Günthardt-Goerg M, Matyssek R, Scheidegger C, Keller T (1993) Differentiation and structural decline in the leaves and bark of birch (Betula pendula) under low ozone concentrations. Trees 7:104–114

Güsewell S (2004) N:P ratios in terrestrial plants: variation and functional significance. New Phytologist 164:243–266

Hacke UG, Plavcová L, Almeida-Rodriguez A, King-Jones S, Zhou W, Cooke JEK (2010, aug) Influence of nitrogen fertilization on xylem traits and aquaporin expression in stems of hybrid poplar. Tree Physiology 30(8):1016–1025. https://doi.org/10.1093/treephys/tpq058

Hacket-Pain AJ, Friend AD, Lageard JGA, Thomas PA (2015) The influence of masting phenomenon on growth-climate relationships in trees: Explaining the influence of previous summers' climate on ring width. Tree Physiology 35(3):319–330. https://doi.org/10.1093/treephys/tpv007

Harvey HP, van den Driessche R (1997) Nutrition, xylem cavitation and drought resistance in hybrid poplar. Tree Physiology 17:647–654

Harvey HP, van den Driessche R (1999) Nitrogen and potassium effect on xylem cavitation and water use efficiency in poplars. Tree Physiology 19:943–950

Hicks DA (1978) Diagnosing Vegetation Injury Caused by Air Pollution. Springfield: U.-S. Department of Commerce, National Technical Information Center

Holuša J, Lubojacký J, Čurn V, Tonka T, Lukášová K, Horák J (2018) Combined effects of drought stress and Armillaria infection on tree mortality in Norway spruce plantations. Forest Ecol Manag 427:434–445. https://doi.org/https://doi.org/10.1016/j.foreco.2018.01.031

Hopf S, Tresch S, Belyazid S, Sverdrup H, Augustin S, Kurz D, Rihm B, Braun S (2023, oct) Dendrochemical indicators of tree rings reveal historical soil acidification in Swiss forest stands. Dendrochronologia 81(July):126099. https://doi.org/10.1016/j.dendro.2023.126099

ICP Forests (2016) Manual on methods and criteria for hamonized sampling, assessment, monitoring and analysis of the effects of air pollution on forests. Hamburg: Johann Heinrich von Thünen Institute

ICP Forests (2022) Manual on methods and criteria for hamonized sampling, assessment, monitoring and analysis of the effects of air pollution on forests. Part XIV. Hamburg: Johann Heinrich von Thünen Institute

IPCC (2021) Climate Change 2021 The Physical Science Basis Working Group I Contribution to the Sixth Assessment Report of the Intergovernmental Panel on Climate Change Summary for Policymakers. Climate Change 2021: The Physical Science Basis: 3949

Jany JL, Martin F, Garbaye J (2003) Respiration activity of ectomycorrhizas from Cenococcum geophilum and Lactarius sp. in relation to soil water potential in five beech forests. Plant Soil 255(2):487–494

Jarcuska B, Mihal I, Cicak A, Tsakov H (2013) Beech bark necrosis: partitioning the environmental and spatial variation of the damage severity in Central and South-Eastern Europe. Ann Forest Res 56(2):317–338

Jarvis PG (1976) The interpretation of the variations in leaf water potential and stomatal conductance found in canopies in the field. Philos Trans Royal Soc Lond 273:593–610

Johnson R, Vishwakarma K, Hossen MS, Kumar V, Shackira AM, Puthur JT, Abdi G, Sarraf M, Hasanuzzaman M (2022) Potassium in plants: Growth regulation, signaling, and environmental

stress tolerance. Plant Physiol Biochem 172:56–69. https://doi.org/10.1016/j.plaphy.2022.01.001
Jönsson AM, Ingerslev M, Raulund-Rasmussen K (2004) Frost sensitivity and nutrient status in a fertilized Norway spruce stand in Denmark. Forest Ecol Manag 201(2-3):199–209. https://doi.org/10.1016/j.foreco.2004.06.030
Kahle HP, Karjalainen T, Schuck A, Agren G, Kellomäki S, Prietzel J, Rehfuess KE, Spiecker H (2008) Causes and consequences of forest growth trends in Europe: Results of the recognition project. Research Report 21. Number 21. Leiden: Brill
Karlsson P, Braun S, Broadmeadow M, Elvira S, Emberson L, Gimeno B, Le Thiec D, Novak K, Oksanen E, Schaub M, Uddling J, Wilkinson M (2007, apr) Risk assessments for forest trees: The performance of the ozone flux versus the AOT concepts. Environ Pollut 146(3):608–616. https://doi.org/10.1016/j.envpol.2006.06.012
Karnosky DF, Percy K, Xiang B, Callan B, Noormets A, Mankovska B, Hopkin A, Sober J, Jones W, Dickson RE, Isebrands JG (2002) Interacting elevated CO2 and tropospheric O3 predisposes aspen (Populus tremuloides Michx.) to infection by rust (Melampsora medusae f. sp. tremuloidae). Global Change Biol 8:329–338
Karnosky DF, Werner H, Holopainen T, Percy K, Oksanen T, Oksanen E, Heerdt C, Fabian P, Nagy J, Heilman W, Cox R, Nelson N, Matyssek R (2007) Free-air exposure systems to scale up ozone research to mature trees. Plant Biol 9(2):181–190
Kint V, Aertsen W, Campioli M, Vansteenkiste D, Delcloo A, Muys B (2012, nov) Radial growth change of temperate tree species in response to altered regional climate and air quality in the period 1901–2008. Climatic Change 115(2):343–363. https://doi.org/10.1007/s10584-012-0465-x
Klesse S, Wohlgemuth T, Meusburger K, Vitasse Y, von Arx G, Lévesque M, Neycken A, Braun S, Dubach V, Gessler A, Ginzler C, Gossner M, Hagedorn F, Queloz V, Samblás Vives E, Rigling A, Frei E (2022, dec) Long-term soil water limitation and previous tree vigor drive local variability of drought-induced crown dieback in Fagus sylvatica. Sci Total Environ 851: 157926. https://doi.org/10.1016/j.scitotenv.2022.157926
Kubiak K, Zólciak A, Damszel M, Lech P, Sierota Z (2017) Armillaria pathogenesis under climate changes. Forests 8(4). https://doi.org/10.3390/f8040100
Lehto T, Zwiazek JJ (2011, feb) Ectomycorrhizas and water relations of trees: a review. Mycorrhiza 21(2):71–90. https://doi.org/10.1007/s00572-010-0348-9
Love DM, Sperry JS (2018, jul) In situ embolism induction reveals vessel refilling in a natural aspen stand. Tree Physiol 38(7):1006–1015. https://doi.org/10.1093/treephys/tpy007
LRTAP C (2017) Mapping critical levels for vegetation. Revised Chapter 3 of the UNECE manual on methodologies and criteria for modelling and mapping critical loads and levels and air pollution effects, risks and trends.
Lux D, Leonardi S, Müller J, Wiemken A, Flückiger W (1997) Effects of ambient ozone concentration on contents of non-structural carbohydrates in young Picea abies (L.) Karst. and Fagus sylvatica L. New Phytologist 137:399–409
Maier-Maercker U, Koch W (1991) Experiments on the control capacity of stomata of Picea abies (L.) Karst. after fumigation with ozone and in environmentally damaged material. Plant Cell Environ 14:175–184
Manion PD (1981) Tree disease concepts. Prentic Hall, Englewood Cliffs, NJ
Manning WJ, von Tiedemann A (1995) Climate change: potential effects of increased atmospheric carbon dioxice (CO2), ozone (O3) and ultraviolet-B (UV-B) radiation on plant diseases. Environ Pollut 88(2):219–245
Matyssek R, Karnosky DF, Wieser G, Percy K, Oksanen E, Grams TEE, Kubiske M, Hanke D, Pretzsch H (2010) Advances in understanding ozone impact on forest trees: Messages from novel phytotron and free-air fumigation studies. Environ Pollut 158(6):1990–2006
Mayer P, Brang P, Dobbertin M, Zimmermann S (2004) Sturmschäden auf sauren Standorten häufiger. Informationsblatt Forschungsbereich Wald 2004(16):1–4

McDowell NG, Beerling DJ, Breshears DD, Fisher RA, Raffa KF, Stitt M (2011, oct) The interdependence of mechanisms underlying climate-driven vegetation mortality. Trends Ecol Evolut 26(10):523–532. https://doi.org/10.1016/j.tree.2011.06.003

McDowell NG, Sevanto S (2010, apr) The mechanisms of carbon starvation: how, when, or does it even occur at all? New Phytologist 186(2):264–266. https://doi.org/10.1111/j.1469-8137.2010.03232.x

McLaughlin SB, Wullschleger S, Sun G, Nosal M (2007) Interactive effects of ozone and climate on water use, soil moisture content and streamflow in a southern Appalachian forest in the USA. New Phytologist 174(1):125–136

Meining S, Wilpert KV, Schäffer J, Schröter H (2008) Waldzustandsbericht der Forstlichen Versuchs- und Forschungsanstalt Baden-Württemberg. Forstliche Versuchs- und Forschungsanstalt Baden-Württemberg (FVA), Freiburg

MeteoSwiss (2021) Wärmstes Jahrzehnt seit Messbeginn

Mills G, Pleijel H, Braun S, Büker P, Bermejo V, Calvo E, Danielsson H, Emberson L, IG Fernández, Grünhage L, Harmens H, Hayes F, Karlsson PE, Simpson D (2011) New stomatal flux-based critical levels for ozone effects on vegetation. Atmospheric Environ 45(28). https://doi.org/10.1016/j.atmosenv.2011.06.009

Mills G, Wagg S, Harmens H (2013) Ozone pollution: Impacts on ecosystem services and biodiversity. Centre Ecol Hydrol:1–108

Mrazek F (1983) Erhöhung der Widerstandskraft von Kiefernbeständen gegen den Blauen Kiefernprachtkäfer (Phaenops cyana F.) durch mineralische Düngung. Beiträge für die Forstwirtschaft 17(3):122–123

Müller M, Olsson PO, Eklundh L, Jamali S, Ardö J (2022) Features predisposing forest to bark beetle outbreaks and their dynamics during drought. Forest Ecol Manag 523:120480. https://doi.org/10.1016/j.foreco.2022.120480

Müller-Haubold H, Hertel D, Leuschner C (2015, sep) Climatic drivers of mast fruiting in European beech and resulting C and N allocation shifts. Ecosystems 18(6):1083–1100. https://doi.org/10.1007/s10021-015-9885-6

Nabuurs GJ, Lindner M, Verkerk PJ, Gunia K, Deda P, Michalak R, Grassi G (2013) First signs of carbon sink saturation in European forest biomass. Nature Climate Change 3(9):792–796. https://doi.org/10.1038/nclimate1853

Nellemann C, Thomsen MG (2001) Long-term changes in forest growth: Potential effects of nitrogen deposition and acidification. Water Air Soil Pollut 128(3-4):197–205. https://doi.org/10.1023/A:1010318800180

Neycken A, Scheggia M, Bigler C, Lévesque M (2022, sep) Long-term growth decline precedes sudden crown dieback of European beech. Agricultural Forest Meteorol 324:109103. https://doi.org/10.1016/j.agrformet.2022.109103

Nihlgård B (1985) The ammonium hypothesis - an additional explanation to the forest dieback in Europe. Ambio 14(1):1–8

Nussbaumer A, Gessler A, Benham S, de Cinti B, Etzold S, Ingerslev M, Jacob F, Lebourgeois F, Levanic T, Marjanović H, Nicolas M, Ostrogović Sever MZ, Priwitzer T, Rautio P, Roskams P, Sanders TG, Schmitt M, Šrámek V, Thimonier A, Ukonmaanaho L, Verstraeten A, Vesterdal L, Wagner M, Waldner P, Rigling A (2021) Contrasting resource dynamics in mast years for European beech and oak—a continental scale analysis. Front Forests Global Change 4. https://doi.org/10.3389/ffgc.2021.689836

Nussbaumer A, Waldner P, Apuhtin V, Aytar F, Benham S, Bussotti F, Eichhorn J, Eickenscheidt N, Fabianek P, Falkenried L, Leca S, Lindgren M, Manzano Serrano MJ, Neagu S, Nevalainen S, Pajtik J, Potočić N, Rautio P, Sioen G, Stakėnas V, Tasdemir C, Thomsen IM, Timmermann V, Ukonmaanaho L, Verstraeten A, Wulff S, Gessler A (2018) Impact of weather cues and resource dynamics on mast occurrence in the main forest tree species in Europe. Forest Ecol Manag 429. https://doi.org/10.1016/j.foreco.2018.07.011

Oddo E, Inzerillo S, Grisafi F, Sajeva M, Salleo S, Nardini A (2014, aug) Does short-term potassium fertilization improve recovery from drought stress in laurel? Tree Physiol 34(8): 906–913. https://doi.org/10.1093/treephys/tpt120

Pääkkönen E, Holopainen T, Kärenlampi L (1997) Differences in growth, leaf senescence and injury, and stomatal density in birch (Betula pendula Roth.) in relation to ambient levels of ozone in finland. Environ Pollut 96(2):117–127

Paar U, Guckland A, Dammann I, Albrecht M, and Eichhorn J (2011) Häufigkeit und Intensität der Fruktifikation der Buche. AFZ-DerWald 66(6):26–29

Paoletti E (2005) Ozone slows stomatal response to light and leaf wounding in a Mediterranean evergreen broadleaf, Arbutus unedo. Environ Pollut 134(3):439–445

Paoletti E, Contran N, Bernasconi P, Gunthardt-Goerg MS, Vollenweider P (2009) Structural and physiological responses to ozone in Manna ash (Fraxinus ornus L.) leaves of seedlings and mature trees under controlled and ambient conditions. Sci Total Environ 407(5):1631–1643

Pathak J, Ahmed H, Kumari N, Pandey A, Rajneesh, Sinha RP (2020, jun) Role of calcium and potassium in amelioration of environmental stress in plants. Protective chemical agents in the amelioration of plant abiotic stress, pp 535–562. Wiley. https://doi.org/10.1002/9781119552154.ch27

Pearson M, Mansfield TA (1993) Interacting effects of ozone and water stress on the stomatal resistance of beech (Fagus sylvatica L.). New Phytologist 123:351–358

Pfeiffer TZ, Stolfa I, Zanic M, Pavicic N, Cesar V, Lepedus H (2013) Oxidative stress in leaves of two olive cultivars under freezing conditions. Acta Biologica Hungarica 64(3):341–351

Pickles BJ, Simard SW (2017) Chapter 18 - mycorrhizal networks and forest resilience to drought. In: Johnson NC, Gehring C, Jansa JBTMMoS (eds.) Mycorrhizal mediation of soil: fertility, structure, and carbon storage, pp 319–339. Elsevier. https://doi.org/10.1016/B978-0-12-804312-7.00018-8

Plavcová L, Hacke UG (2012, nov) Phenotypic and developmental plasticity of xylem in hybrid poplar saplings subjected to experimental drought, nitrogen fertilization, and shading. J Exp Botany 63(18):6481–6491. https://doi.org/10.1093/jxb/ers303

Plavcová L, Hacke UG, Almeida-Rodriguez AM, Li E, Douglas CJ (2013, jan) Gene expression patterns underlying changes in xylem structure and function in response to increased nitrogen availability in hybrid poplar. Plant Cell Environ 36(1):186–199. https://doi.org/10.1111/j.1365-3040.2012.02566.x

Powles SB (1984, jun) Photoinhibition of photosynthesis induced by visible light. Annu Rev Plant Physiol 35(1):15–44

Pretzsch H, Block J, Dieler J, Gauer J, Göttlein A, Moshammer R, Schuck J, Weis W, Wunn U (2014) Nährstoffentzüge durch die Holz- und Biomassenutzung in Wäldern. Teil 1: Schätzfunktionen für Biomasse und Nährelemente und ihre Anwendung in Szenariorechnungen. Allgemeine Forst- und Jagdzeitung 185(11/12):261–285

Puhe J (1994) Die Wurzelentwicklung der Fichte (Picea abies (L.) Karst.) bei unterschiedlichen chemischen Bodenbedingungen. Dissertation Universität Göttingen, Göttingen

Reich PB, Oleksyn J, Wright IJ (2009) Leaf phosphorus influences the photosynthesis-nitrogen relation: A cross-biome analysis of 314 species. Oecologia 160(2):207–212. https://doi.org/10.1007/s00442-009-1291-3

Reuss JO, Cosby BJ, Wright RF (1987) Chemical processes governing soil and water acidification. Nature 329:27–32

Rihm B, Künzle T (2019) Mapping Nitrogen Deposition 2015 for Switzerland. Technical Report on the Update of Critical Loads and Exceedance, including the years 1990, 2000, 2005 and 2010. Technical report, Meteotest, Bern

Rihm B, Thimonier A, Albrecht S, Waldner P (2013) Berechnung der Deposition basischer Kationen für Wälder, Provisorische Depositionskarten für Ca, Mg, K, Na und Cl. Technical report, Meteotest

Roth T, Tresch S, Du E, Braun S (2022) Hierarchical change-point regression models including random effects to estimate empirical critical loads for nitrogen using Bayesian Regression Models (brms) and JAGS. MethodsX 9:101902. https://doi.org/10.1016/j.mex.2022.101902

Samuelson LJ, Kelly JM (1996) Carbon partitioning and allocation in northern red oak seedlings and mature trees in response to ozone. Tree Physiology 16:853–858

Sardans J, Penuelas J, Coll M, Vayreda J, Rivas-Ubach A (2012) Stoichiometry of potassium is largely determined by water availability and growth in Catalonian forests. Funct Ecol 26(5): 1077–1089

Schindler U (1951) Das Buchensterben; bisherige Veröffentlichungen und Auswertungen der staatlichen Unterlagen. Forstarchiv 22:109–119

Schuldt B, Buras A, Arend M, Vitasse Y, Beierkuhnlein C, Damm A, Gharun M, Grams TE, Hauck M, Hajek P, Hartmann H, Hiltbrunner E, Hoch G, Holloway-Phillips M, Körner C, Larysch E, Lübbe T, Nelson DB, Rammig A, Rigling A, Rose L, Ruehr NK, Schumann K, Weiser F, Werner C, Wohlgemuth T, Zang CS, Kahmen A (2020, jun) A first assessment of the impact of the extreme 2018 summer drought on Central European forests. Basic Appl Ecol 45: 86–103. https://doi.org/10.1016/j.baae.2020.04.003

Schulla J (2019) Model description WaSiM-ETH. Hydrology Software Consulting J. Schulla, Zürich

Schweizerischer Bundesrat (2022) Umweltbericht 2022. Schweizerischer Bundesrat, Bern

Seidl R, Müller J, Hothorn T, Bässler C, Heurich M, Kautz M (2016) Small beetle, large-scale drivers: How regional and landscape factors affect outbreaks of the European spruce bark beetle. J Appl Ecol 53(2):530–540. https://doi.org/10.1111/1365-2664.12540

Senser M, Kloos M, Lütz C (1990) Influence of soil substrate and ozone plus acid mist on the pigment content and composition of needles from young Norway spruce trees. Environ Pollut 64(3-4):295–312. https://doi.org/10.1016/0269-7491(90)90052-E

Shukla P, Skea J, Slade R, van Diemen R, Haughey E, Malley J, Pathak M, Pereira JP (2019) IPCC Technical Summary 2019 Technical report, Intergovernmental Panel on Climate Change

Singh DK, Sale PWG, Pallaghy CK, McKenzie BM (2000) Phosphorus concentrations in the leaves of defoliated white clover affect abscisic acid formation and transpiration in drying soil. New Phytologist 146:249–259

Smith S, Read D (2008) Mycorrhizal symbiosis. Academic Press

Spellmann H, Sutmöller J, Meesenburg H (2007) Risikovorsorge im Zeichen des Klimawandels. AFZ-Der Wald 62(23):1246–1249

Spiecker H, Mielikäinen R, Köhl M, Skorgsgaard JP (1996) Growth trends in European forests. Springer, New York

Sutton MA, Simpson D, Levy PE, Smith RI, Reis S, van Oijen M, de Vries W (2008) Uncertainties in the relationship between atmospheric nitrogen deposition and forest carbon sequestration. Global Change Biol 14(9):2057–2063. https://doi.org/10.1111/j.1365-2486.2008.01636.x

Suz LM, Bidartondo MI, van der Linde S, Kuyper TW (2021) Ectomycorrhizas and tipping points in forest ecosystems. New Phytologist 231(5):1700–1707. https://doi.org/10.1111/nph.17547

Sverdrup H and P. Warfvinge (1993) The effect of soil acidification on the growth of trees, grass and herbs as expressed by the (Ca+Mg+K)/Al ratio. Technical report, Lund University, Department of Chemical Engineering II, Lund

Talkner U, Meiwes KJ, Mindrup M, Nagel R (2011) Bewertung des Ernährungszustands eines Fichtenforsts: Möglichkeiten des DRIS (Diagnosis and recommendation integrated system). Forstarchiv 82:82–91

Tamm CO (1991) Nitrogen in terrestrial ecosystems. Ecol Stud 81:1–115

Thomas RQ, Canham CD, Weathers KC, Goodale CL (2010) Increased tree carbon storage in response to nitrogen deposition in the US. Nature Geosci 3:13–17

Tresch S, de Witte LC, Hopf SE, Schindler C, Rihm B, Braun S (2025, jun) Drought and N deposition impact roots and ectomycorrhizal colonisation of European beech down to deep soil layers. Appl Soil Ecol 210:106055. https://doi.org/10.1016/j.apsoil.2025.106055

Tresch S, Roth T, Schindler C, Hopf SE, Remund J, Braun S (2023, sep) The cumulative impacts of droughts and N deposition on Norway spruce (Picea abies) in Switzerland based on 37 years of forest monitoring. Sci Total Environ 892:164223. https://doi.org/10.1016/j.scitotenv.2023.164223

Ulrich B (1986) Die Rolle der Bodenversauerung beim Waldsterben: Langfristige Konsequenzen und forstliche Möglichkeiten. Forstwissenschaftliches Centralblatt 103:421–435

Ulrich B (1991) An ecosystem approach to soil acidification. In: Ulrich B, Sumner ME (eds) Soil acidity, pp 28–79. Springer, Berlin

Urli M, Porté AJ, Cochard H, Guengant Y, Burlett R, Delzon S (2013) Xylem embolism threshold for catastrophic hydraulic failure in angiosperm trees. Tree Physiology 33(7):672–683. https://doi.org/10.1093/treephys/tpt030

Van der Heyden D, Skelly J, Innes J, Hug C, Zhang J, Landolt W, Bleuler P (2001) Ozone exposure thresholds and foliar injury on forest plants in Switzerland. Environ Pollut 111(2):321–331

van der Linde S, Suz LM, Orme CDL, Cox F, Andreae H, Asi E, Atkinson B, Benham S, Carroll C, Cools N, De Vos B, Dietrich HP, Eichhorn J, Gehrmann J, Grebenc T, Gweon HS, Hansen K, Jacob F, Kristöfel F, Lech P, Manninger M, Martin J, Meesenburg H, Merilä P, Nicolas M, Pavlenda P, Rautio P, Schaub M, Schröck HW, Seidling W, Šrámek V, Thimonier A, Thomsen IM, Titeux H, Vanguelova E, Verstraeten A, Vesterdal L, Waldner P, Wijk S, Zhang Y, Žlindra D, Bidartondo MI (2018, jun) Environment and host as large-scale controls of ectomycorrhizal fungi. Nature 558(7709):243–248. https://doi.org/10.1038/s41586-018-0189-9

Veresoglou SD, Peñuelas J, Fischer R, Rautio P, Sardans J, Merilä P, Tabakovic-Tosic M, Rillig MC (2014) Exploring continental-scale stand health - N : P ratio relationships for European forests. New Phytologist 202(2):422–430. https://doi.org/10.1111/nph.12665

Verschuren L, De Mil T, De Frenne P, Haneca K, Van Acker J, Vandekerkhove K, Van den Bulcke J (2023, dec) Heading for a fall: The fate of old wind-thrown beech trees (Fagus sylvatica) is detectable in their growth pattern. Sci Total Environ 903:166148. https://doi.org/10.1016/j.scitotenv.2023.166148

Vitasse Y, Bottero A, Cailleret M, Bigler C, Fonti P, Gessler A, Lévesque M, Rohner B, Weber P, Rigling A, Wohlgemuth T (2019, nov) Contrasting resistance and resilience to extreme drought and late spring frost in five major European tree species. Global Change Biol 25(11): 3781–3792. https://doi.org/10.1111/gcb.14803

Vitousek PM, Porder S, Houlton BZ, Chadwick OA (2010) Terrestrial phosphorus limitation: mechanisms, implications, and nitrogen-phosphorus interactions. Ecol Appl 20(1):5–15

Wallander H (2000) Use of strontium isotopes and foliar K content to estimate weathering of biotite induced by pine seedlings colonised by ectomycorrhizal fungi from two different soils. Plant Soil 222(1-2):215–229. https://doi.org/10.1023/a:1004756221985

Wellburn FAM, Wellburn AR (1994) Atmospheric ozone affects carbohydrate allocation and winter hardiness of Pinus halepensis Mill. J Exp Botany 45:607–614

Wittig VE, Ainsworth EA, Naidu SL, Karnosky DF, Long SP (2009) Quantifying the impact of current and future tropospheric ozone on tree biomass, growth, physiology and biochemistry: a quantitative meta-analysis. Global Change Biol 15(2):396–424

Wolf S, Paul-Limoges E (2023, oct) Drought and heat reduce forest carbon uptake. Nature Commun 14(1):6217. https://doi.org/10.1038/s41467-023-41854-x

WSL and BUWAL (2001) Lothar. Der Orkan 1999. Ereignisanalyse. WSL Swiss Federal Research Institute, Birmensdorf, Bern

Zweifel R, Sterck F, Braun S, Buchmann N, Eugster W, Gessler A, Häni M, Peters RL, Walthert L, Wilhelm M, Ziemińska K, Etzold S (2021, sep) Why trees grow at night. New Phytologist 231(6):2174–2185. https://doi.org/10.1111/nph.17552

# Drivers of Forest Degradation and Biodiversity Changes in Mongolian Forest-Steppes During the Holocene

**Julia Unkelbach**

## Contents



**Abstract** This review analyses forest-steppe vegetation dynamics in western and north-central Mongolia during the Holocene with respect to the extent of natural and anthropogenic drivers of vegetation change and deforestation. Under the influence of the highly continental mountain climate, the environmental conditions for the establishment of forest and forest-steppe vegetation are harsh. The spatial distribution of forest and steppe is controlled by several climatic and topographic factors. Human activities such as tree logging and grazing by nomadic livestock are another important factor limiting the distribution of forested areas. Based on selected pollen, macro-charcoal, and coprophilous fungi concentrations from two example forest-steppe sites, one in a region with high anthropogenic influence and the second with very low anthropogenic impact, general changes in the forest and steppe vegetation

Communicated by Christoph Leuschner

J. Unkelbach (✉)
Department of Palynology and Climate Dynamics, Albrecht-von-Haller-Institute for Plant Sciences, University of Göttingen, Göttingen, Germany
e-mail: julia.unkelbach@biologie.uni-goettingen.de

 
Progress in Botany (2026) 86: 269–284, https://doi.org/10.1007/124_2025_87,
Published online: 25 April 2025

over the last 9,570 cal yr BP were reconstructed. Based on the presented multi-proxy data, it can be concluded that during the early and mid-Holocene, it appears most likely that the vegetation dynamics in both regions are defined by supra-regional climate variability. Whereas during the late Holocene, small-scale variations, regional climatic trends, and the local human activities may have had a greater impact on the forest-steppe ecosystems of Mongolia than large-scale climate patterns.

**Keywords** Forest-steppe ecosystems, Human impact, Mongolia, Multi-proxy approach, Vegetation dynamics

## 1 Introduction

Mongolia covers a total area of around 1,600 million hectares. Today, only a small part is still forested (17.5 million ha, 11.2%), which is distributed between the closed boreal forest regions in the north and the adjacent open forest-steppe areas in the western and central montane parts of the country (Tsogtbaatar 2007). Due to Mongolia's continental location and the great distance to the water resources of the oceans, water availability and high altitude are the main limiting factors for the establishment of forest vegetation. In the forest-steppe, the spatial distribution of forest and steppe is controlled by several further climatic and topographic factors: temperature, precipitation, evaporation as well as topographic position and relief (e.g., Dulamsuren and Hauck 2008; Hais et al. 2016; Klinge et al. 2021). Human activity is another important factor influencing the characteristics of the forest-steppe area. Human disturbance through logging for timber and fuelwood is massively reducing the tree populations of the forest stands. In addition, in most forest-steppe areas, large-scale grazing by nomadic livestock is preventing the regeneration of tree populations, particularly in small forest patches or at the forest margins (Dulamsuren et al. 2014; Khishigjargal et al. 2013). Due to their small numbers, wild herbivores seem to play only a subordinate role in this regard (Klinge et al. 2021).

Since the mid-twentieth century, the vulnerable forest-steppe ecosystem has also been threatened by global warming and its various side effects (Batima et al. 2005). Increased drought stress causes forest regeneration and wood production to further decline (e.g., Dulamsuren et al. 2009; Gunin et al. 1999). In addition, forest fires, which have been increasing in frequency and intensity especially over the last decades, have destroyed large areas of woodland. In 1996 and 2002, the two most extensive fire events in Tarvagatai Nuruu National Park (Northern Khangai Mountains) have affected 2.36 million ha and 80,000 ha, respectively, of forested land (Ariuntsetseg et al. 2008; Dulamsuren et al. 2014; Tsogtbaatar 2007). This is a concerning trend, as the livelihood of Mongolia's rural and mostly nomadic population is still strongly dependent on natural resources (Lkhagvadorj et al. 2013a).

The aim of this review is to identify the main factors influencing the proportional composition and the structural characteristics of the forest-steppe vegetation in

Mongolia. Using two example study sites, one in a forest-steppe region with comparatively high anthropogenic influence and the second with very low anthropogenic impact, the changes in the ratio of forest to steppe vegetation over the Holocene are reconstructed based on pollen concentrations. The focus is on the mid and especially the late Holocene. Regarding further parameters (coprophilous fungal spores and charcoal particles) and beyond, various driving factors of the changes in vegetation will be compared and then discussed. Which drivers predominate? Do they change over time?

## 2 Reconstructing the Dynamics of Mongolian Forest-Steppe-Ecosystems

### 2.1 Geographical Setting

Mongolia is located in Central Asia and borders Russia to the north and China to the south. With an area of 1.6 million square kilometers but only 3.5 million inhabitants, Mongolia is the most sparsely populated country in the world. Mongolia is a diverse landscape with several mountain ranges as well as wide flat plains and the Gobi Desert. The high mountains include the Mongolian Altai in the west, the Khangai in the center, Khentei and Sayan in the north, and the Gobi Altai in the south. In the south and east of Mongolia, extensive plains and a low relief characterize the landscape. The distribution of vegetation is similarly diverse, with zones ranging from alpine vegetation to steppe and desert (see Fig. 1). The forest-steppe zone

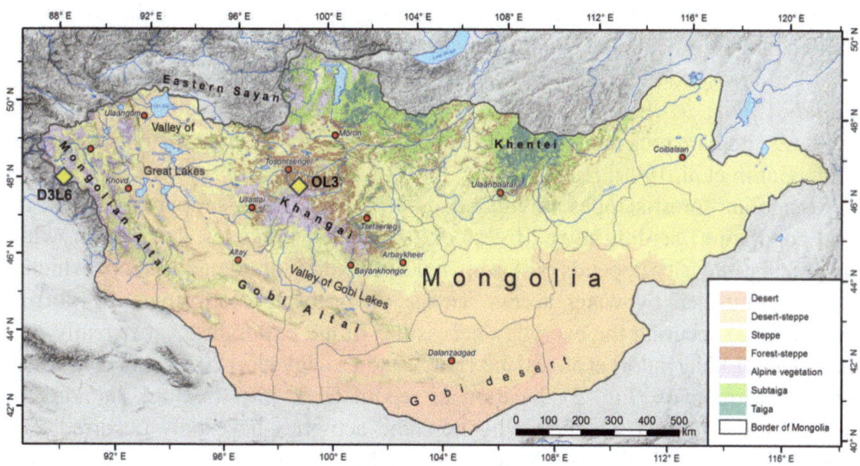


**Fig. 1** Overview map showing the vegetation zones of Mongolia with the location of the two investigated palynological records in the forest-steppe areas in the Altai Mountains (D3L6) and in the northern Khangai Mountains (OL3). (Modified from Klinge et al. (2018))

covers 375.000 $km^2$ (Wallis de Vries et al. 1996) which is around 23% of the total Mongolian territory.

The climate in the forest-steppe regions in north-western and north-central Mongolia is highly continental due to their position within Central Asia and their montane elevation. Mean annual temperatures are −0.9°C for Ulgii in the Mongolian Altai and −5.9°C for Tosontsengel in the northern Khangai. Mean annual precipitation mainly follows a west-east directed gradient and ranges from 150 to 500 mm (Schneider et al. 2021; AM Online Projects 2024). Low-pressure cells brought by the prevailing westerlies cause most of the annual precipitation to occur in summer (Batima et al. 2005). In winter, the Siberian High is predominant, which usually leads to rather dry conditions during this time. The cold winter climate favors the occurrence of discontinuous permafrost, which mainly forms in upper mountain areas, valley bottoms, and partially on mountain slopes.

The Mongolian forest-steppe forms the transition zone between the southern border of the boreal forest of Central Asia and the dry Gobi Desert. The forest-steppe is highly sensitive to climate change and any form of anthropogenic land-use (Hauck and Lkhagvadorj 2013; Khansaritoreh et al. 2017; Lkhagvadorj et al. 2013b; Yang et al. 2016). In the Mongolian mountain forest-steppe, there is an altitudinal zonation of vegetation controlled by elevation, relief, precipitation, and solar radiation (e.g., Beket and Knapp 2012; Gunin et al. 1999). Above 2,800 m a.s.l., the slopes are dominated by alpine tundra, followed by dry steppe vegetation mainly consisting of grasses and sedges, and the valleys below 2,200 m a.s.l. are covered by open grasslands of shrubby species, grasses, sedges, and forbs. Forest patches do only occur on north-facing slopes through a complex interplay of more water availability due to the presence of permafrost and less evapotranspiration due to reduced solar radiation at elevations between around 1,700 and 2,500 m a.s.l. Siberian larch (*Larix sibirica*) is the most dominant tree species, accounting for 75–80% of the total boreal forest and forest-steppe of Mongolia. Other occurring conifers are *Pinus sibirica*, *Pinus sylvestris*, *Picea obovata*, and *Abies sibirica*. In disturbed areas, coniferous forests are enriched by *Betula platyphylla* and *Populus tremula* (MNE 2012; Tsogtbaatar 2004). Azonal vegetation types occur sporadically, e.g., on bare rock areas or in and around freshwater habitats (Ariuntsetseg et al. 2008; Gunin et al. 1999; Hilbig 1995).

Mongolian forest-steppes are inhabited by pastoral nomads (approx. 30% of the total population) herding mixed livestock (sheep, goats, horses, cattle, yak), while the grazing intensity depends on the spatial distribution of nomadic camps which is usually controlled by water access. Hence, damages to biodiversity by grazing activities and securing the everyday necessities of the pastoralists vary greatly on a local scale (Lkhagvadorj et al. 2013a; Hauck and Lkhagvadorj 2013). In the northern Khangai Mountains (Fig. 2), location of record OL3, infrastructure for nomadic pastoralists is limited which is why nomadic activities have only occurred very sparsely in the area in the past. Whereas in the Mongolian Altai (Fig. 3), location of record D3L6, traditional land-use, e.g., animal husbandry and grazing by nomadic pastoralists as well as hunting activities are tolerated and are historically common (Beket and Knapp 2012).

**Fig. 2** View of the forest-steppe ecosystem in the northern Khangai area with typical forest-steppe distribution and dead trees due to the impact of recent forest fires. (Photo: H. Behling, 2018)

**Fig. 3** View of the forest-steppe ecosystem in the Mongolian Altai with typical forest-steppe distribution and representative density of nomadic settlements. (Photo: H. Behling, 2016)

### 2.2 Materials and Methods

For a comparative analysis of temporal and spatial patterns of vegetation change in both forest-steppe areas with different characteristics of anthropogenic impact, our two longest palynological and paleoecological records from each area have been selected.

The first study area (record OL3) is located in the Travagatai Nuruu NP, northern Khangai Mountains, in north-central Mongolia (Fig. 1). The NP was established in 2000, and its summits reach up to 3,100–3,200 m a.s.l. Fieldwork was conducted in 2018 within the valley of the Tegsh river, 50 km southwest of Tosontsengel. Lacustrine sediment core OL3 (48°18′37.97″N, 98°0′38.48″E, 2,012 m a.s.l.) was taken from Olgi Lake, a closed groundwater-fed lake comprising an area of approximately 300 m × 700 m. Core OL3 is 238 cm long, divided into three overlapping 1 m-sections.

The second study area (record D3L6) is located in the Altai Tavan Bogd National Park (NP), Altai Mountains, in north-western Mongolia (Fig. 1). The NP, established in 1996, is situated between 2,000 and 4,000 m a.s.l. Fieldwork was conducted in 2016 south of Dayan Nuur, located 110 km southwest of Ölgii. Lacustrine sediment core D3L6 (48°16′22.18″N, 88°54′48.50″E, 2,450 m a.s.l.) was taken from a small circular lake of around 150 m in diameter and 2.5 m max. water depth. Core D3L6 is 94 cm long, divided into two core sections.

For palynological analyses, 187 subsamples of 0.5 $cm^3$ taken in 1-cm or 2-cm intervals from cores D3L6 and OL3 were processed using standard treatment protocols after Faegri et al. (1990). *Lycopodium clavatum* spores were added as an exotic marker for concentration calculations (pollen sum per $cm^3$). Pollen and non-pollen palynomorph (NPP) identification were carried out based on the reference collection of Mongolian pollen and spores at the Department of Palynology and Climate Dynamics (University of Göttingen), the NPP database (Shumilovskikh et al. 2022), and other relevant literature (Beug 2004; Murad 2011). For macro-charcoal (particles >125 μm) analysis, 187 subsamples of 0.5 $cm^3$ were taken at the same intervals. Subsamples were processed based on the method established by Stevenson and Haberle (2005). In each subsample, all macro-charcoal particles were counted using a binocular stereomicroscope. For each subsample, macro-charcoal concentrations were calculated. TILIA and TGView version 2.0 (Grimm 2004) were used for visualization of palynological and charcoal data.

Information on age-depth modelling and more detailed sedimentological and methodological descriptions can be found in Unkelbach et al. (2019, 2021).

## 3 Results

### 3.1 Northern Khangai

The summarized multi-proxy dataset of core OL3 from the forest-steppe zone in the northern Khangai (Fig. 4) covers the period from about 9,570 cal yr BP to present.

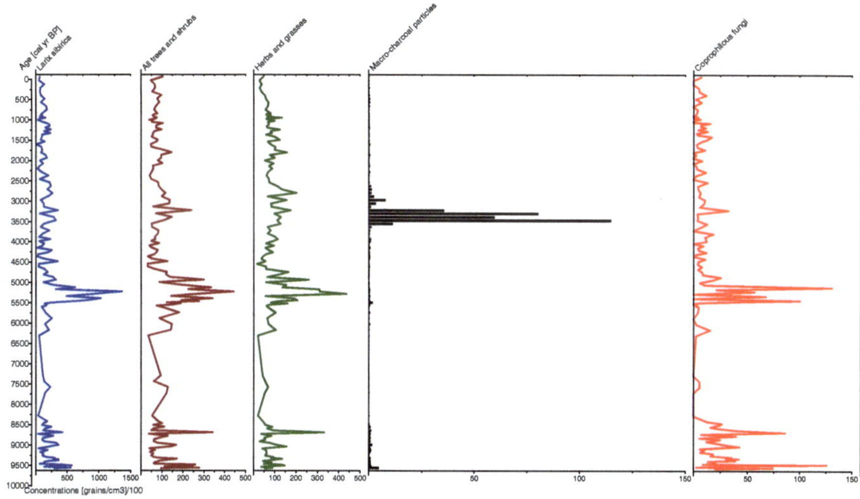


**Fig. 4** Multi-proxy diagram of selected pollen (grains/cm$^3$), macro-charcoal (particles/cm$^3$), and NPPs (spores/cm$^3$) for core OL3 from the forest-steppe in the northern Khangai Mountains

Based on the progression of the pollen and NPP records, the dataset can be divided into four sections: (1) from 9,570 to 8,500 cal yr BP, (2) from 8,500 to 5,500 cal yr BP, (3) from 5,500 to 4,500 cal yr BP, and (4) from 4,500 cal yr BP to present. The lowermost section from 9,570 to 8,500 cal yr BP is characterized by a high variability in all three pollen concentrations shown in the diagram. *Larix sibirica* concentrations range from 420 to 5,600 pollen grains per cm$^3$, the total tree and shrub pollen concentrations from 3,700 to 34,400 grains per cm$^3$, whereas the herb and grass concentrations vary between 28,900 and 156,300 grains per cm$^3$. Macro-charcoal concentrations are at a comparatively high level of between 20 and 460 particles per cm$^3$. Coprophilous fungal spore concentrations vary in accordance with the pollen concentrations between 300 and 8,600 spores per cm$^3$. In the subsequent section (2) from 8,500 to 5,500 cal yr BP, herb and shrub pollen, macro-charcoal, and coprophilous fungal spore concentrations decrease markedly to their lowest values. *Larix sibirica* and the overall tree and shrub pollen concentration decrease only slightly, still ranging from 420 to 2,600 and 3,400 to 19,100 grains per cm$^3$, respectively. The subsequent short period (3) from 5,500 to 4,500 cal yr BP is characterized by a marked renewed increase in pollen and coprophilous fungal spore concentrations. All four records reach their peak values between 5,220 and 5,280 cal yr BP. Macro-charcoal concentrations remain on a low level. In the uppermost section from 4,500 cal yr BP to present, pollen and coprophilous fungal spore concentrations decrease again, resulting in rather stable values with only minor variations. In the trees and shrubs and coprophilous fungal spore records, there are minor peaks at around 3,220 cal yr BP. This falls into a more extended period with markedly higher fire intensity between 3,630 and 2,880 cal yr BP, as can be seen from the significant increase in the macro-charcoal concentration up to 11,500 particles per cm$^3$.

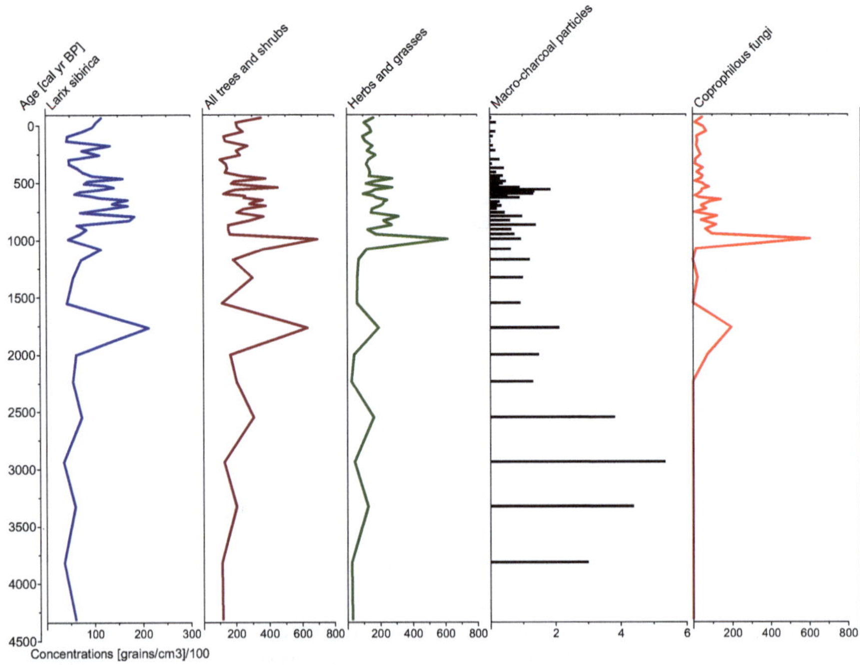


**Fig. 5** Multi-proxy diagram of selected pollen (grains/cm$^3$), macro-charcoal (particles/cm$^3$), and NPPs (spores/cm$^3$) for core D3L6 from the forest-steppe in the Mongolian Altai

## 3.2 *Mongolian Altai*

The summarized multi-proxy record of core D3L6 from the forest-steppe ecosystem in the Mongolian Altai (Fig. 5) comprises the period from around 4,310 cal yr BP to present. The lower core section from 4,310 to 1,000 cal yr BP is characterized by high tree and shrub pollen concentrations between 11,500 and 36,700 grains per cm$^3$. There is one major peak of 63,400 pollen grains per cm$^3$ at around 1,770 cal yr BP. In this period, the concentrations of herb and grass pollen are comparably low ranging from 25,300 to 193,100 grains per cm$^3$. Macro-charcoal concentrations are between 62 and 534 particles per cm$^3$, whereas their peak is at around 2,940 cal yr BP. There is no input of coprophilous fungal spores until 2,000 cal yr BP. From 2,000 to 1,000 cal yr BP, the concentration varies on a low level between 17 and 215 spores per cm$^3$.

In the upper core section from 1,000 cal yr BP to present, the overall concentrations of pollen grains increase, whereas the concentrations of grass and herb pollen as well as *Larix sibirica* increase more than the total tree and shrub pollen concentrations. *Larix sibirica* pollen concentrations vary between 4,400 and 18,400 grains per cm$^3$, the concentrations of all trees and shrubs slightly decrease to around 10,700–45,300 grains per cm$^3$, and the herb and grass pollen concentrations increase to values ranging from 95,500 to 315,200 grains per $cm^3$. The latter two concentrations show a distinct peak at around 990 cal yr BP with 69,500 and 616,200 grains per $cm^3$, respectively, while the concentration of *Larix sibirica* pollen grains is rather stable at this point. In the period from 1,080 cal yr BP to present, macro-charcoal concentrations decrease to 2–186 particles per $cm^3$, showing their maximum at around 560 cal yr BP. The input of coprophilous fungal spores increases markedly, varying between 1,200 and 14,200 spores per $cm^3$. The coprophilous fungal spore record also has a clear peak at 990 cal yr BP with 60,200 spores per $cm^3$.

## 4 Discussion

The multi-proxy data obtained from records D3L6 and OL3 provide different aspects of past environmental changes in the Mongolian forest-steppe ecosystem during the Holocene regarding vegetation and climate dynamics as well as human disturbance due to fire, tree logging, and grazing activities. These different data sets allow a reconstruction of long- and short-term regional and local changes, and to distinguish between natural climatic factors and anthropogenic influences.

### *4.1 Vegetation Changes in the Forest-Steppe Ecosystems During the Holocene*

In the northern Khangai Mountains, the multi-proxy data (Fig. 4) suggest the presence of *Larix sibirica* forest stands with an understory of open-habitat shrubs and steppe communities from 9,570 to 8,500 cal yr BP, distributed in higher proportions than the present vegetation structure. After 8,500 cal yr BP, steppe and open-habitat shrub vegetation expanded at the cost of *Larix sibirica* forests. From 5,500 to 4,500 cal yr BP, there was a sharp simultaneous increase in the input of arboreal and non-arboreal pollen, possibly indicating a somewhat denser vegetation cover. After 4,500 cal yr BP, forest and steppe proportions returned to their initial fractions. However, the proportion of steppe vegetation remains higher than at the beginning of the recorded period, indicating a reduction of forest cover in the northern Khangai.

From 4,310 to around 2,500 cal yr BP, the forest-steppe landscape in the Mongolian Altai was mainly dominated by steppe communities, whereas a slightly lower representation of forest can be inferred compared to the modern setting. According to the increasing representation of *Larix sibirica* and other arboreal pollen, the forest started to expand after 2,500 cal yr BP until around 1,000 cal yr BP. The period from 1,000 to present is characterized by major vegetation changes. One aspect is the greater variability of the individual pollen concentrations, but the greater change is the opening of the forests and a significant increase in the proportion of steppe vegetation.

The vegetation changes in the two areas are fairly different over the last 4,500 years. While the palynological reconstructions indicate surprisingly stable long-term dynamics in the northern Khangai, the vegetation in the Mongolian Altai changed significantly more during this period.

## 4.2 Natural Impact Factors

At the end of the early Holocene, from 9,570 to 8,500 cal yr BP, the reconstructed vegetation distribution from the northern Khangai in central Mongolia with larger quantities of pollen grains deposited suggests rather moist and warm conditions with a higher precipitation compared to the modern setting. Overall decreasing pollen concentrations, especially of *Larix sibirica*, reveal that the forested areas started to retreat slightly after 8,500 cal yr BP, indicating a climatic cooling and less precipitation. During the mid-Holocene, there was a short period of more favorable climatic conditions, as suggested by the marked increases in all pollen concentrations, but particularly among the trees and shrubs from 5,500 to 4,500 cal yr BP. In the northern Khangai, the period after 4,500 cal yr BP is characterized by stable climatic conditions. A slight decrease of forest stand area, and a renewed expansion of steppe vegetation may indicate lower temperatures, while rather humid conditions still prevailed.

In western Mongolia, the period from 4,310 to 2,500 cal yr BP was also characterized by rather cold, but slightly drier conditions according to the palynological data from the Mongolian Altai (Fig. 5). As indicated by the lower representation of *Larix sibirica* trees in the area, there was presumably slightly less annual precipitation compared to the modern setting. In the period from 2,500 to 1,000 cal yr BP, an increase in temperature and precipitation featured the expansion and density of forests and steppe areas. Around 1,000 cal yr BP, significant changes in the vegetation distribution of forest and steppe occurred in the Mongolian Altai. One theory could be a regional climate change toward more continental climate conditions. Probably, the extreme temperature differences between summer and winter developed around this time. However, in contrast to the results of Andreev et al. (2007) from the Russian Altai, there is no evidence for the Medieval Warm Period around 750 to 550 cal yr BP or the Little Ice Age from 310 to 110 cal yr BP in record D3L6.

In the northern Khangai, all vegetation changes described in Sect. 4.1 remain at a low level in contrast to the findings from the Mongolian Altai, and also in comparison with climatic reconstructions from western-central Mongolia by Tian et al. (2013, 2014) at Khuisiin and Bayan Nuur, and they are also partially disagreeing with the findings of Brugger et al. (2018) at Tsambagarav Mountain, Sun et al. (2013) at Achit Nuur, Wang et al. (2009, 2011) at Ugii Nuur, and Fowell et al. (2003) at Telmen Nuur. Renewed vegetation changes due to increasing temperature and precipitation after around 2,500 cal yr BP are not recorded in the pollen data of northern Khangai. Generally low pollen concentrations and minor variations in the proportions of forest and steppe confirm that climatic factors did not change on a larger scale in central Mongolia over the recorded period, limiting possible impacts on the vegetation. Overall, our findings of an initially moist and warm period from 9,570 to 8,500 cal yr BP, followed by increasingly arid and cooler conditions from 8,500 to 5,500 cal yr BP, interrupted by a shorter phase of more humid and slightly warmer conditions from 5,500 to 4,500 cal yr BP, mainly correspond to the climatic dynamics reconstructed from pollen data at Telmen Nuur nearby the northern Khangai study site. Fowell et al. (2003) suggest that changes in the solar insolation may be the driving factor for temperature oscillations in the area, whereas changes in the atmospheric circulation control precipitation variations. Pollen and diatom data from Ugii Nuur and Bayan Nuur, both located in the Valley of Great Lakes in between the Altai and Khangai Mountains, documented cold and rather humid climate conditions from around 8,100 to 5,800 BP, an increase in temperature and precipitation from 5,800 to 3,000 BP, and repeated cold and humid conditions from 3,000 BP to present (Tian et al. 2014; Wang et al. 2009, 2011). These reconstructed climate dynamics differ largely from our reconstructions in terms of rainfall or water availability, especially in the northern Khangai area. This could be explained by the location of the site north of the Khangai mountain range and most likely outside of the Asian summer monsoon's area of influence during the mid- and late Holocene (Fowell et al. 2003; Tarasov and Harrison 1998).

## *4.3 Impact of Fire and Grazing*

Comparison of the macro-charcoal and the tree/shrub and herb/grass pollen concentrations of the same site permits a rough evaluation of the influence of fire on the overall distribution of forest and steppe in both study areas. There can be decennial- to centennial-scale (short-term) or millennial-scale (long-term) changes in the vegetation composition following a fire event. In addition to the frequency of fires, their intensity, extent, and location are equally important factors in assessing their influence on changes in forest composition and structure (Kuosmanen et al. 2014; Ryan 2002).

At site D3L6 in the Mongolian Altai, slightly higher macro-charcoal concentrations are reconstructed after declines in the *Larix sibirica* and overall tree/shrub pollen concentrations, at least before 1,500 cal yr BP and around 600 cal yr BP (Fig. 5). This correlation is probably linked with larger quantities of dry wood after a possible climate-related decline of forest stands (Brugger et al. 2018; Umbanhowar et al. 2009; Unkelbach et al. 2019). For the Northern Khangai, a comparison of pollen and macro-charcoal concentrations suggests little to no linkages between the forest-steppe distribution and the occurrence of fire, even after the major fire event at around 3,000 cal yr BP (Fig. 4). The fire event, despite its intensity, might have been of a rather local character and thus has not affected the regional forest and steppe vegetation distribution or the ecosystem in general that much.

Even though *Larix sibirica* is classified as a pyrophytic species (Kharuk et al. 2019; Tsvetkov 2004), the increased fire activity between 3,630 and 2,880 cal yr BP (Northern Khangai) and over the last 1,000 cal yr BP (Mongolian Altai) does not provide evidence that it has increased the local *Larix* populations or altered the local vegetation composition markedly.

In the NPP assemblage, coprophilous fungi are the most robust indicators of grazing activity. These types of fungi only grow on the dung of wild herbivores, such as various ungulates or rabbits, or the livestock of nomadic pastoralists. Especially since the colonization of the region in the mid- and late Holocene, their examination enables inferences to be derived about the extent of local nomadic settlement activity and grazing pressure on vegetation (e.g. Davies 2019; Davis and Shafer 2006; Baker et al. 2013).

In the northern Khangai, the distribution of coprophilous fungi suggests a presence of a significant number of large herbivores within the investigated area since the beginning of the recorded period from around 9,570 cal yr BP. Indeed, from 9,570 to 8,500 cal yr BP, coprophilous fungal spores occurred more frequently than during the mid- and late Holocene. This abundance can be attributed to wildlife activities in the area during that time and can be a valuable indication of how high their values can be under little to no human influence.

After a period of presumably very little wildlife grazing activity in the area, there was a marked increase in the concentration of coprophilous fungi during the introduction of pastoralism on Mongolian territory during the Bronze Age (5,500–-3,300 cal yr BP) from 5,500 to 5,000 cal yr BP (Fernandez-Gimenez 1999). After 5,000 cal yr BP and during the period of the expansion of mobile pastoralism in Mongolia during the early Iron Age (2,800–2,400 cal yr BP), the concentration of coprophilous fungal spores remains on a very low level. These relations suggest a rather low and limited human impact on the vegetation dynamics in the northern Khangai area, especially during the late Holocene. Most likely, the dispersal of coprophilous fungal spores might still be attributed to wild animals more often than to the livestock of mobile pastoralists. In this period, more distinct vegetation changes would have been expected due to a general increase in nomadic activities in Mongolia, at least over the last 3,000 years (e.g., Fernandez-Gimenez 1999; Fernandez-Gimenez et al. 2017). The very small variations in vegetation composition and the proportions of forest and steppe reconstructed over the last 4,500 cal yr BP are therefore very likely the result of the very low human influence. Hence, our data suggests a mainly natural, climate-driven fire history for the Northern Khangai region. However, it should be stated that the dataset only displays a rather local signal, and the small total number of charcoal particles counted per sample must also be considered.

The coprophilous fungi patterns from the Mongolian Altai region provide an entirely different scenario. The marked peak of coprophilous fungi coincides with the abrupt decrease of arboreal pollen and the expansion of steppe communities starting from 1,000 cal yr BP. Thus, in contrast, wild animals appear to be of only minor importance for vegetation dynamics in the Mongolian Altai. The multi-proxy dataset suggests an intensification of human land-use, possibly induced by slightly wetter conditions. At around 1,000 cal yr BP, an increase in the number of cattle and an expansion of tree logging for firewood are the likely factors for the observed decline of natural forests despite favorable climatic conditions. This interpretation is supported by the almost simultaneous changes seen in our proxy data. If those shifts had been caused solely by climatic signals, NPPs would have responded more quickly than the forests surrounding the lake. Thus, our findings from the Mongolian Altai demonstrate a strong relationship between changes in the distribution of forest and steppe, and the increased impact on the landscape by pastoral nomads in the area during the period of Chinese invasions into Mongolia and increased nomadic movement starting at around 1,400 cal yr BP (Huang et al. 2018; Unkelbach et al. 2020). Tree logging, trampling, and grazing of the animals diminished the number of saplings, reduced the forested areas, and interfered with their recovery (Hauck and Lkhagvadorj 2013; Khishigjargal et al. 2013). Instead, ruderal herbaceous plants became more dominant.

## 5 Conclusions

The aim of this paper is to review the regional and local forest-steppe vegetation dynamics in western and north-central Mongolia over the last 9,570 cal yr BP with special emphasis on the extent of natural (climatic) or anthropogenic (hunting, grazing, tree logging) drivers of vegetation change and deforestation.

Based on the presented multi-proxy data from the northern Khangai and the Mongolian Altai, it can be concluded that during the early and mid-Holocene, it appears most likely that the vegetation dynamics in both regions is defined by the supra-regional climate variability. In the northern Khangai region, which was and is still historically only sparsely populated by nomads, this trend seems to have continued in the late Holocene.

Whereas in the Mongolian Altai, markedly larger and more densely forested areas occurred during the mid-Holocene that were strongly reduced after around 1,000 cal yr BP. Intensified local grazing activities might have caused the intense forest reduction and steppe expansion at around 1,000 cal yr BP. Those changes lead to contradictive reconstructions of climate signals under concurrent occupation of the area by nomads. Hence, during the late Holocene, small-scale site variability, regional climatic trends, and the local human activities may have a greater impact on the forest-steppe ecosystems of Mongolia than large-scale climate patterns. Especially over the last millennium, the ever-increasing human disturbance seems to outweigh climatic influences on the vegetation.

## References

AM Online Projects (2024) Climate Bayan-Ölgii. https://en.climate-data.org/asia/mongolia/bayan-oelgii-2275. Accessed 23 Dec 2024

Andreev AA, Pierau R, Kalugin IA et al (2007) Environmental changes in the northern Altai during the last millennium documented in Lake Teletskoye pollen record. Quat Res 67:394–399

Ariuntsetseg L, Boldgiv B, Golde A (2008) Main habitats and floristic diversity in the Tarvagatai Nuruu National Park – general view. Mong J Biol Sci 6:11–20

Baker AG, Bhagwat SA, Willis KJ (2013) Do dung fungal spores make a good proxy for past distribution of large herbivores. Quat Sci Rev 62:21–31. https://doi.org/10.1016/j.quascirev.2012.11.018

Batima P, Natsagdorj L, Gombluudev P et al (2005) Observed climate change in Mongolia. AIACC Work Pap 12:1–25

Beket U, Knapp HD (2012) Protection of the natural and cultural heritage of the Mongolian Altai. Explor Biol Resour Mong 12:335–352

Beug HJ (2004) Leitfaden der Pollenbestimmung für Mitteleuropa und angrenzende Gebiete, vol 1. Pfeil, München

Brugger SO, Gobet E, Sigl M et al (2018) Ice records provide new insights into climatic vulnerability of central Asian forest and steppe communities. Glob Planet Change 169:188–201

Davies AL (2019) Dung fungi as an indicator of large herbivore dynamics in peatlands. Rev Palaeobot Palynol 271:104108. https://doi.org/10.1016/j.revpalbo.2019.104108

Davis OK, Shafer DS (2006) Sporormiella fungal spores, a palynological means of detecting herbivore density. Palaeogeogr. Palaeoclimatol. Palaeoecol. 237:40–50. https://doi.org/10.1016/j.palaeo.2005.11.028

Dulamsuren C, Hauck M (2008) Spatial and seasonal variation of climate on steppe slopes of the northern Mongolian mountain taiga. Grassl Sci 54(4):217–230. https://doi.org/10.1111/j.1744-697X.2008.00128.x

Dulamsuren C, Hauck M, Bader M et al (2009) Water relations and photosynthetic performance in Larix sibirica growing in the forest-steppe ecotone of northern Mongolia. Tree Physiol 29(1):99–110. https://doi.org/10.1093/treephys/tpn008

Dulamsuren C, Khishigjargal M, Leuschner C et al (2014) Response of tree ring width to climate warming and selective logging in larch forests of the Mongolian Altai. J Plant Ecol 7(1):24–38. https://doi.org/10.1093/jpe/rtt019

Faegri K, Kaland PE, Krzywinski K (eds) (1990) Textbook of pollen analysis. Wiley, Oxford

Fernandez-Gimenez ME (1999) Sustaining the steppes: a geographical history of pastoral land use in Mongolia. Geogr Rev 89(3):315–342

Fernandez-Gimenez ME, Venable NH, Angerer J et al (2017) Exploring linked ecological and cultural tipping points in Mongolia. Anthropocene 17:46–69

Fowell SJ, Hansen BCS, Peck JA et al (2003) Mid to late Holocene climate evolution of the Lake Telmen Basin, North Central Mongolia, based on palynological data. Quat Res 59:353–363

Grimm E (2004) Tilia and TGView 2.0. Illinois State Museum, Research and Collections Center, Springfield

Gunin PD, Vostokova EA, Dorofeyuk NI et al (1999) Vegetation dynamics of Mongolia. Kluwer Academic Publishers, Dordrecht

Hais M, Chytrý M, Horsák M (2016) Exposure-related forest-steppe: a diverse landscape type determined by topography and climate. J Arid Environ 135:75–84. https://doi.org/10.1016/j.jaridenv.2016.08.011

Hauck M, Lkhagvadorj D (2013) Epiphytic lichens as indicators of grazing pressure in the Mongolian forest-steppe. Ecol Indic 32:82–88. https://doi.org/10.1016/j.ecolind.2013.03.002

Hilbig W (1995) The vegetation of Mongolia. SPB Academic Publishing, Amsterdam

Huang X, Peng W, Rudaya N et al (2018) Holocene vegetation and climate dynamics in the Altai mountains and surrounding areas. Geophys Res Lett 45:6628–6636. https://doi.org/10.1029/2018gl078028

Khansaritoreh E, Dulamsuren C, Klinge M et al (2017) Higher climate warming sensitivity of Siberian larch in small than large forest islands in the fragmented Mongolian forest steppe. Glob Chang Biol 23(9):3675–3689. https://doi.org/10.1111/gcb.13750

Kharuk VI, Ranson KJ, Petrov IA et al (2019) Larch (*Larix dahurica* Turcz) growth response to climate change in the Siberian permafrost zone. Reg Environ Chang 19:233–243. https://doi.org/10.1007/s10113-018-1401-z

Khishigjargal M, Dulamsuren C, Lkhagvadorj D et al (2013) Contrasting responses of seedling and sapling densities to livestock density in the Mongolian forest-steppe. Plant Ecol 214(11): 1391–1403. https://doi.org/10.1007/s11258-013-0259-x

Klinge M, Dulamsuren C, Erasmi S et al (2018) Climate effects on vegetation vitality at the treeline of boreal forests of Mongolia. Biogeosciences 15:1319–1333. https://doi.org/10.5194/bg-15-1319-2018

Klinge M, Dulamsuren C, Schneider F et al (2021) Geoecological parameters indicate discrepancies between potential and actual forest area in the forest-steppe of Central Mongolia. For Ecosyst 8: 55. https://doi.org/10.1186/s40663-021-00333-9

Kuosmanen N, Fang K, Bradshaw RH et al (2014) Role of forest fires in Holocene stand-scale dynamics in the unmanaged taiga forest of northwestern Russia. Holocene 24(11):1503–1514. https://doi.org/10.1177/0959683614544065

Lkhagvadorj D, Hauck M, Dulamsuren C et al (2013a) Pastoral nomadism in the forest-steppe of the Mongolian Altai under a changing economy and a warming climate. J Arid Environ 88:82–89. https://doi.org/10.1016/j.jaridenv.2012.07.019

Lkhagvadorj D, Hauck M, Dulamsuren C et al (2013b) Twenty years after decollectivization: mobile livestock husbandry and its ecological impact in the Mongolian forest-steppe. Hum Ecol 41(5):725–735. https://doi.org/10.1007/s10745-013-9599-3

MNE (2012) Forest resources of Mongolia. Mongolia Ministry of Environment and Green Development, Government Agency of Forests, Ulan Bator

Murad W (2011) Late quaternary vegetation history and climate change in the Gobi Desert, South Mongolia. Dissertation, University of Goettingen

Ryan KC (2002) Dynamic interactions between forest structure and fire behavior in boreal ecosystems. Silva Fenn 36(1):13–39. https://doi.org/10.14214/sf.548

Schneider F, Klinge M, Brodthuhn J et al (2021) Hydrological soil properties control tree regrowth after forest disturbance in the forest-steppe of central Mongolia. Soil 7:563–584

Shumilovskikh LS, Shumilovskikh ES, Schlütz F et al (2022) NPP-ID – Non-Pollen Palynomorphs Image Database as a research and educational platform. Veg Hist Archaeobotany 31:323–328. https://doi.org/10.1007/s00334-021-00849-8, http://non-pollen-palynomorphs.uni-goettingen.de. Accessed 15 Dec 2024

Stevenson J, Haberle S (2005) Palaeoworks technical papers 5. Macro Charcoal Analysis: a modified technique used by the Department of Archaeology and Natural History. Department of Archaeology & Natural History, Research School of Pacific & Asian Studies, Australian National University, Canberra, pp 1–8

Sun A, Feng Z, Ran M et al (2013) Pollen-recorded bioclimatic variations of the last ~ 22,600 years retrieved from Achit Nuur core in the western Mongolian Plateau. Quat Int 311:36–43

Tarasov PE, Harrison SP (1998) Lake status records from the former Soviet Union and Mongolia: a continental-scale synthesis. Palaeoklimaforschung 25:115–130

Tian F, Herzschuh U, Dallmeyer A et al (2013) Environmental variability in the monsoon-westerlies transition zone during the last 1200 years: lake sediment analyses from central Mongolia and supraregional synthesis. Quat Sci Rev 73:31–47

Tian F, Herzschuh U, Telford RJ et al (2014) A modern pollen-climate calibration set from central-western Mongolia and its application to a late glacial-Holocene record. J Biogeogr 41:1909–1922

Tsogtbaatar J (2004) Deforestation and reforestation needs in Mongolia. For Ecol Manag 201:57–63

Tsogtbaatar J (2007) Forest rehabilitation in Mongolia. In: Lee DK (ed) Keep Asia green, vol 2, "Northeast Asia", IUFRO World Series. IUFRO, Vienna, pp 91–116

Tsvetkov PA (2004) Pyrophytic properties of the larch *Larix gmelinii* in terms of life strategies. Russ J Ecol 35(4):224–229. https://doi.org/10.1023/b:ruse.0000033790.50606.54

Umbanhowar CE, Shinneman ALC, Tserenkhand G et al (2009) Regional fire history based on charcoal analysis of sediments from nine lakes in Western Mongolia. Holocene 19(4):611–624. https://doi.org/10.1177/0959683609104039

Unkelbach J, Kashima K, Enters D et al (2019) Late Holocene (Meghalayan) palaeoenvironmental evolution inferred from multi-proxy-studies of lacustrine sediments from the Dayan Nuur region of Mongolia. Palaeogeogr Palaeoclimatol Palaeoecol 530:1–14. https://doi.org/10.1016/j.palaeo.2019.05.021

Unkelbach J, Kashima K, Punsalpaamuu D et al (2020) Decadal high-resolution multi-proxy analysis to reconstruct natural and human-induced environmental changes over the last 1350 cal. yr BP in the Altai Tavan Bogd National Park, Western Mongolia. Holocene 30(7): 1016–1028. https://doi.org/10.1177/0959683620908662

Unkelbach J, Dulamsuren C, Klinge M et al (2021) Holocene high-resolution forest-steppe and environmental dynamics in the Tarvagatai Mountains, north-central Mongolia, over the last 9570 cal yr BP. Quat Sci Rev 266:107076. https://doi.org/10.1016/j.quascirev.2021.107076

Wallis de Vries MF, Manibazar N, Dügerlham S (1996) The vegetation of the forest-steppe region of Hustain Nuruu, Mongolia. Vegetatio 122:111–127

Wang W, Ma Y, Feng Z et al (2009) Vegetation and climate changes during the last 8660 cal a BP in central Mongolia, based on a high resolution pollen record from Lake Ugii Nuur. Chin Sci Bull 54(9):1579–1589

Wang W, Ma Y, Feng Z et al (2011) A prolonged dry mid-Holocene climate revealed by pollen and diatom records from Lake Ugii Nuur in central Mongolia. Quat Int 229:74–83

Yang Y, Wang Z, Li J et al (2016) Comparative assessment of grassland degradation dynamics in response to climate variation and human activities in China, Mongolia, Pakistan and Uzbekistan from 2000 to 2013. J Arid Environ 135:164–172. https://doi.org/10.1016/j.jaridenv.2016.09.004

MIX
Papier aus verantwortungsvollen Quellen
Paper from responsible sources
FSC® C105338

If you have any concerns about our products,
you can contact us on
**ProductSafety@springernature.com**

In case Publisher is established outside the EU,
the EU authorized representative is:
**Springer Nature Customer Service Center GmbH**
**Europaplatz 3, 69115 Heidelberg, Germany**

Printed by Libri Plureos GmbH
in Hamburg, Germany